EINSTEIN BEFORE ISRAEL

EINSTEIN BEFORE ISRAEL

Zionist Icon or Iconoclast?

Ze'ev Rosenkranz

PRINCETON UNIVERSITY PRESS

PRINCETON AND OXFORD

Published by Princeton University Press,
41 William Street, Princeton, New Jersey 08540
In the United Kingdom:
Princeton University Press, 6 Oxford Street, Woodstock, Oxfordshire OX20 1TW
All Rights Reserved

Jacket photograph: *Famous Zionists Arrive in the United States*, courtesy of Bettman/Corbis

Library of Congress Cataloging-in-Publication Data

Rosenkranz, Ze'ev.
Einstein before Israel : Zionist icon or iconoclast? / Ze'ev Rosenkranz.
p. cm.
Includes bibliographical references and index.
ISBN 978-0-691-14412-2 (hardcover : alk. paper)
1. Einstein, Albert, 1879–1955. 2. Einstein, Albert, 1879–1955—Political
and social views. 3. Zionism. I. Title.
QC16.E5R673 2010
320.54095694092—dc22   2010036902

British Library Cataloging-in-Publication Data is available

This book has been composed in Janson Text and Franklin Gothic
Printed on acid-free paper. ∞
press.princeton.edu
Printed in the United States of America

1 3 5 7 9 10 8 6 4 2

To my mother, Ruth
and
my children, Naomi and Daniel

CONTENTS

ILLUSTRATIONS

PREFACE

Whether they realize it or not, historians choose the topics they write about for very subjective reasons. I can trace the roots of my interest in Zionism back to that heady time in June 1967 when, on the verge of turning six years old, I became aware of a special sensation in my family's home in Melbourne, Australia. It had something to do with Israel and a war, and with how "we" had won. To my late father, who had been born in Vienna and had survived the Holocaust in China, and for whom the Zionist-Revisionist youth group Betar had been a main focal point of identification in the bleak and perilous reality of the Jewish ghetto in Shanghai, the Six Days' War was akin to a miracle. My father was not impervious to the general euphoria among Western Jewry caused by the swift Israeli victory. He was not one of the very few who had the foresight to realize what the Israeli occupation would eventually entail for the Palestinian inhabitants of the conquered territories, or indeed for Israeli society itself.

The seeds of my own enthusiasm for Israel and Zionism possibly stemmed from that exciting time. Later, after I had myself become an emigrant when my father decided to return to his native Vienna and uprooted our family, my perception of Zionism took on a slightly more sophisticated character. I guess I was around fifteen or sixteen when I first read Leon Uris's *Exodus* (in German) and identified with Ari Ben Canaan's exploits against the British and the Arabs. A year or two later I started listening to the shortwave English-language service, the Voice of Israel, and imagined how life would be so much better for me in Israel. After all, growing up Jewish and a foreigner in post-Holocaust Vienna was a very good way to become a Zionist. However, my Zionism took on a different character from my father's. Coming of age in social-democratic "Red" Vienna as I did, progressive ideals clearly left their mark on me, and my Zionism sprouted a left-wing, dovish hue. This development was no doubt due in part to my adolescent rebellion against my father's Zionist Revisionism.

After I emigrated to Israel in 1981, I joined the Peace Now student cell at the Hebrew University. Barely off the boat, and not yet fully conversant in Hebrew, I handed out leaflets for upcoming demonstrations and was told by some passersby to "go back to America"—this was presumably where all the wacko lefties came from (although it was well-known that many right-wing fanatics also originated there). Perhaps unsurprisingly, my youthful idealism did not fare that well in the harsh, cynical reality of Israeli society. My belief in Zionism was subject to continuous erosion as the years passed—the first war in Lebanon, the slaying of Peace Now demonstrator Emil Grunzweig, the discovery of the Jewish underground, the apathy toward Palestinian suffering among middle-of-the-road Israelis, Rabin's "break-their-bones" policy during the first Intifada, his own assassination, and many more events all did their share to make me realize that the Zionist vision had gone awry. Just as Vienna had been a good place to become a Zionist, Israel was an excellent place to become a post-Zionist.

It is perhaps ironic that for someone who has been in the Einstein business for twenty-two years, first as curator of his personal papers in Jerusalem and for the past eight years as one of the editors at the Einstein Papers Project at Caltech, Albert Einstein was not one of my childhood heroes. As a pacifist, I was deeply impressed by his dictum that "the next world war will be fought with sticks and stones" (I later learned that this was only an apocryphal statement). Shortly after I took up my position at the Albert Einstein Archives in Jerusalem, the electronics giant Sanyo launched a massive advertising campaign that depicted a very non-PC Einstein lookalike with slanted eyes and included the slogan, "Sanyo—the Einstein of Japan." I came to see this motto as quite indicative of Israelis' attitude toward Einstein—at the time, he seemed to be much more Japan's Einstein than Israel's Einstein. Yes, connotations of genius and smartness were prevalent in Israel, as they were elsewhere. But his image had become depoliticized, all his radicalism had been depleted, and his grandfatherly figure was as cuddly as a plush animal. In fact, if you mentioned Einstein's name, more often than not, people would think you were referring to the popular Israeli pop star Arik Einstein, who even figured in a widespread ad campaign for an Israeli bank dressed up as an Albert Einstein lookalike. And when I was invited to present a lecture on Einstein and the Hebrew University at a seminar on the history of that institution, I was taken aback when one of the participants attacked Einstein for being critical of the university's executive. He believed Einstein had no right to voice such criticism from the comfort of his study in Berlin.

Thinking over these vignettes, I would often ask myself why Einstein's very public role on behalf of Zionist causes and as a towering figure of Jewish moral authority had largely been forgotten by Israeli society. Was it because Einstein had never settled in Israel? Was it because his opinions on Zionist and Israeli policies were too controversial? If he had immigrated to Israel, would he have also been told to go back where he came from?

Pasadena, California, April 2010

ACKNOWLEDGMENTS

I would like to express my deep appreciation to Ingrid Gnerlich, Peter Dougherty, Debbie Tegarden, and Marjorie Pannell at Princeton University Press for their astute guidance and assistance throughout the publication process. I am especially indebted to Moshe Zimmermann, my PhD adviser at the Richard Koebner Center for German History at the Hebrew University, for his knowledgeable, kind, and patient support throughout the writing of my thesis, on which this book is based. I also owe a very special debt to Diana Buchwald, my supervisor at the Einstein Papers Project, for her generous support of this effort and her sagacious comments on my manuscript. Furthermore, I would like to thank the members of my PhD committee at the Hebrew University, Yaron Ezrahi, Steven Aschheim, and Dan Diner, for their helpful remarks. Warm thanks are also due to the reviewers of the manuscript of this book for their perceptive and valuable comments.

I am especially grateful to my colleagues at the Einstein Papers Project Jeroen van Dongen and József Illy, Moshe Sluchovsky of the Hebrew University, and Barbara Wolff of the Albert Einstein Archives for their insightful comments on early versions of the manuscript; to Ofer Ashkenazi of the Hebrew University for helpful and enjoyable discussions on Einstein, pacifism, and Zionism; to Rosy Meiron of the Einstein Papers Project for assistance with the translation of French texts; to Issachar Unna of the Hebrew University for help in deciphering some complex Hebrew texts; to Roni Gross and Chaya Becker of the Albert Einstein Archives and Daphne Ireland of Princeton University Press for their assistance with copyright issues; to Rochelle Rubinstein and Simone Schlaichter of the Central Zionist Archives for help with initial research queries; to Rachel Misrati of the National Library of Israel, Anat Banin of the Central Zionist Archives, and Christine Nelson of the Pierpont Morgan Library for their friendly assistance with photograph inquiries; and to Sandy Garstang,

Shady Peyvan, and the rest of the staff at the interlibrary loan department at the Millikan Library at Caltech for their kind and efficient assistance. I am also particularly indebted to Sara Palmor and Nurit Lifshitz for their untiring efforts in assisting me with my archival and library research.

All of these individuals have helped to improve this book in some significant way. However, any faults are entirely mine.

EINSTEIN BEFORE ISRAEL

INTRODUCTION

"The Zionists are shameless and importunate; I have a hard time adopting the appropriate position in each instance, considering that I am, of course, well-disposed to the cause."[1] This is how Albert Einstein defined his relationship to the Zionist movement in a letter to his close friend and fellow physicist Paul Ehrenfest in the spring of 1922. By then he had been associated with Zionism for more than three years, and had even crossed the Atlantic and toured the United States for the first time as part of a Zionist delegation. Einstein's exasperation with the demands of this ideological movement is patently clear in his statement. Yet his sympathy to the cause is also decidedly evident.

This quotation is indicative of the delicate relationship between Einstein and the Zionist movement, as well as of his own internal struggle as to how he should relate to a political cause that made demands on his time and energy and for which he felt considerable affinity. Yet how far did Einstein's commitment to Jewish nationalism extend? Was he, indeed, a fully fledged Zionist? Did he accept all the tenets of Zionist ideology, or did he maintain a certain distance from the movement and its leaders? How did Einstein become affiliated with a political organization that strove to establish a Jewish state in the British Mandate for Palestine? How much of his time and effort did this most prominent of scientists dedicate to Zionism, and what activities was he prepared to undertake on its behalf? What kind of relationships did he establish with the prominent German and international Zionist leaders of the movement? Which Zionist projects was he specifically interested in? How did he reconcile his support for

1

Zionism with his well-known disdain for all forms of nationalism? What type of political entity did he envisage for the Jewish and Arab populations in Palestine, and how did he react to the mounting inter-ethnic violence in the region? How was he perceived by the leaders and the rank and file of the Zionist movement? Did he assume the role of a Zionist icon or iconoclast? These are the central issues dealt with in this book.

This study tackles one topic that is highly controversial, Zionism, and another that has largely been forgotten, Einstein's affiliation with Jewish nationalism. Prior to 1967, the consensus among historians of Zionism was to stress its "epic struggle for survival and supremacy against implacable, antisemitic and murderous enemies."[2] The incisive divisions in Israeli society in the aftermath of the Six Days' War also brought the first cracks in that historical consensus.[3] The public dispute over Zionism spread to the international arena and perhaps reached an all-time high in 1975, when the UN passed the controversial "Zionism is Racism" resolution.[4] The contentious debate on Zionist and Israeli history has intensified even more since the late 1980s. Work by a group of Israeli scholars termed the "new" or "post-Zionist" historians has stimulated acrimonious discussion of Zionist history in general and the Israeli government's treatment of its Arab minority in particular. These publications have created a sharp division between those "who insist on discussing Zionism and Israel, warts and all" and those "who prefer more traditional and sympathetic interpretations of Israel's stance."[5]

Because of this very emotional public debate and the ongoing Arab-Israeli conflict, it is well-nigh impossible to discuss any topic related to Zionism with complete impartiality. Nevertheless, I have tried to approach this specific historical case study with a maximum amount of objectivity. I am well aware of the potential for exploiting my conclusions for propagandistic purposes. If this study finds that Einstein was indeed a Zionist, will that enhance the legitimacy of Zionism? Conversely, if I conclude that he was not a Zionist, will that be a moral victory for those who oppose Zionism? Even though I concur with those who do not believe that value-free history actually exists,[6] I will leave the propaganda to others, focus on the historical sources, and strive to be as objective as possible.

This book also explores what Einstein's association with the Zionist movement says about his relationship to his Jewish identity in general. How did he see himself as a Jew and as a member of the Jewish people? How did he view German Jewry, Western Jewry, and the Jews from Eastern Europe? What was his impression of the Jewish masses he encountered in the United States and of the new Jewish community in Palestine? Which political, social, and cultural developments that affected his fellow coreligionists did he view favorably, and which did he disapprove of?

This book also examines Einstein's Zionism in the context of his political outlook and his general views on nationalism. How did Einstein define nationalism? What were his opinions on German nationalism, and what impact did these attitudes have on his perception of Zionism? How did Einstein feel in Germany in light of both his status as a member of the academic establishment and as a target of anti-Semitic attacks? How did his opinions on Zionism and nationalism fit in with his general political worldview?

This book deals with Einstein's association with the Zionist movement during the European period of his life. Specifically, it covers the years between 1919 and 1933. The year 1919 saw Einstein burst onto the world stage as an international figure following the verification of his general theory of relativity by British astronomers. After Hitler's rise to power in January 1933, Einstein decided never to return to Germany, and by the end of that year he had emigrated to the United States and settled in Princeton, New Jersey.

Einstein hardly figures in general histories of Zionism. Such works as do mention him in passing have referred to him as one of the "eminent people outside the orbit of Zionism" recruited for the Zionist cause,[7] as a supporter of Chaim Weizmann during his 1921 U.S. tour,[8] as a "participant" in the Zionist Federation of Germany and as being present at the Constituent Assembly of the Jewish Agency in August 1929,[9] and as an opponent of the creation of a Jewish state and a supporter of a binational solution in the Middle East.[10] Even in works specifically dedicated to Zionism in Germany, references to Einstein are rare: one major work fails to mention him entirely;[11] another cites a visit by Weizmann to Einstein in 1925, a speech given by him at a rally on behalf of the Palestine Foundation Fund the following year, and his role in the internal dispute within the German Zionist Federation over the most suitable response to the outbreak of violence in Palestine in August 1929.[12] These scant references testify to Einstein's marginal role in Zionist historiography.

To date, there have been no monographs on Einstein's affiliation with Zionism. Two collections dedicated exclusively to Einstein's writings on Zionism have been published: *About Zionism*, which was edited by the veteran Zionist official Leon Simon in 1930, and *Einstein on Israel and Zionism: His Provocative Ideas about the Middle East*, which was compiled by the New York journalist Fred Jerome in 2009.[13] In addition, an anthology of Einstein's political writings includes a substantial section titled "Anti-Semitism and Zionism, 1919–1930."[14] However, none of these collections offers an extensive analysis of Einstein's affinity with the Zionist movement.

In general, Einstein's association with Zionism has been examined in two very different types of publications: biographical studies, which deal inter alia with this topic as part of a comprehensive discussion of Einstein's life and work, and scholarly articles, which either pertain exclusively to this topic or deal with it as part of an analysis of Einstein's Jewish identity. The majority of the biographical studies have been written either by colleagues or close associates of Einstein or, in more recent years, by journalists. A handful of German Jewish and American Jewish historians have written brief articles on the subject of Einstein and Zionism, yet have usually based their conclusions on limited published material. No extensive studies of Einstein's Jewish identity or of his affinity to Zionism have been authored by professional historians.

In these works, Zionism's appeal for Einstein has been attributed to various causes: to the primordial bond he allegedly felt with the Jewish people,[15] to his concern for the dire plight of Eastern European Jewry in the aftermath of World War I and Zionism's plan to offer them a refuge in Palestine,[16] to the rise of anti-Semitism in Germany during and immediately after the war,[17] to his disdain for the "undignified mimicry" of his fellow coreligionists in Berlin and his rejection of assimilation as a solution to the "Jewish problem,"[18] to his antipathy toward German nationalism,[19] to his sense of "solidarity with outsiders as an outsider,"[20] and to the beneficial effect Zionism would presumably have on restoring pride and self-respect to Western Jewry.[21]

Widely varying theories have also been proposed to define the nature of Einstein's role in the Zionist movement. At one end of the spectrum he has been ascribed Messiah-like qualities in "preaching" nationalism to the Jewish masses[22] and has been described as a "spiritual head" of the Jewish people.[23] He has been defined as a "Zionist leader"[24] and as "an ardent advocate of Zionism."[25] A prominent head of the movement termed him a Zionist, although admitting there were definite limits to his Zionism because of his idiosyncratic personality.[26] A recent biographer dubbed him "the Wandering Zionist," yet simultaneously claimed that "his allergy to nationalism kept him from being a pure and unalloyed Zionist."[27] Others have viewed Einstein's Zionism as akin to the cultural Zionism of the ideologue Ahad Ha'am.[28] A recent article defines Einstein's Zionism as a "syncretistic private Zionism, which is not subsumed to any political strategy or organizational discipline."[29] And in the abovementioned recent anthology of Einstein's writings on Zionism, he is seen as having "mixed feelings about Zionism."[30] At the other end of the spectrum, Zionism and Palestine have been viewed as being "only peripheral concerns" for Einstein;[31] his advocacy of Zionism has been described as "limited,"[32] and the notion that he was a Jewish nationalist has even been dismissed as "ridiculous."[33] He has also been characterized as a "sympathizing non-Zionist."[34]

These previous studies on Einstein's relationship to Jewish nationalism are not extensive and often make generalizations based on very limited source material. Consequently, ambiguities and contradictions in Einstein's positions have often been overlooked. In contrast, in this book I am committed to avoiding such generalizations and intend to subject the available materials to a meticulous examination. Such a careful study of Einstein's writings, correspondence, and third-party materials should help illuminate the ambiguities and contradictions in his attitudes, beliefs, and actions in regard to his support for Zionism. Moreover, I am of the opinion that such a detailed study will reveal how Einstein could sometimes change his mind on various issues literally from one day to the next.

Consequently, this study constitutes a major departure from previous studies on Einstein and Zionism in a number of highly significant ways: its analysis is both far deeper, delving into the major emotional, social, and cultural factors that led to Einstein's subsequent affinity for the Zionist movement, and also far more meticulous, examining certain key issues in his relationship to the movement with great regard to details, nuances, and ambiguities. Moreover, the analysis is grounded in a vast plethora of archival material, much of which has not previously been utilized for a study of Einstein and Zionism, and at the same time it is far more expansive than extant studies in both examining the major factors that contributed to Einstein's mobilization on behalf of Zionism from his earliest years till his young adulthood and in offering an extensive analysis of Einstein's interaction with the Zionist movement from his induction into the movement until the end of his Berlin years.

In a wider context, this work provides a reinterpretation and substantial expansion of the Jewish and Zionist aspects of Einstein's biography. It contributes to the demythologization of Einstein by juxtaposing the real historical figure of the prominent physicist with his public image. It also adds to the general reconstruction of Einstein as an authentic individual and helps rescue him both from the popular myth, in which he is merely perceived as a genius, and from the narrow focus of historians of science, to whom only his contributions to science appear noteworthy. It should also enhance our understanding of why Zionism appealed to those who, like Einstein, were essentially outsiders vis-à-vis the movement.

In my analysis of Einstein's relationship to the Zionist movement, I have drawn amply from the rich historical sources housed at the Albert Einstein Archives at the Hebrew University in Jerusalem and with the Einstein Papers Project at the California Institute of Technology in Pasadena. I have delved into critical materials held at the Central Zionist Archives in Jerusalem and the personal papers of major Zionist figures such as Chaim Weizmann and Louis D. Brandeis. I have also availed myself of important sources at official German repositories. In addition, I have made extensive use of the press coverage of Einstein's Zionist-related activities and of the memoirs of central figures with whom he was in contact as a consequence of his involvement in Zionism.

Nevertheless, this study makes no claim to be exhaustive or all-inclusive. Because of the sheer mass of available materials relevant to Einstein's involvement in the Zionist movement during the period under discussion, I could not scrutinize every single aspect of his engagement with the same level of intensity. Therefore, I have consciously chosen specific events and issues I believe to be crucial in elucidating Einstein's involvement in the Zionist movement during his European years. I preface my analysis of his association with Zionism with an introductory chapter that surveys the decisive factors in Einstein's family background, childhood, adolescence, and early adulthood that, in my opinion, led to the radical ideological transformation in his position vis-à-vis Jewish nationalism in the aftermath of World War I. This beginning is followed by an inquiry into Einstein's mobilization and induction into the Zionist movement in 1919 and his first actions on behalf of Zionism. I then examine Einstein's first trip to the United States, in the spring of 1921, when he visited as a member of a Zionist delegation and unwittingly played a key role in the explosive conflict between Zionist leaders Chaim Weizmann and Louis D. Brandeis over funding for the settlement of Palestine. Einstein's tour of the land of his forefathers in the winter of 1923 is considered next; I focus on his inaugural lecture at the future site of the Hebrew University and his reception by both the Jewish and Arab communities and his British and Zionist hosts. I then turn to Einstein's role in the establishment and early development of his pet Zionist project, the Hebrew University in Jerusalem. This is followed by an examination of Einstein's reaction to the mounting violence in the burgeoning Arab-Jewish conflict in Palestine. His break with the executive body of the Hebrew University in 1928 initially left him severely disillusioned over the future course of the nascent institution, but eventually reforms were put into place that would guarantee his renewed support. In a final chapter, I bring together all the relevant evidence and present my conclusions about Einstein's association with the Zionist movement during his European years.

CHAPTER 1

"A VIVID SENSE OF STRANGENESS"

Einstein's Path to the Zionist Movement

"Physical attacks and verbal abuse on the way to school were common, yet mostly not that malicious. Yet they sufficed to consolidate in the child a vivid sense of strangeness."[1] This is how Einstein described his first experience of anti-Semitism at the Petersschule elementary school in Munich. In the spring of 1920, merely a few months after he had burst upon the world stage, Paul Nathan, the editor of Berlin's most prominent liberal newspaper, asked the newly famous celebrity for a brief article on his views on anti-Semitism. This topic was a particularly sensitive one for Einstein to deal with at the time, as he had recently become the target of anti-Semitic attacks at his own lectures at the University of Berlin. Despite being inundated with requests from journalists during this period, Einstein obliged and delivered a piece that revealed his public views on the status of Jews in Germany and his opinions on the issues of nationalism, assimilation, and the origins of anti-Jewish prejudice.[2]

In his reply to Nathan, Einstein explained that he had decided not to include in his essay a description of his school days in Munich as, in his opinion, "these experiences are not particularly meaningful for those not involved." Yet writing the article had clearly brought back memories of his plight as the school's only Jewish pupil. In the interim, thirty-five years had passed, and from that unique perspective he recalled his brush with anti-Semitism at the hands of some of his classmates in a somewhat assuaging tone. However, the verbal and physical abuse had obviously been sufficiently significant to be remembered despite all the years that had since elapsed. Most important, however, his letter reveals how Einstein conceptualized the impact the early experience of anti-Semitism had on him as an impressionable and vulnerable child. In his mind, these episodes had led to a deep sense of alienation.

But what was the context in which he experienced this early form of ethnic prejudice? And how did the "vivid sense of strangeness" affect Einstein's subsequent views on the status of the individual within the body politic and on his own ethnic identity as a Jew in Germany?

Albert Einstein's ancestors originated in the region of Swabia in the southern German state of Württemberg, one of the most long-established and densely populated regions in Germany in which rural Jews resided. The importance of legitimizing the Einstein family's identity as German and deeply rooted in the country is apparent from one of the first claims in his sister Maja's biography, that "all his known ancestors" had been German citizens. Maja also states that through intramarriage among the Jewish families of southern Germany, the Einsteins were related to most of the other Jewish families of the region, especially in Württemberg and Bavaria.[3] Indeed, village Jews were often "related several times over," and this was definitely the case with the Einsteins. Thus, they conformed to the patterns of intramarriage widely prevalent among village Jews in Germany. The consequence of these patterns of endogamy was that the rural Jews in any given region were "like an extended clan."[4]

Einstein's grandparents came from similar regional yet markedly different socioeconomic and cultural backgrounds. His paternal grandparents, Abraham Ruppert Einstein and Hindel Helene Moos, had Jewish first names and German middle names and remained in Buchau, a small village southwest of Ulm, all their lives.[5] In contrast, his maternal grandparents, Julius Dörzbacher and Jette Bernheimer, were known by their German first names and were born in Jebenhausen, a village southeast of Stuttgart.[6] However, they followed the trend common among the young elite of village Jewry by leaving their rural surroundings and moving to the town of Cannstatt near Stuttgart.

Julius established a prosperous business there and adopted the more German-sounding surname of Koch.[7] Thus, Einstein's maternal grandparents were considerably more upwardly mobile and modern and far more ambitious in their aspirations to conform to the culture of the German educated middle class than his paternal grandparents.

As a consequence, Einstein's parents, Hermann Einstein and Pauline Koch, were actually situated at different places on the continuum that led from the traditional existence of village Jewry toward urbanization, modernity, and *embourgeoisement*.[8] Hermann was born in Buchau in 1847, the fifth of seven siblings. Pauline was born in Cannstatt in 1858, the youngest of three siblings.[9] Even though they were only eleven years apart in age, there was, in effect, somewhat of a generational gap between them. Indeed, from a socioeconomic perspective, Einstein's father had more in common with his father-in-law than with his own spouse. Like Julius Koch, Hermann became a merchant and moved to an urban area—first to Stuttgart, then to Ulm, where he married Pauline.[10] As was the case with other village Jews of the time, Hermann's move from his rural surroundings was motivated by both marriage and the establishment of a business.

By all accounts, Einstein's parents had very differing personalities. Hermann Einstein was introspective and not suited to being a businessman, particularly as he found it difficult to make quick decisions, owing to a tendency toward excessive meticulousness. His initially successful business ventures in electrotechnology eventually suffered repeated failures. In contrast, Pauline had "a warm and caring personality," "seldom gave free reign to her emotions," and, following the collapse of Hermann's business enterprises, adapted herself "with difficulty, but with understanding" to the family's reduced living conditions.[11]

As members of the Jewish minority in the Wilhelmine Empire, Einstein's parents were subject to the "dual condition" of Diaspora Jewry, which has been defined by one historian of German Jews as an "adaptation to the circumstances of the non-Jewish environment, coupled with a sustained Jewish identity."[12] In this context, a unique German Jewish subculture was created that differed in important respects both from the narrow constraints of traditional Jewish society and from the majority German culture. Each Jewish family created its own idiosyncratic version of the German Jewish subculture, and this was no different with Einstein's parents.

Hermann's and Pauline's specific version of the German Jewish subculture into which their first son, Albert, was born was markedly influenced by the rural and small-town attitudes they had acquired in their families of origin. And, like other rural Jews, they brought these attitudes with them when they moved to the city.

In regard to religious observance, it seems that Hermann and Pauline abandoned the specific traditional piety of village Jewry. Like other rural Jews, the physical separation from village life had made the traditional practices of Judaism seem meaningless to them once they migrated to urban surroundings. This was especially the case with Hermann, for whom Jewish religious practices were an "ancient superstition."[13] However, certain vestiges of Jewish practices prevailed, though in a secular form. The most pronounced of these was charity to the poor: a needy Eastern European Jewish student was invited for a weekly meal every Thursday evening.[14] In this, Einstein's parents adhered to a common rural Jewish tradition, even though they did not observe this ritual on a Friday night, thereby divesting the practice of its religious content. This tradition also expressed the rural Jewish value of pan-Jewish solidarity, a "sense of [a] shared fate with their brethren."

Perhaps the most crucial characteristic of the specific subculture prevalent in the Einstein household was the pervasive sense of a familial Jewish identity and of a potent, all-encompassing Jewish milieu in which the family was situated. Similar to practices among other rural Jews, this was a consequence of intramarriage and migration patterns and was reinforced by joint business ventures. Thus the Einstein nuclear family constituted only one small segment of a much wider Einstein-Koch-Moos-Bernheimer clan, the members of which were related to each other several times over. The resulting sense of Jewish identity may not have been explicit and was possibly even unacknowledged, yet it was nevertheless inescapable and would have had a strong impact on young Albert.

This almost incestuous family atmosphere was complemented by a certain wariness of the German majority. Even though Hermann apparently "admired" the new Wilhelmine Empire, its rulers, and its culture, simultaneously he was also "afraid of the Prussians."[15] Yet the family seems to have been more tolerant of their more immediate Gentile neighbors, their fellow Bavarians. Similar to other rural Jews, they exhibited a high degree of interdenominational tolerance: among village Jews it was common for pupils to attend Christian public schools and receive private Jewish tutoring at home. Indeed, as we will see, Hermann and Pauline actually went one step farther and allowed young Albert to receive Catholic religious instruction.

Thus, like other members of the German Jewish minority, the Einsteins clearly maintained a "Jewish familial 'inside'" and a "German 'outside'"—a condition defined by one German historian as the "situative ethnicity" of the German Jews.[16]

Gender roles in the Einstein family were apparently well-defined and traditional in nature, similar to those in the rural Jewish family: the father's role was to earn a living and support the family financially; the mother's role was to maintain familial harmony, uphold the family's "moral values," and create the "bourgeois domesticity" required of families seeking embourgeoisement.[17] However, Einstein's parents' very different personalities and the subsequent progressive deterioration in their financial circumstances placed severe strains on the family's well-being.

From a socioeconomic perspective, the Einsteins belonged to the Jewish property-owning bourgeoisie.[18] However, culturally, their aspirations derived to a greater extent from the German-Jewish educated middle classes. They strived to inculcate in their children what have been described by historians of German Jewry as the two core values of the ersatz religion of the Jewish educated middle class, cultivation (*Bildung*) and respectability (*Sittlichkeit*).[19] Biographers have noted the importance of German classical literature for Hermann and of classical music for Pauline. And Einstein's mother's emotional re-straint (as noted by her daughter) was undoubtedly an essential pre-requisite for inculcating her children with a "respectable" lifestyle.[20]

When Albert Einstein was born in 1879 in Ulm, his parents' first act in his German Jewish enculturation[21] consisted of the names they bestowed on him. Their choice of Albert as a first name, which was common among the ruling dynasty of the Hohenzollern, and of Abra-ham as a middle name, in memory of Einstein's paternal grandfather, testified to their dual cultural allegiances and to their acculturationist aspirations.

When Albert was one year old, the family moved to Munich in pursuit of Hermann's ambition to establish a more lucrative business enterprise. In the Bavarian capital, Einstein grew up in an extended household that encompassed the families of both Hermann Einstein and his brother (and business partner) Jakob. Pauline's widowed fa-ther also lived with them.[22] Thus, the typical residential patterns of rural Jewry were emulated by the Einsteins in their new urban sur-roundings and played a significant role in their children's immersion in an all-encompassing Jewish milieu.

After receiving private tuition at home, Einstein entered elemen-tary school in second grade at the age of six, thereby exhibiting a fair degree of precociousness. A year earlier, Einstein's enculturation as a bona fide member of the Jewish cultivated middle class got well under way when he began to learn to play the violin. Einstein's enrolment at the Catholic Petersschule primary school clearly constituted a dra-matic transformation for him. He suddenly transitioned from his small, overly intimate family environment to a very large school at-tended by two thousand pupils. He immediately had to contend with the status of being an outsider, both as a new pupil among second-graders and as the only Jewish pupil among his seventy classmates.[23]

According to later accounts, Einstein had a difficult time with the strict discipline enforced at the Petersschule. Perhaps it was the cor-poral punishment used to teach multiplication tables that led him to subsequently describe the teachers as "staff sergeants."[24] These difficulties at an early age have been seen (by Einstein and biographers alike) as precursors of his future defiance of authority.[25] They also possibly demonstrate that Einstein began to conceptualize the German educational system in military terms. However, it is not possible to ascertain whether Einstein actually viewed his school experiences in such terms at the time or whether this was merely a later projection back into the past. Furthermore, the common perception of Einstein as being consistently in defiance of authority is in need of serious reexamination.

As we saw at the beginning of this chapter, Einstein's unique status as the only Jewish pupil in his class soon brought about his first recorded encounters with anti-Semitism. In elementary school, he was struck by "the children's remarkable awareness of racial characteristics."[26] And in his abovementioned article on anti-Semitism and assimilation, Einstein presented his view of the process by which Jewish pupils like him became aware of the ways in which they were different from their Gentile peers:

> Soon after the Jewish child begins attending school, it realizes that it distinguishes itself from the other children and that it is not treated as an equal by its peers. This quality of being different is based on one's descent; it is not at all based merely on religious affiliation or certain idiosyncracies of tradtion.[27]

Thus, Einstein held both racial and religious anti-Semitism responsible for the discriminatory attitudes he was exposed to at school. Yet more important than his fellow pupils' perception of ethnic differences between Jews and Gentiles was Einstein's own employment of such distinctions. His classmates were not the only ones to utilize racial stereotypes, specifically racial physiognomy, to distinguish between their social in-group and their out-group; Einstein did so too. This is extremely important for Einstein's emerging ethnic self-concept as a Jew, which was obviously based on physiognomy and ethnic origin and not on religious identity. Such distinctions "based on observable, biological features" have been seen by political psychologists as typical of the very early development of ethnic identity among minority-group children.[28] In any case, Einstein's fledgling social identity as a Jew became intricately linked with his social role as an outsider. Yet we should keep in mind that we only have Einstein's later descriptions of his perceptions of himself as a child; we have no contemporary sources of his actual perceptions at the time.

The other main venue of Einstein's early encounter with anti-Semitism was the Catholic religious instruction classes he attended. Even though Einstein later characterized the teacher body at his elementary school as being "liberal and not making any distinctions between the various religious denominations,"[29] this does not seem to have extended to the school's religious instruction teacher. According to various biographies, on one occasion this teacher brought a large nail to class and informed his pupils that "this was the nail with which the Jews had crucified Jesus." All the pupils turned to Einstein and he "felt embarrassed." However, rather than mere embarrassment, it is far more likely that he felt excruciating shame at such a humiliating experience, and this may well have had a crucial impact on his self-esteem at precisely the time when, developmentally, approval by his peers would have been of great importance to him.[30] However, the effects of Catholic religious instruction may not have been all negative. It also seems to have contributed substantially to Einstein's later open-mindedness toward other denominations. In his biography of Einstein, his son-in-law Rudolf Kayser claimed that simultaneous instruction in both Judaism and Catholicism led Einstein to perceive "the sameness of all religions."[31]

Einstein's parents could have sent him for Jewish religious instruction at a nearby school yet apparently were not interested in doing so, thus demonstrating once again their acculturationist tendencies. However, neither did they seem content with his being instructed only in Catholic liturgy and lore. Therefore, they arranged for a distant relative to give young Albert private lessons in Judaism at home.[32] Perhaps this was a compromise between his parents: we know that Hermann was fervently antireligious; we do not know what Pauline's stance was on these issues. In any case, asking a relative to give Albert tuition in Judaism had the additional advantage of ensuring that the

Figure 1. Einstein's class photo from the Luitpold Gymnasium, Munich, 1889

Jewish enculturation he received would remain as close as possible to
what his parents were familiar with.

Einstein commenced secondary school at the age of nine, when
he began attending the Luitpold-Gymnasium in Munich. This transition brought with it several major changes. His encounter with Judaism took on a more formal character, as he now began to participate in Jewish religious instruction. His impression of the militaristic
character of the German educational system was reinforced at the
gymnasium. He perceived his teachers as "leaning predominantly towards the character of lieutenants . . . in the sense of the *Vormärz*," a
reference to a particularly reactionary period in Central European
history.[33]

Perhaps the most dramatic change brought about by Einstein's
transition to secondary school was that he ceased to be the only Jewish
pupil. His encounter with other Jewish students propelled a further
stage in the development of his evolving Jewish identity: the "vivid
sense of strangeness" was transformed into something new, a sense of
solidarity, which compensated for the hostility encountered at school:
"The sense of strangeness very easily leads to a certain animosity, espe-

cially when there are several Jewish children in the class, who then, quite naturally, close ranks and gradually form a small, closely knit community."[34] These critical experiences seem to have had a crucial impact on the formation of Einstein's stereotypes of his own Jewish minority in-group and the majority German out-group.[35] With regard to his in-group, the sense of solidarity, which was based both on a common ethnic origin and on shared experiences, contributed to Einstein's identification with the Jewish minority. This would most likely have also strengthened the tendency toward Jewish ethnocentrism, which his family had already inculcated in him. With regard to the out-group, Einstein clearly did not identify with the representatives of the German majority culture. Thus, his experiences at the gymnasium reinforced both the rudiments of Jewish in-group social cohesion and an accompanying alienation from the social identity of the majority out-group. Consequently, if Einstein internalized the German majority groups' negative stereotypes of the Jewish minority group, it would seem that he did so only to a limited extent during this period. This is noteworthy, as recent research has shown that minority children often adopt such negative stereotypes of their own ethnic group.[36] Einstein's rejection for the most part of such negative stereotypes paved the way for the subsequent development of his Jewish identity.

The Jewish pupils at the gymnasium shared another experience that may have enhanced their sense of in-group social cohesion: Jewish religious instruction. However, the increase in solidarity may have been achieved in an opposite fashion from that intended by the religious instruction teachers. The lessons were not characterized by a great deal of discipline, and the teachers were not held in high regard. Indeed, there was a substantial amount of "skylarking," which Einstein felt contrite about decades later.[37] Consequently, rather than increasing Einstein's respect for Judaism or for Jewish male role models, the religious instruction lessons achieved the opposite effect and eventually reinforced the antireligious stance inculcated by his parents.

Nevertheless, around the age of eleven Einstein went through a brief religious phase. Various factors have been cited to explain this transformation. Both the anti-Semitism he experienced at school and his solitary pursuits in the "rustic" garden adjacent to the family home have been proposed as contributing to this temporary turn to religion. Einstein's exposure to Jewish religious instruction and a passing youthful rebellion against his parents' antireligious attitudes have also been cited as precipitating this radical change.[38] Einstein himself (in a much later description) attributed this phase to a rejection of materialism: he viewed religion as "the first means of escape" from "the horror of the drive" for material pursuits to which everyone is condemned "through the existence of one's stomach." In his own mythopoeia of his youth, Einstein significantly came to see this phase as a "lost religious paradise of youth" which was "the first attempt to liberate myself from the shackles of the 'merely personal', from an existence which was dominated by wishes, hopes and primitive feelings."[39] This has been seen by one biographer as a "projection of ideas pertaining to his mature age into his youth," and I would have to agree.[40] However, it does reveal a tendency in Einstein (at least in his later years) to abnegate the importance of his deepest needs and emotions. Indeed, it would seem that Einstein's temporary religiosity was a coping mechanism he employed to deal with the vicissitudes of puberty, which were exacerbated by his transition to middle school.

According to Einstein himself, his religious phase found "an abrupt end at the age of twelve." His contact with an impoverished Polish Jewish medical student named Max Talmey, who came to partake of a family meal every Thursday evening, has been widely credited with having brought about the major transition in Einstein's allegiance from religion to science. It was particularly the perusal of popular works of science and philosophy encouraged by Talmey that led to this dramatic transformation. As Talmey himself later pointed out, it was ironic that "this conversion was, indirectly at least, the consequence of the only religious custom his parents observed, namely, to host a poor Jewish student for a weekly meal."[41] Einstein described his turn away from religion and toward science as follows:

Out yonder there was this huge world, which exists independently of us human beings and which stands before us like a great, eternal riddle, at least partially accessible to our inspection and thinking. . . . The mental grasp of this extra-personal world within the frame of our capabilities presented itself to my mind, half consciously, half unconsciously, as a supreme goal. . . . The road to this paradise was not as comfortable and alluring as the road to the religious paradise; but it has shown itself reliable, and I have never regretted having chosen it.[42]

This fascinating passage elucidates that it was the nonpersonal and objective nature of the "extrapersonal world" that tempted Einstein away from the seductive inveiglement of religiosity. Like religion, science could provide him with an enticing way of escaping from what he obviously perceived as the humdrum and common nature of his most profound feelings and desires. Moreover, unlike religion, science had the added crucial bonus of being able to stimulate the intellectual curiosity of the budding cerebral adolescent. Unfortunately, we do not have any sources from this period in Einstein's life that would provide authentic evidence of his deep-seated need to avoid dealing with what he apparently perceived as the daunting and forbidding emotions faced by every adolescent. All the accounts of his striking turn from religion to science derive from much later.

However, we do know of one concrete consequence of the turn to science. As far we know from the historical sources available to us, as a result of the end of his religious phase, Einstein did not celebrate his Bar Mitzvah. Even though he did attend preparatory lessons, we have no evidence that he actually celebrated this major Jewish rite of passage.[43]

It was during this stage in Einstein's life that science began to fulfill another critical role: a realm in which he could bond with like-minded individuals:

The contemplation of [the extrapersonal world] beckoned as a liberation, and I soon noticed that many a man whom I have learned to esteem and to admire had found inner freedom and security in its pursuit. . . . Similarly motivated men of the present and the past, as well as the insights they had achieved, were the friends who could not be lost.[44]

This passage demonstrates clearly that science was not only quickly becoming an emotional refuge for Einstein on an individual level but that the social contacts Einstein was to forge in the realm of science (and Talmey was the first of such contacts outside his immediate family)[45] would come to play a crucial role in providing a highly significant reference group with which he could bond.

As mentioned previously, the only noteworthy vestige of religious practices preserved in the Einstein home was the provision of a weekly meal for a needy Eastern European student, albeit in a form in which the tradition was stripped of its religious and ritualistic content. Einstein's weekly contact with Talmey, which extended over a period of five years, when he was between the ages of ten and fifteen, had a major influence on the impressionable youth. As far as we know, it was his first significant encounter with an *Ostjude*, a representative of Eastern European Jewry.[46] Talmey's association with the Einstein family occurred against the background of a very large influx of Eastern European Jews into Germany. In the period between 1880 and 1910, the number of Ostjuden in the Wilhelmine Empire grew from 15,000 to 80,000 and their share of the Jewish population increased from 2.7 percent to 12.8 percent.[47]

The intense contact with Talmey also had a fundamental impact on Einstein's emerging Jewish identity. He was exposed to an entirely different ethnic Jewish identity than that he encountered in his own family. However, he may well have sensed the (perhaps implicit) affinities between his parents' rural Jewish background and Talmey's Eastern European Jewish identity. Both of these Jewish subcultures were shunned by the urban German Jewish bourgeoisie, and this may well have heightened the appeal of the representatives of Eastern European Jews for the impressionable descendant of village Jews.[48]

Talmey has been seen by one biographer as a "substitute father in a spiritual and intellectual sense" for Einstein.[49] However, this ignores the momentous psychological element in their relationship. As his mentor, Talmey filled a crucial emotional role for the young Einstein. Indeed, he created an internal intellectual and emotional refuge for the inquisitive adolescent without Einstein's even having to leave his familial abode. Moreover, as Talmey was only ten years older than Einstein, he may well have seemed far more authentic to him than his materialistic and middle-aged father. In ethnic terms, this may have led Einstein to deduce that Eastern European Jewry were a far more authentic (and vibrant) Jewish subgroup than his Western Jewish parents.[50] Ironically, it was Talmey the Ostjude who strengthened Einstein's internalization of the German Jewish ideal of cultivation (*Bildung*). Perhaps this is what his parents had wanted, yet the outcome was significantly different from what they had originally planned. And it was Talmey, a medical student (and proto-scientist), who was the first of those abovementioned "[s]imilarly motivated men" whom Einstein learned "to esteem and to admire" at this vital stage in his emotional development. Furthermore, the stark contrast between his negative perception of the rigid German educational system and his obviously favorable learning experience with Talmey's informal instruction apparently led Einstein to associate Ostjuden in general with academic pursuits in a most positive manner. Thus, the adolescent Einstein rejected both the ineffectual bourgeois materialism of his father and the uncharismatic conformist Judaism of his religious instruction teachers. Instead, he chose the vibrant and authentic intellectualism of Max Talmey.

At the age of fifteen, Einstein experienced a dramatic upheaval. With the failure of the family's electrotechnical business, Hermann and his brother uprooted their families and moved to Italy. However, it was decided to leave Einstein in Munich with relatives to finish his schooling. This was motivated by the concern that adapting to the Italian school system would be too difficult at such a late stage in Einstein's education.[51]

However, his parents seem to have been oblivious to the potential emotional impact their move to Italy could have on young Albert. Half a year after his family's departure he suffered a severe emotional crisis. After a major confrontation with his class teacher, he decided to withdraw from school and follow his family to Italy. The most pressing motivation for Einstein to leave Germany suddenly, however, was likely to avoid later military service. According to the German citizenship law, males could emigrate till their completed seventeenth year; otherwise they would be declared deserters.[52]

Yet when Einstein arrived in Milan, he went one step farther than merely leaving his country of birth; he renounced his German citizenship altogether. He subsequently remained stateless for five years until he adopted Swiss citizenship in 1901. Indeed, Einstein renounced not only his national identity at this time. In the Danube district register of persons released from Württemberg citizenship in 1896, Einstein was listed as being "without religious affiliation."[53] As far as we know, this was the first time that Einstein described his religious allegiance in this manner; additional instances would follow in subsequent years. Thus, in one fell swoop, Einstein renounced both his national and his denominational identities. This constituted a rejection of both his official Germanness and his official Jewishness. Yet it was not a repudiation of his ethnic identity as a Jew, rather a rejection of his denominational identity as a member of the Jewish community. On the other hand, it also represented a repudiation of his bourgeois identity. He would later describe his experience of the Jewish bourgeoisie during this period as follows: "For the members of the Jewish bourgeois circles whom I became acquainted with in my younger years, with their luxuriousness and deficient sense of community, offered nothing which seemed of any value to me."[54] This simultaneous rejection of his parents' identity on three fronts—as denominationally Jewish, as nationally German, and as socially bourgeois—could well be viewed as the adoption of what social scientists call an "oppositional identity."[55] We will see that Einstein attempted to adhere to this identity for quite a long time.

Einstein spent roughly nine months in Italy with his family, from December 1894 to October 1895. He then took the entrance examination for the Zurich Polytechnic, yet failed in the linguistic and historical subjects. In a display of bravado ostensibly devoid of emotion, which, as we will see, would become typical for Einstein, he later

claimed that his failure had been "completely justified."[56] Yet this must
have been quite a painful experience for him at the time. Einstein was
advised to attend the last year of Swiss high school. If he completed
the final year successfully, he would be eligible to commence his studies at the ETH. He therefore enrolled in the cantonal school in Aarau
in the Swiss canton of Aargau to complete his secondary education.
This was facilitated by a friend of the Einstein family, the German
Jewish banker and social reformer Gustav Maier, who arranged for
Einstein to lodge with his close friend and ideological associate Jost
Winteler, the history teacher at the cantonal school.[57] Einstein felt a
close affinity for Winteler and his family and was greatly impressed by
their political and religious liberalism.[58] Maier and Winteler were
both instrumental in founding the Swiss Society for Ethical Culture
in 1896. Indeed, some scholars have pointed out that the ideological
progressivism of this society had a notable impact on Einstein's subsequent political views.[59]

The city of Aarau and the canton of Aargau were known for their
tolerant attitude toward Jews: they had granted them emancipation
relatively early.[60] In this liberal atmosphere, Einstein enjoyed his sojourn with his hosts. It seems that the Wintelers even constituted ersatz parents for the young student; tellingly, he referred to them as
"Papa" and "Mama" in his subsequent correspondence with them.
The liberal-minded Swiss school "thoroughly destroyed" Einstein's
prejudice against German-speaking secondary schools that had been
instilled in his early adolescence; here there was no military atmosphere or adoration of authority figures. Thus, Einstein was presented
with another positive model for a scholarly environment at an age in
which he was still impressionable. This time it was a formal framework (unlike Talmey's informal mentorship). The Wintelers also provided Einstein for the first time with a Gentile social milieu at home.
It is important to note that as the Wintelers were Protestants, their
home offered Einstein a different kind of Gentile environment than
he had been acquainted with in Catholic-dominated Munich. This
experience no doubt reinforced Einstein's universalistic tolerance of

non-Jewish social groups and would have countered (to some extent) his aforementioned tendency toward Jewish ethnocentrism.

In early October 1896, Einstein received his *Matura* school-leaving certificate from the cantonal school. Later that month he moved to Zurich and began studying toward a diploma that would entitle him to teach mathematics and physics at the secondary school level.[61]

Einstein's period at the Zurich Polytechnic brings with it a highly important change in the historical source material available to us. With a large increase in his own correspondence, we no longer have to rely on biographical (and autobiographical) material to reconstruct the various aspects of the formation of his Jewish identity and political socialization that helped carve his path to the Zionist movement. Consequently, from his student years on, we are able to draw on contemporary sources, which have a far greater degree of authenticity.

Just as Einstein was beginning his studies, his father's second business venture failed, and his parents' financial situation became precarious. Throughout his degree work Einstein received little financial support from his family and had to drastically restrict his expenses. In fact, his parents' desperate situation led to feelings of worthlessness and despondency on Einstein's part: "After all, I am nothing but a burden to my family. . . . It would indeed be better if I were not alive at all." The only thought that kept him from despair, he later claimed, was that he had not allowed himself any pleasures apart from his studies.[62] This provides us with an insight into another reason for the importance of his studies to Einstein—during this period, they became an utmost existential necessity for him.

Einstein's status at the Zurich Polytechnic was a complex one. As a German and a Jew, he was doubly an outsider. Perhaps this was a critical factor in determining that his most important relationship during his studies was with his fellow classmate, Mileva Marić, who, as a Serbian woman from the Austro-Hungarian Empire, was even more of an outsider than Einstein. As for the socialization that Einstein experienced at the Polytechnic, Maja claimed her brother never felt any urge to participate in the "rambunctious student life" customary at German-speaking universities. It is difficult to ascertain whether this was indeed the case. Einstein may have felt an aversion to such practices because of their overly German nature, or alternatively, he may have felt uncomfortable admitting such behavior to his sister. For at least one future occasion we have evidence that Einstein may have been anything but a teetotaler. Visiting Munich in the spring of 1911, he and Mileva sent a postcard written in very shaky handwriting to a Swiss friend that stated, "Both of us, alas, dead drunk under the table." There is a distinct possibility that Einstein was not actually drunk and that this was merely a lark among friends. In any case, the police detective report that was prepared in 1900 as one of the requirements for his naturalization as a Swiss citizen stated that Einstein was an "abstainer."[63]

Einstein evidently showed no interest in student associations, Jewish or otherwise, during his studies at the Polytechnic. Indeed, such associations were not particularly prominent at that institution during the period of Einstein's studies. Yet this is highly significant in light of Einstein's subsequent attraction to the Zionist movement. The student associations in Germany that advocated Zionism or Jewish nationalism played a central role in the formation of the future German Zionist leadership.[64] Yet this was not an experience that Einstein was exposed to because he studied in Switzerland. Perhaps more important, neither was he confronted with the virulent anti-Semitism common among German students of the period. Einstein's lack of exposure to anti-Semitism in academia during this period certainly distinguished him from most of his contemporary Jewish colleagues at German universities, who went on to become leaders of the Zionist movement. Later on, I explore how this affected his subsequent relationship with official Zionism.

His parents' worsening financial circumstances during his studies in Zurich had a deep impact on Einstein. His father's fate was perceived by him as a warning of the devastating effect dire economic conditions could have on a vulnerable individual. After the second failure of the family business in Pavia in July 1896, the seventeen-

year-old Albert begged his father not to start another business. However, his pleading was of no avail, and Hermann established a third
electrical company in Milan that same year, this time without the
business acumen of his brother Jakob. Two years later he was forced to
liquidate this company as well, at a great financial loss. According to
Maja, this failure and the ensuing financial worries led to their father
contracting severe heart disease, from which he died in 1902 at the
age of fifty-five.[65] Albert was twenty-three years old at the time.

Einstein continued to cope with these adversities by finding solace
in his studies. Soon after the beginning of his time at the Polytechnic,
we find further confirmation of the emotional importance for him of
intellectual activities. For the first time, this evidence derives from a
contemporary source and not from a much later one. Just before his
move to Zurich he had become romantically involved with Marie
Winteler, one of the daughters of his hosts in Aarau. At the termina-
tion of this liaison he wrote to her mother, Pauline, "Strenuous intellectual work and the observation of God's nature are the angels that
will guide me through all the tribulations of life in a conciliatory, for-
tifying and yet unrelentingly strict manner."[66]

To some extent, Einstein's Swiss years were a moratorium in the
development of his Jewish identity. He undertook a number of actions
that interpolated more distance between himself and official forms of
Judaism. This was also the case in respect to his German identity, as
we saw earlier.

The most significant move away from his Jewish background was
his marriage to his Serbian classmate at the Polytechnic, Mileva Marić.
Describing his mother's dramatically negative reaction to being told
the news that Mileva was "his woman," he did not mention explicitly
that she rejected his betrothed because Mileva was not Jewish. In fact,
Einstein's parents had welcomed his previous liaison with the Wintelers' daughter, Marie.[67] Yet Pauline Einstein openly objected to Mileva
being older than her beloved son and to her "bookishness," and—in a
statement that illustrated her allegiance to the cultural values of the
educated German Jewish middle classes—claimed that her son's prospective wife was not suitable to be accepted into a "respectable fam-

ily."[68] Perhaps it was Mileva's Serbian background that distressed Einstein's parents the most. His marrying a Serbian national was also a further rejection of his Germanness, not just of his Jewishness.

It is important to note that Einstein's distancing himself from official Jewishness during this period was not accompanied by a simultaneous social aloofness from other Jews. On the contrary, some of Einstein's most crucial relationships, especially his close male friendships, were with fellow members of the ethnic group with which he clearly continued to feel a close affinity: his devoted friend Michele Besso, his classmate at the Polytechnic Marcel Grossmann, and a fellow member of the discussion group "Olympia Academy," Maurice Solovine, to name the most important of his circle.

Around this time, Einstein began to refer to himself half-mockingly as "the valiant Swabian." In reference to a possibly unpleasant confrontation with his parents in regard to his relationship with Mileva during his upcoming visit to Milan, Einstein claimed, "yet the valiant Swabian is not afraid."[69] Thus, he utilized in a playful manner an imagined identity that he directed against his parents. Perhaps in his mind he was the authentic "valiant Swabian" and they were the unauthentic bourgeois Jews who were trying to deny their village Jewish roots.

Einstein received his diploma as an expert teacher in mathematics from the Zurich Polytechnic in July 1900. Of the four graduates in his year, Einstein was the only one not to secure a position as a university assistant, the first level in the academic hierarchy. Einstein later recalled he was not liked by his professors because of his "independence" and therefore was not offered a position. It is not clear whether this was a veiled reference to his Jewishness, but he did not explicitly mention his ethnic origin as a reason for not obtaining a position as an assistant at the Polytechnic upon completion of his studies. Einstein began work on his doctoral thesis under the supervision of Polytechnic professor Heinrich Weber. However, he soon found that he could not continue working with Weber owing to a clash of personalities.[70]

In the spring of 1901, Einstein embarked on a desperate search for a position as an assistant to a number of physics professors in the entire German-speaking region, yet his efforts were to no avail. This unsuccessful search left Einstein quite depressed. He was firmly convinced that it was his former professor Weber who had wielded influence behind the scenes and caused him to fail to secure a position.[71] While visiting his parents in Milan, Einstein now stated his plan to focus his search in Italy: "First of all, one of the main obstacles does not apply here, i.e., anti-Semitism, which would be as unpleasant as it is obstructive in the German[-speaking] countries, and secondly, I have a fair amount of patronage here."[72] This comment reveals that he was well aware of the anti-Semitism at German-speaking universities and was profoundly affected by it. Mileva was also very aware of the role that ethnic prejudice played in sidetracking Einstein's academic career: "And yet it is not very likely that he will obtain a secure position any time soon; you know that my sweetheart has a very wicked tongue and is a Jew into the bargain."[73]

Perhaps the most revealing document of the extent of Einstein's deep despair during this period came from his father's hand. After Einstein had sent Leipzig physicist Wilhelm Ostwald two requests for a position without receiving a reply, Hermann wrote to inform him of Einstein's fruitless search for a position as an assistant:

> My son is profoundly unhappy with his current lack of a position and the idea that his career has been derailed and that he will no longer be able to make the [necessary] connexions is becoming more and more entrenched in him by the day. Moreover, he is oppressed by the thought that he is a burden on us, as we are of modest means.

Hermann proceeded to ask Ostwald to write his son "if possible, a few words of encouragement, so that he might recover his joy for life and creativity."[74] However, no such letter from Ostwald is extant.

Even though he eagerly wanted to pursue a life of the intellect, Einstein found that he urgently needed to earn a livelihood, especially as Mileva was now pregnant with their first child. Consequently, he was prepared to compromise on the nature of the work he would accept. While waiting for a possible position at the Swiss Patent Office in Bern (which had been arranged with the assistance of his friend

Marcel Grossmann's father), Einstein took on temporary positions as a high school teacher.[75] In the autumn of 1901 he resumed work on his dissertation, this time under the supervision of University of Zurich professor Alfred Kleiner. At the end of the year, he learned that he had been accepted for the position at the Patent Office. Einstein was elated:

> I am giddy with joy when I think of it. . . . We will remain students (horribile dictu) as long as we live and won't give a damn about the world. But we will never forget that we owe everything to good old Marcelius, who tirelessly looked out for me. Also, I will always assist gifted youth wherever this is in my power, I am undertaking this as a solemn oath.[76]

This is an exceptionally significant statement by Einstein, and it came at an important juncture in his life, as he prepared to make the transition from a quasi-nomadic existence as an itinerant substitution teacher to an established pater familias with a fixed monthly income. Yet he clearly did not want to settle into this new lifestyle too quickly; hence his promise to Mileva that they would remain students, at least as far as their social ethos was concerned. Two weeks later Einstein would make a further comment to Mileva that plainly revealed what he feared a settled existence would entail: "Until you become my dear little woman, we will zealously do scientific work together so that we don't turn into old philistines, alright? My sister seemed to be such a philistine to me. You must never become like that, that would be terrible for me."[77] The reference to his sister, Maja, is quite indicative, as it implies that it was his family of origin that he perceived as being Philistine in character. Yet we can also see that in his perception, the role of science for him had undergone a transformation. It was no longer merely an emotional solace (as we saw previously) but the only way (for him) to escape a staid bourgeois existence. These fears were obviously especially heightened at this period of his life when he was on the brink of taking up a salaried position and the possibility this afforded him of settling down with Mileva. His admonition to her also seems to be a desperate attempt to hang on to an oppositional identity, which first manifested itself in his adolescence.

Yet of even greater significance for our purposes is Einstein's "sol-emn oath" that he will lend his assistance to "talented young men." Later I will examine the impact Einstein's own experiences as a desperate young academic had on his subsequent involvement in the educational projects of the Zionist movement.

Einstein's letter of application for the position at the Swiss Patent Office reveals a further twist in his evolving political and ethnic identity. In his transition to a more settled existence, he reconfirmed his official identity as being of German origin and a Swiss citizen, yet made no mention of his Jewish identity: "I am the son of German parents, but I have been living in Switzerland without interruption since the age of sixteen. I am a citizen of the city of Zurich."[78]

Einstein and Mileva's daughter, who is referred to as "Lieserl" in their correspondence, was born in Mileva's home town of Novi Sad (which at the time was located in the Hungarian part of the Habsburg Empire) in January 1902. Half a year later, Einstein took up the position at the Patent Office. Having secured a permanent position, he felt he could marry Mileva in January 1903.[79]

The following years brought more setbacks for Einstein in his aspirations for an academic career. He withdrew his first attempt at submitting his thesis to the University of Zurich in 1902. His second attempt at completing his dissertation on molecular theory was successful in 1905. However, according to Maja, the publication of his two revolutionary articles on special relativity that same year was followed by an "icy silence" from the scholarly community. A crucial breakthrough in his formal academic career only came in 1907, when Einstein was twenty-eight: he received his postdoctoral lecture qualification (Habilitation) as a private lecturer (Privatdozent) at the University of Bern. As he lectured for only two hours a week, he held on to his job at the Patent Office for another two years.[80]

A letter to Marcel Grossmann reveals what Einstein perceived as obstacles to advancing his academic career. He wanted to apply for a teaching position at the engineering school in Winterthur, yet he was concerned that his lack of Swiss German and his "Semitic appearance" would jeopardize his application. This provides us with further evidence of Einstein's dual outsider status in Switzerland as a German and as a Jew. Recent literature on Swiss anti-Semitism has stressed its character as discreet and xenophobic rather than overtly racist.[81] Einstein's letter also presents us with proof that, at least by this stage, his perception of the alleged physiognomic traits of Jews was well established.

Einstein's physical appearance during this period also seems to have been an important factor in his position in Bern. In her unpublished biographical sketch of her brother, Maja related that as a private lecturer, he was so absorbed in his work that he neglected his external appearance. Herself a student at the University of Bern at the time, she once asked the beadle which hall her brother was lecturing in and received an astounded reaction:

> "What, the . . . Ruski is your brother?" And he was close to bestowing a far less flattering epithet on the Russian. As Albert Einstein had never learned to converse in the widely accepted Swiss dialect, in spite of his long-time sojourn [there], the shabby-looking private lecturer could be regarded by the beadle as one of the despised Russians.[82]

Thus, Einstein's outsider status had heightened to an even greater extent: he was now mistaken for an Ostjude, one of the Jewish immigrants from Eastern Europe.

The most crucial sea change in Einstein's academic career came in the autumn of 1909, when he received a call as an associate professor at the University of Zurich.[83] It was now certain that his future in academia was secured; it was merely a matter of how lucrative a position he could obtain. During the next five years he would switch locations and positions three times, each time taking on a more prestigious appointment, until he finally moved to Berlin to take up a position that met all his requirements. In doing so, Einstein exhibited a great amount of self-interest, which to my mind should be seen as proof that the prominent physicist was far less naive than is popularly held to be the case.

After a very brief tenure in Zurich, he received a call to be a full professor at the German University of Prague in 1911. Here the issue of his religious affiliation at first constituted a stumbling block. Initially, Einstein believed that the Austrian authorities did not accept the philosophical faculty's proposal "due to my Semitic descent."[84] For our purposes, it is important that he attributed the initial hitch in his appointment to his ethnic origin. This provides us with further proof that there was no hiatus in Einstein's self-identification as a Jew during his Swiss years.

Yet it was actually the fact that he was registered as being "without religious affiliation" in Switzerland that led to difficulties with the appointment. When the Austro-Hungarian authorities in Vienna made it clear to Einstein that, as a professor of the Habsburg Empire, he had to declare his allegiance to a recognized denomination, he resignedly declared himself as belonging to the Jewish faith. The next year, Einstein described in the typically cavalier tone of his early years how meaningless this step had been for him: "To return to the bosom of Abraham—that was nothing. A signed piece of paper."[85] It is particularly intriguing that Einstein employed a phrase from the New Testament ("the bosom of Abraham") to describe his return to adopting an official Jewish identity. It was clearly not a major issue for Einstein to "return to the fold," because he viewed this bureaucratic step as merely a formality pertaining to his denomination. He was not actually returning to his Jewish identity because, as we have seen, he had never abandoned it. From a developmental point of view this episode was highly significant, as it reveals Einstein's willingness to compromise (as well as a fair degree of expediency); his oppositional identity, which manifested itself so strongly in his younger years, was now apparently mellowing. He even counterposed himself to his close friend Paul Ehrenfest, who was prepared to sacrifice an advance in his academic career rather than declare himself Jewish: "he adamantly insists on remaining without a religious affiliation."[86]

Einstein's brief sojourn in Prague also brought with it his first recorded encounter with members of the Zionist movement. Yet he was not influenced by them ideologically or impressed by them, for that matter—quite the opposite. Four years later, Einstein made some very disparaging remarks about the Zionists he had met in Prague in reference to a book he had recently read by Max Brod:

> Incidentally, I think I made this man's acquaintance in Prague. He may belong to a small circle which is infested by philosophy and Zionism and which was loosely affiliated with the university philosophers, a small band of seemingly medieval and unworldly people, with whom you have become familiar on reading the book.[87]

Einstein was most likely referring to the Prague Zionist association Bar Kochba, which was strongly influenced by both cultural Zionist Ahad Ha'am and Jewish philosopher Martin Buber. According to Brod, he met Einstein in the literary salon of the Prague socialite Bertha Fanta. In a subsequent memoir, philosopher Hugo Bergmann, another prominent member of Bar Kochba, claimed that Einstein had never discussed Zionism with him during his sojourn in Prague.[88]

Yet at least one Einstein biographer, Philipp Frank, who was Einstein's successor in Prague, has argued that despite this indifference (or even antipathy) to Zionism during his period in Prague, it was in the Czech capital that Einstein's interest in wider Jewish affairs was stimulated. Frank mentions that Emil Nohel, Einstein's assistant in Prague, aroused Einstein's "interest in the relation between the Jews and the world around them." In light of the complex situation in which the Jews of Prague found themselves between the fervent adherents of German and Czech nationalism, this certainly seems a distinct possibility. Einstein may have felt a special affinity to Nohel, as he was originally a village Jew from Bohemia. Einstein's brief sojourn in Prague also saw the beginnings of his awareness of political affairs in general: he noted what seemed to him to be the "considerable" animosity between Germans and Czechs in the city.[89]

The Einsteins did not feel at home in Prague. Unlike in Zurich, they felt quite "foreign." In the conflict between the Czechs and the Germans, it seems that Einstein identified with the German-speaking population, at least culturally. He complained that most of the locals did not know German and displayed hostility toward the Germans. Not long after their arrival, he stated the wish to get away from "semi-barbaric Prague." This statement is intriguing, as it reveals an anti-Slav sentiment in Einstein, which is slightly baffling, as he was married to a Serbian. In any case, he had clearly not yet discovered the appeal of Eastern Europe. After receiving an appointment as full professor at his alma mater, the Zurich Polytechnic, in January 1912, Einstein wrote to a friend, "All of us are very glad that we are returning to Europe."[90] Thus, he clearly did not define Prague as belonging to Europe, or at least to the Europe he felt comfortable in.

His departure from Prague was an occasion for Einstein to reconfirm the importance for him of his social and ethnic ties to his fellow Jews: "I am only sorry that I have to take my leave of my colleague Pick, with whom I have become very good friends. Sangue non é aqua, as the Italians say!"[91] This was a reference to Georg Pick, who held a chair in mathematical physics at the German University and had been instrumental in Einstein's securing an appointment there. The Italian expression Einstein used can be translated as "blood is thicker than water," thus again stressing the ethnic definition of his Jewish identity at this time.

Einstein's emerging political awareness, first in evidence during his period in Prague, continued to evolve after his return to Zurich to take up the position at the Polytechnic. When tensions rose between Austria and Serbia over access to the Adriatic, Einstein wrote, "If only the Austrians will remain calm; a conflict with Austria would be bad for the Serbians, even if they were victorious."[92] Perhaps having left Prague behind him, Einstein could now feel more sympathy for the Slavs and for the nation from which his wife hailed.

By July 1913, Einstein had been offered a lucrative position in Berlin as a member of the prestigious Prussian Academy of Sciences. In his characteristically wry and tongue-in-cheek style, he expressed his anticipation that this would entail a further elevation in his social status: "At Easter I am moving to Berlin as an academician, virtually as a living mummy." The transition to Berlin was motivated by two factors: Einstein's marriage had seriously deteriorated following the beginning of his affair with his cousin Elsa, and he eagerly desired to be at the hub of European science among his esteemed colleagues. This is confirmed by his statement to Elsa: "You can hardly imagine how much I am looking forward to the spring, primarily because of you, but also because of Haber and Planck."[93]

Einstein moved to Berlin in late March 1914. Soon afterward he took his first semi-public stance on the issue of anti-Semitism in rejecting an invitation from the vice president of the Imperial Academy of Sciences in St. Petersburg to visit Russia: "I am reluctant to travel without necessity to a country in which my ethnic comrades are being persecuted in such a brutal manner." This was most likely a reference to the notorious blood libel trial of the Ukrainian Jew Menahem Mendel Beilis, which had recently been held in Kiev.[94] More important, it constituted the first recorded time Einstein used the term "ethnic comrades" (*Stammesgenossen*) to describe his fellow Jews, thus reconfirming the ethnic nature of his perception of Jewish identity.[95]

However, at the time of the outbreak of World War I, Einstein was mainly preoccupied with his own family turmoil. Following a further deterioration in his marriage, a separation was agreed upon in July 1914. Mileva and their two sons left Berlin and returned to Zurich just days before the outbreak of war.[96]

Einstein's close friend, Michele Besso, later surmised he may have played a role in the deterioration of Einstein's marriage to Mileva by advocating a stronger connection to one's Jewish identity: "and perhaps as a result of my defense of Judaism and the Jewish family, some of [the guilt] lies with me that your family life took such a turn and that I had to bring Mileva back from Berlin to Zurich."[97] What part Besso actually played in Einstein's move to Berlin and his subsequent separation from Mileva we cannot know for sure. Yet it is intriguing to note that Einstein's first public remarks on behalf of his "ethnic comrades" coincided with his taking up residence in Berlin, inter alia as a consequence of his liaison with his cousin Elsa.

Mid-August saw Einstein's first substantive remark on the war: "Europe in its madness has now embarked on something truly incredible. At such times one realizes to what sorrowful species of brutes one belongs. I am maundering along in my tranquil musings and feel only a mixture of pity and revulsion."[98] This powerful comment reveals several crucial elements in Einstein's reaction to the war, some of which would remain constant throughout its course. He clearly perceived the war as an act of insanity, and this view would only intensify as the war dragged on. Moreover, he attributed responsibility for the war to Europe as a whole; at this stage he did not blame any one side in the conflict. In addition, the war reinforced his tendency toward misanthropy. And finally, he viewed the events from an emotional distance, not as a participant.

When viewed in the more general context of German Jewish reactions to the start of hostilities, Einstein's early dissident views were extraordinary. He did not share the hopes of the urban Jewish middle classes, who viewed the internal German truce (*Burgfrieden*) proclaimed by Wilhelm II as a golden opportunity for an integration of the Jewish minority into German society.[99] Indeed, this issue does not seem to have been on his agenda. Even though Einstein's views on the war were not shared by many German Jews, there were a few German-speaking intellectuals, such as Walter Benjamin, Gershom Scholem, Arthur Schnitzler, Karl Kraus, and Sigmund Freud, who did perceive the war as a disaster for humanity.[100] There were also left-wing politicians such as Hugo Haase, Rosa Luxemburg, and Gustav Landauer who opposed the war from the beginning. However, there are no indications that Einstein was consciously aware of their opposition.

Einstein's antiwar agenda was not explicitly political at this stage. Rather, it was the disruption in ties between scholars of the various countries involved in the conflict that Einstein found the most irksome consequence of the war in its initial stages: "The international catastrophe weighs heavily on me as a person with such international affinities. Living in this 'great era,' it is difficult to grasp that one belongs to this insane, degenerate species which credits itself with free will. If only there were an island somewhere for the benevolent and the sober-minded! That is where I too would want to be a fervent patriot."[101] Here we detect for the first time in Einstein an explicit longing for an isolated and sheltered environment in which like-minded individuals could live and flourish.

In the meantime, however, he felt deep disappointment when some of his closest scientific colleagues in Berlin, such as Fritz Haber, Max Planck, and Walther Nernst, signed the pro-war manifesto *To the Civilized World* (*An die Kulturwelt*). Despite this obvious setback, Einstein found solace in distinguishing between scientists and science (*Wissenschaft*) per se: "I love science twice as much in these times when I feel so painfully for the emotional aberrations of almost all my fellow human beings and for the sorrowful consequences thereof. It is as if a malicious epidemic had addled their brains!"[102] A few days after the publication of the Manifesto of the 93, as it became known, Einstein's associate Georg Nicolai drew up a counter-manifesto called *Manifesto to the Europeans* (*Aufruf an die Europäer*). The statement was circulated among many professors yet cosigned only by Nicolai, Einstein, and two other academics, and was not published until 1917.[103]

At a time when his closest German colleagues had disappointed him by supporting the war, Einstein seems to have found some refuge not only in science itself but also in developing new ties with fellow Jews. He had recently met Władisław Natanson, a professor of theoretical physics at the University of Cracow, who was spending a year in Berlin: "He is a Polish Jew, grew up in Russia and is now 50 years old. I quickly took a liking to him as I rarely do with people, blood runs thicker than water!" Of note, he used the same Italian phrase, "sangue non è aqua," he had employed a few years previously to stress what he perceived as the crucial importance of a common ethnic origin with Natanson and attributed his feeling of affinity with him to that joint ancestry. The importance of this encounter for Einstein in the general context of his social life in Berlin can also be discerned in his description of Natanson's role once he had returned to Cracow: "As long as you were here, you were my favorite Berliner; now I very much miss our relaxed contact with each other."[104] This statement is noteworthy for two reasons. It illustrates Einstein's very real estrangement from his colleagues who were authentic Berliners (or at the very least native Germans). And Einstein's continued adherence to the crucial significance of his ethnic Jewish identity can be seen as confirmation of the persistence theory in political sociology, which claims that "identification with an original nationality group" remains stable over the life span.[105] This certainly seems to have been the case with Einstein, even though by this time he had already migrated five times between the ages of fifteen and thirty-three.[106]

At this stage in the internecine conflict, Einstein did not believe in concrete actions to establish an international framework of those who opposed the war: "I am now starting to feel comfortable in the insane present-day turmoil, in conscious detachment from all things which preoccupy the deranged public at large. Why should one not enjoy life as one of the staff members of the insane asylum?" In the same letter, he went on to state that he thought the French writer and pacifist Romain Rolland was an optimist because he, unlike Einstein, believed in an international organization of all the "sane staff members."[107]

Einstein's separation from his first family and the experience of the war both underscored the importance for him of friendship, and in particular of male friendship. To his close friend Heinrich Zangger, who fulfilled the role of father confessor for Einstein during this period, he wrote:

> As difficult as the separation from them [i.e., from his sons Hans Albert and Eduard] is for me, it was a matter of life and death. . . . In these times one realizes that the only thing really worth striving for in this world is the friendship of splendid and independent people who are not swayed by every piece of printed rubbish to adopt the most perverse emotional attitudes.[108]

This insight was possibly the motivation for Einstein to seek out new contacts with like-minded individuals in a more formal framework: a few weeks later, he joined a political organization for the first time and became a member of the pacifist New "Fatherland League" (Bund "Neues Vaterland"). This led the following month to his first open political action: he cosigned the Delbrück-Dernburg declaration, which opposed Germany's annexationist policy.[109]

As the war dragged on and the initial militaristic enthusiasm waned, Einstein became more encouraged than he had been by the views of his fellow scientists and mathematicians, whom he now described as being "strictly internationally-minded." This was clearly a consequence of some of them (e.g., Max Planck) regretting their erstwhile decision to sign the pro-war Manifesto of the 93. The historians and the philologists, on the other hand, were (in Einstein's opinion) "mostly chauvinistic hotheads." The fact that highly educated individuals could subscribe to "narrow-minded nationalistic sentiment" was "a bitter disappointment" for him. Yet his disillusionment was not limited to developments in Germany: his regard for the "politically more advanced states" had decreased considerably, as, in his opinion, they were all ruled by oligarchs who had the press and might in their hands.[110] This reveals a growing political awareness on Einstein's part, even though at the same time he described himself as "being absolutely inexperienced and incompetent in political affairs."[111] In contrast to theories of the persistence of political attitudes throughout the course of one's life that I referred to above, Einstein's continuing politicization in his middle adulthood can be seen as a prime example of what some political sociologists have termed "lifelong openness." Proponents of this theory recognize that "change during adulthood is normal." Discontinuities within adulthood (such as emigrating to another country) are seen as playing a major role in contributing to changes in political attitudes in this stage of one's life.[112]

Contrary to his earlier beliefs, Einstein was now prepared to participate in an international pacifist organization, and joined the Dutch Anti-War Council. He did so, though, as a Swiss citizen and not as a German.[113] Indeed, Einstein's denial of his official ties to Germanness continued during this period. Asked by the Berlin Goethe-Association for a contribution to a "patriotic memorial book," Einstein defined his relationship to official state entities as such: "The state to which I belong as a citizen does not play the slightest role in my emotional life; I view the affinity to a state as a business matter, akin to one's relations with a life insurance company."[114] This passage reveals Einstein's desire to keep his relationship to the state devoid of any patriotic emotions. As we have seen, the only patria he was willing to acknowledge at this stage was the international one of science.

There was another domain he would rather have kept devoid of emotion: his relationship with his sons. After he received another less than friendly letter from Hans Albert, he wrote, "Come, dear old friend, Lady Resignation, and sing me your old familiar song, so that I can continue to spin quietly in my corner." To his mind, his influ-

ence on Hans Albert was limited to "intellectual and aesthetic matters."[115] Here (and elsewhere) it is clear that Einstein felt greatly
challenged (if not overwhelmed) by the emotional demands of others, and in particular of his closest family members. This was another
reason why science took on such a monumental role in his life. It also
partly explains why he would soon be attracted to an organizational
framework that, in his mind, would not demand too much emotionally from him.

The need for such an institutional framework may have seemed
even more urgent to Einstein in light of his perceived diminished scientific output: "With age, one gradually becomes more and more sta-
tionary and lacking in imagination, as is fitting. In one's place there are
young ones in whose brains the tracks are not so worn out." Yet he was
also deeply concerned about the effect of the war on the younger generation: "Over here much is stagnating as a consequence of the long
war. The younger people are going under in this evil treadmill."[116]

Meanwhile, the year 1916 saw the infamous census of the Jews
(*Judenzählung*) in the German army carried out by the Prussian Ministry of War. This census was seen by the Jewish community as the
culmination of anti-Semitic attacks, in light of devastating battles on
the Western front.[117] Yet there is no mention of the census in Einstein's correspondence or writings.

Soon afterward, at a low point in the war, Einstein seemed to start
losing hope that Europe as a whole had much of a future:

> The war does not seem to want to end at all anymore before the whole of
> Europe is destroyed. But the people don't deserve any different. . . . Why don't
> we have an analogue to the medieval monasteries, a refuge for people who
> want to withdraw from all worldly dealings, while renouncing certain things
> commonly held to be worth striving for? Could not such an international in-
> stitution with cultural objectives be established?[118]

At this stage, Einstein still obviously hoped that such a refuge would be
universalist in nature. Yet it is significant that what he longed for during this period was a sanctuary, and that it have a cultural objective.

By the end of 1917, Einstein was wondering whether it would not be preferable for Europe to self-destruct: "Would it not be beneficial for the world if decadent Europe were to destroy itself entirely? . . . I seriously believe that the Chinese ought to be held in higher esteem than we and hope that their healthy proliferation will outlast the extinction of our 'toughened' comrades."[119] Accompanying these increasingly anti-European sentiments were indications that Einstein began to see himself as a non-European: in a letter to Zangger, he referred to himself as an "indolent Oriental,"[120] an obvious reference to his Jewish origin.

In the meantime, Einstein's despair over his sons grew even greater. Informed by Zangger of Eduard's pulmonary tuberculosis, he lashed out at Mileva, seeing his son's illness as a "well-deserved punishment" for fathering offspring with a "physically and morally inferior person." We do not know whether he believed she was "inferior" because she was a Serbian Gentile. Yet he did not think his own genetic input should be "held in such high esteem" either. He simply did not believe he had fathered "valuable offspring," thus implying that neither Hans Albert, whom he regarded as obstinate, nor Eduard, who was seriously ill, were "valuable" human beings.[121] The further implication was that the only humans who did have some worth were those who were healthy and could be productive intellectually. And as in earlier times of crisis, Einstein found solace in science to cope with emotional difficulties:

> I console myself with the fact that one will also live on through the fruits of one's labor. The happy awareness of having thus been effective in such a productive and liberating manner, and lastingly so, is a consolation for me which nothing can destroy. Knowing this will enable me to bear whatever sad experiences I have with my children.[122]

Yet his hopes that a panacea could be found among young *German* academics were dashed by their fervent nationalism. In contrasting their role to that of two representatives of his own generation, he wrote, "the younger ones are considerably worse than they [i.e., Max

Planck and Wilhelm von Waldeyer-Hartz] are. I am convinced that some kind of intellectual epidemic is involved." In his opinion, critical minds were not in abundance in academia during the war: "Only very uncommon independent individuals can elude the pressure of prevailing opinions. There seems to be no such person at the Academy."[123] Yet when conditions worsened even more (in Einstein's eyes) following Russia's withdrawal from the war in December 1917, when Germany was able to focus exclusively on the Western front. Einstein found that "[e]ven the habitual escape into physics does not always help.[124]

The dramatic increase in mortality as a consequence of the war affected Einstein deeply. Some of his "best Academy brethren" had died, yet a worse loss for him than the deaths of those who had been killed in battle was the death of the "exceptional and still young [Marian] *Smoluchowski*," who had succumbed to dysentery. Einstein defined him as another "war casualty." He obviously felt that if the young and talented were dying, there was no hope for a productive future: "We've simply reached that age [he was thirty-eight] when one gradually substitutes dignity for labor."[125] The issue of dignity, especially in a Jewish context, would become of signal importance to Einstein.

The war and a prolonged bout of abdominal ailments, which worsened greatly during this period, heightened Einstein's sense of his own mortality: "But remember that the most important thing besides the demands of the moment is that my boys are reasonably provided for in the event of my untimely death."[126]

Einstein's distinction between "valuable" and "unvaluable life" within the context of his own family was further solidifed in the last months of the war. Upon hearing that Otto Heinrich Warburg—"one of Germany's most talented and promising younger biologists"—was near the front lines, he wrote to him and asked, "Is this not madness? Can't this post of yours out there not be filled by an unimaginative average person of the type that comes 12 to the dozen? Is it not more important than all that big scuffle out there that valuable people stay alive?"[127] Thus, Einstein revealed the true meaning of "valuable" in his opinion: a life was most valuable if it contributed to the advancement of science. We will encounter this elitism again, yet it appears here in its purest form, literally as an issue of life or death. This passage also reveals the limits of Einstein's much-touted humanism.[128]

Meanwhile, discriminatory anti-Semitic measures by the German government continued to gather momentum. The closure of the German border (*Grenzschluss*) against Eastern European Jewish refugees in April 1918 was another act that affected German Jewry during the war,[129] yet there is no documentation of Einstein's views on this development.

However, it cannot be doubted that Einstein was aware of these events. Half a year before the armistice, Einstein received his first (extant) letter from the German Zionist Federation. It pertained to the Ostjuden and was related to their cultural and humanitarian efforts on behalf of their eastern brethren. Einstein was invited to attend a discussion with Rabbi Markus Braude from the Great Synagogue in Lodz, who was visiting Berlin to raise interest in the establishment of a Jewish teachers' seminary.[130] It is not known whether Einstein attended the meeting.

Just before the end of the war, Einstein received a request that forced him to define the extent of his political radicalism. Yet despite the politicization he had undergone during the war years, he rejected an initiative by leftist writer Kurt Hiller that called for the emergence of intellectuals as a new political force. He used his Swiss citizenship as an excuse not to participate, while also admitting that the proposal did not seem realistic to him.[131] Even as he distanced himself from this explicitly left-wing initiative, however, he continued to be perceived as sympathetic to progressive causes. Two days after the armistice he wrote to his mother, "Among the academicians, I am some kind of high-placed Red."[132]

With the end of the devastating world war, a brand-new era was dawning in Europe. Einstein was also on the brink of a radically new period in his life: he was about to become an international celebrity. He was also on the verge of undergoing a profound transformation in his views on Jewish nationalism—a political ideology for which he had only had contempt just a few years previously. Einstein's experiences during the first forty years of his life now made him susceptible to become affiliated with the Zionist movement, a cause he felt could play a large role in helping him overcome that pervasive "sense of strangeness" he had first encountered as a vulnerable and impressionable child.

CHAPTER 2

A DIFFERENT KIND OF NATIONALISM

Einstein's Induction and Mobilization into
the Zionist Movement

In late February 1919, three months after the end of the Great War, Einstein was walking home after attending a lecture organized by the German Zionist Federation. As he strolled through the streets of Berlin, he was accompanied by the prominent Zionist propagandist Kurt Blumenfeld. In an article published thirty-five years later, Blumenfeld remarked that it was during this encounter that he noticed a "transformation" in Einstein in regard to his attitude toward Jewish nationalism. According to this subsequent account, Einstein, who had previously been wary of all particularistic strivings among the Jews, now declared he was prepared to become engaged on behalf of the Zionist cause.

Blumenfeld was highly satisfied with this dramatic turnabout in Einstein's views. He had been working toward this goal for the past several months. Einstein had expressed strong disdain for the adherents of Jewish nationalism just two and a half years earlier.[1] But on that winter evening in the streets of Berlin, the Zionists landed a major coup: one of the most highly respected German Jewish intellectuals declared his support for their cause.

The transformation in Einstein's views on Jewish nationalism occurred during a turbulent period both in German politics and society and in his personal life. The end of World War I brought about tumultuous changes in the German political scene: the abdication of the Kaiser and the proclamation of the new Weimar

46

Figure 2. German Zionist leader Kurt Blumenfeld in the early 1920s

Republic. There was also extreme political instability, with revolutionary uprisings, right-wing putsch attempts, and political assas-
sinations.[2] At the same time, Einstein's personal life was hardly less
turbulent: in February 1919 his divorce from Mileva Marić was fi-
nalized. Even though Swiss law stipulated that he was not to remarry
for two years, Einstein waited merely four months before wedding
his cousin, Elsa Einstein. In addition, his mother suffered a recur-
rence of abdominal cancer in the autumn of 1918, and her illness
entered its terminal stage. Einstein's younger son, Eduard, was seriously ill with a pulmonary ailment. His financial situation was also
precarious during this period, both because of the unfavorable exchange rate between the Swiss franc and the German mark and because of his attempts to comply with the monetary terms of the divorce. Finally, Einstein had to cope with his sudden emergence into

worldwide fame and be-coming a prominent public figure, which occurred overnight in early November 1919.[3]

Einstein's views on Jewish nationalism must be understood in the context of his perception of the nationalism of the predominant culture in which he resided, namely, German society. How did Einstein view Germany and the Germans during this period?

First (and most basically), Einstein continued to deny his own German identity. At this point, Einstein clearly viewed the Germans as "them," from whom he could distinguish himself by citizenship (as a Swiss national) and by ethnicity (as a Jew). He even believed that a period of postwar hardship would be beneficial for Germany: "This period will be curative for the inner development of the country because it educates politically and sweeps away the luxury and mollycoddling."[4] He also believed that an economic boycott would be advantageous, as "the last remnants of megalomania and the voracious appetite for power . . . will be extirpated."[5] For him, this was especially true of the German academics, most of whom had supported the war. In the autumn of 1919, he stated that the exclusion of German scholars from international meetings might provide them with "a lesson in humility."[6]

Yet he was obviously conflicted in his feelings about the economic hardships endured by the general German population. On one occasion, he stated, "Ever since things have been going badly for the people here, they appeal to me much more. Misfortune suits mankind immeasurably better than success." On other occasions he was more sympathetic toward their plight.[7]

He was also concerned with the German legacy of the war. He therefore joined a joint German-Belgian commission to investigate the alleged atrocities committed by German troops in Belgium and France during the conflict. This constituted part of his continuing advocacy of internationalism. In this context, reconciliation among intellectuals was one of the most significant issues for him during this period. He was initially optimistic about the prospects for a league of nations and continued his involvement in pacifist organizations, such as the New "Fatherland League," the Council of Intellectual Workers, and the Clarté movement, an association of intellectuals who advocated the establishment of an international federal commonwealth.[8]

In the immediate aftermath of the war, Einstein's hopes for the newly established Weimar Republic swayed between optimism and pessimism. In early December 1919 he decried the tendency among his fellow Germans toward "Prussianization" and expressed feelings of hopelessness. In reaction, he advocated "the political apathy and international mentality of the scholars of previous centuries." Yet by the end of that month he was optimistic again and believed that a "genuine republican attitude" was emerging. After the failed right-wing Kapp-Lüttwitz Putsch, his optimism was dashed again: "Over here confusion reigns, corruption and the dictatorship of the saber. The military murders with impunity, that miserable bunch! The barbarity is horrendous. The government is weak and always tends toward rotten compromises."[9]

The question of whether, in light of these conditions, he would remain in Germany was often on his mind. He repeatedly reassured others (and himself) that he would stay "unless something quite horrible happens." The loyalty he felt he owed to the scientific community in Berlin, especially to his senior colleague Max Planck, was a critical factor: he would remain, as he was "obligated toward the local authorities and colleagues by great gratitude." And it was Planck whom he viewed as being firmly rooted in Germany, whereas he himself did not feel grounded anywhere. He saw Planck as "rooted to his native land with every fiber, like no one else." In contrast, he viewed himself as "completely lacking such sentiments"; he only felt obliged toward "people as such," and not toward a national entity.[10] When asked by his close friends Max and Hedwig Born for his advice as to whether they should move to Göttingen, he cited the reasons for his sense of uprootedness:

> My father's ashes lie in Milan. I laid my mother to rest here a few days ago. I myself have gallivanted around incessantly—a stranger everywhere. My children are in Switzerland under such conditions that it is a major inconvenience

if I want to see them. A person like myself deems it as ideal to feel *at home* anywhere, with his loved ones.[11]

This was the wider personal and ideological context of Einstein's move toward Jewish nationalism. I now turn to the process by which he became involved with the Zionist movement in Germany.

In the previous chapter, we saw that the first recorded contact between Einstein and German Zionists occurred in May 1918 and in the context of an invitation to a discussion on the establishment of a Jewish teachers' seminary in Lodz. The next contact took place in early December 1918, four weeks after the November armistice and the abdication of the Kaiser. To further the "discussions" that had taken place to date, Felix Rosenblüth, a prominent member of the Central Committee of the Zionist Federation of Germany, sent Einstein for approval drafts of two documents that were to be published in the Zionist organ, *Jüdische Rundschau*. One draft was a brief announcement that Einstein had joined the provisional committee for the preparation of a Jewish Congress in Germany. The plan to establish a central organization that would serve as the official representative of German Jewry had just been proposed by the German Zionist Federation to other, non-Zionist Jewish organizations. It was thought that a joint body was needed in preparation for the upcoming peace talks. The second draft was an invitation to a discussion that was to take place on 19 December 1918 in Berlin on the topic "The German Jewish problem." The draft stressed the necessity of a new "policy on the Jews" to guide Jewish delegates at the planned peace conference, who would raise demands that Palestine be recognized as "a region of national settlement for the Jewish people" and that national autonomy be granted to Jews in various countries, especially in light of the recent pogroms in Poland and the rising threat of anti-Semitism in "all countries," including Germany. The Zionist Federation saw the proposed new congress "of all Jews residing in Germany" as the means by which the new policy on the Jews would be introduced. In the draft, the invitation was signed by both Einstein and Federation Central

Committee member and chief ideologue Kurt Blumenfeld. Einstein's name was apparently to be used in the invitation to attract other non-Zionist Jews. In his letter, Rosenblüth stressed that the "relationship of the Jewish notion of nationality to the ideal of internationalism" would be discussed at the planned meeting. Thus, the German Zionists obviously needed to reassure Einstein that his interest in internationalism would not be ignored at the proposed meeting. In a nonextant letter of reply, Einstein apparently consented to participate. However, the German Zionist Federation subsequently decided not to use Einstein's name for the invitation to the discussion as they had not managed to mobilize enough other new names to include in the invitation to be published in the *Jüdische Rundschau*.[12]

It is no coincidence that the first substantial contact between Einstein and the Berlin Zionist Bureau centered on an initiative that was predominantly pan-Jewish in nature rather than Palestinocentric. The German Zionist Federation had already "adopted the concept of Palestine-orientation (*Palästina-Zentrismus*) as a cornerstone of [its] ideology." Indeed, it seems fair to speculate that had the initiative focused instead on Palestine, it would not have earned Einstein's support at this point in time. He would definitely not have been prepared to embrace the German Zionists' Posen Resolution of 1912, which declared that every Jew should "incorporate Palestine into [his] life's program" and which had formed the central tenet of the ideological radicalization the younger German Zionists had brought about within the Zionist Federation.[13]

Following this correspondence with the Berlin Bureau, there is no documentary evidence of contacts between Einstein and German Zionist functionaries until October 1919.[14] However, according to the standard account of Einstein's induction into the Zionist movement provided by Blumenfeld, much happened in the interim. In two later accounts, Blumenfeld described his initial contacts with Einstein and the process by which he fostered the physicist's emerging interest in Zionism. According to Blumenfeld, he and his colleagues believed that with the Balfour Declaration, Zionism had succeeded in establishing a serious political basis. Therefore, it was now time to win over new supporters, especially those who were influential "in the public and cultural spheres." He and Rosenblüth compiled a list of Jewish scholars whom they might interest in Zionism. Rosenblüth proposed Einstein's name. Blumenfeld hesitated at first, as he was not certain of Einstein's "real significance." But he decided to visit Einstein (with Rosenblüth), and this turned out to be "the most significant event of the year 1920." However, this was evidently not the year that this "most significant event" occurred. In his other article on Einstein's induction, Blumenfeld states that his initial contact with Einstein occurred in February 1919.[15] This is much more likely to be closer to the actual date, yet it seems logical that Blumenfeld and Rosenblüth came up with Einstein's name before Rosenblüth first contacted Einstein in December 1918, and not that he contacted him first and they then came up with his name just prior to Blumenfeld's meeting Einstein in February 1919.

According to his memoirs, at their first meeting, Blumenfeld stressed the importance of Zionism's role in strengthening the Jews' "sense of internal security."[16] He apparently realized that this was a primary concern for Einstein. Following their first meeting, Blumenfeld invited Einstein to a lecture in a small circle that same week. There are various candidates for the lecture that Einstein may have attended, but unfortunately, it is not possible to reach a definite conclusion as to which one it was.[17] As described in the opening to this chapter, Blumenfeld allegedly noticed the "transformation" in Einstein in regard to his attitude toward Jewish nationalism while they were walking home. According to Blumenfeld, Einstein stated,

> I am *against* nationalism, but I am *in favor* of the Zionist cause. Today the reason has become clear to me. If a person has both arms and he constantly declares "I have a right arm," then he is a chauvinist. However, if a person lacks a right arm, then he must do everything to replace the missing limb. Therefore in principle, I oppose nationalism. But as a Jew I will support the Jewish-national cause of Zionism from today on.[18]

How accurate this account is (which was recounted approximately thirty-five years after the event and, perhaps tellingly, just one year after Einstein's death), we cannot ascertain. Yet it is clearly an account that stresses Blumenfeld's role in convincing Einstein of the special merits of Zionism. In this it is a self-serving account, and must therefore be regarded with caution.

In any case, it seems that Blumenfeld understood that with this parable, Einstein was making a qualitative and essential distinction between Jewish nationalism and other forms of nationalism. Seen objectively, this is actually a case of special pleading: all the years of Jewish suffering and lack of sovereignty entitled the Jews (in Einstein's view) to a homeland and justified their national aspirations. However, as this would render the Jews just like other nations, their form of nationalism would have to be *qualitatively different* from that of other nations for it to be acceptable to Einstein. The positive aspect of this is that it would commit Jewish nationalism to a higher degree of ethical standards than other nationalisms. Thus, on the one hand, this is an enlightened viewpoint, but it is also another instance of Einstein's tendency toward a Jewish ethnocentrism.

Over the next few months, Blumenfeld held a series of conversations with Einstein, only some of which dealt directly with Zionism. These meetings led Einstein to become interested in Zionism "gradually" (as Blumenfeld puts it), yet he concedes that Zionism was not yet part of Einstein's "worldview." Einstein's own reference to Blumenfeld's role in the transformation he had undergone is revealing, yet vague: a few weeks prior to his death in 1955, he thanked Blumenfeld "retroactively that you helped me to make myself aware of my Jewish soul."[19] From this wording it is clear that Einstein realized that his "Jewish soul" was an existing entity and that Blumenfeld made him conscious of it, and not that Blumenfeld grafted something onto Einstein that had not been there previously. It is also clear that what Blumenfeld had tapped into was primarily an emotional commitment that he had helped to translate into a cognitive realization. It is noteworthy that Einstein makes no mention of Zionism in this letter.

Blumenfeld's account of the process by which he tried to convince Einstein of the merits of Zionism is strikingly similar to Einstein's brief statement of 1955. Blumenfeld admits it was not an easy process.

He found that the method that worked with Einstein was one that he had most likely employed when he was the head of propaganda of the German Zionist Federation: "to attempt to uncover in the person that which is hidden within them and never to attempt to introduce that which is inconsistent with their essence." Blumenfeld felt he could only succeed in convincing Einstein of the merits of the Zionist cause "if I managed to adapt myself to his style in such a fashion that in the end he felt that the formulations were not being introduced by me but rather being created spontaneously by him." Furthermore, because he respected that Zionism and Palestine were peripheral issues for Einstein, he succeeded in gradually winning his trust.

It is clear that there was a very subtle form of manipulation at work here. Blumenfeld knew he had to be cautious to succeed where others, most notably the philosopher Hugo Bergmann, the writer Max Brod, and the artist Hermann Struck, had failed. It is interesting to consider why Blumenfeld (and his method) appealed to Einstein. What made him susceptible to Blumenfeld's art of persuasion? Blumenfeld's charisma was legendary, and Einstein seems to have been swayed by it.[20] But it is clear Einstein would not have been persuaded by the positive aspects of Zionism if he had not developed an interest in the issues addressed by the German Zionists and if some of their solutions had not appealed to him. In this context, it is pertinent to ask whether Blumenfeld fulfilled the role of a guide or mentor in Zionist matters for Einstein (even though Einstein was five years older than Blumenfeld). Were there any similarities between Blumenfeld and Max Talmey, Einstein's mentor in educational and scientific matters? Their backgrounds were very different. Thus, it was most likely to have been their charisma and their willingness to discuss issues that seemed important to Einstein at the time that were the decisive factors in influencing him.

Setting aside Blumenfeld's persuasive narrative, we should consider whether there were any other Zionists Einstein was in contact with even prior to his being contacted by the Berlin Bureau. To date, scholars have not looked into whether any of Einstein's friends or acquaintances played a formal role in the German Zionist Federation before Rosenblüth and Blumenfeld contacted him officially. Intriguingly, the Jewish physician Hans Mühsam was both a delegate to the Zionist Federation from Lübeck and an acquaintance of the Einsteins. While stationed in Lüttich in the dying days of the war, Mühsam wrote a long letter to Einstein merely a few weeks before Rosenblüth contacted him. In his letter, Mühsam expounds on his social Darwinian theories of the survival of the fittest (which Einstein would presumably have found objectionable), yet also offers personal medical advice to Einstein regarding his abdominal ailments. Indeed, we can find confirmation that Hans Mühsam may have functioned as the "missing link" between Einstein and the official functionaries of the Zionist movement in his cousin Paul's memoirs. There we find the bold claim that Hans "initiated . . . Einstein into Zionist thought."[21] However to date, this claim has been overlooked in favor of Blumenfeld's well-known account of his decisive role in Einstein's induction.

The month of February 1919 was an active one for Einstein in general in relation to his interest in Jewish issues. On 23 February 1919 (just after his return from a two-month stay in Switzerland), he participated in a meeting of the Association for the Establishment and Preservation of an Academy for the Science of Judaism (Verein zur Gründung und Erhaltung einer Akademie für die Wissenschaft des Judentums) at the home of the prominent gynecologist and Jewish communal leader Leopold Landau. At the meeting, a course of action for the association was presented. One of the main goals of the association was the revitalization of the Science of Judaism within a Jewish institution as "a consequence of the exclusion of Jewish studies (and scholars) from the German academy." Members of the association included both leading German Jewish communal leaders and prominent German Jewish academics.[22] Both Zionists and non-Zionists participated in the association. We can see what would have appealed to Einstein in this association. Its objective was to assist in compensating for the exclusion of Jewish studies and scholars from German academia, a goal he would have fully identified with. And its focus was on the lofty ideal of science (*Wissenschaft*)—albeit the Science of Judaism and not the exact sciences—which he held in such high esteem. More important, Einstein's participation at this point in time in both the activities of the association and in discussions on Jewish nationalism with prominent Zionist leaders reveals that this was a period in his life in which his Jewish consciousness was being intensely heightened.

The earliest documented *positive* expression of Einstein's views on Jewish nationalism and Zionism reveals his emotional commitment to the ideological cause of Jewish nationalism. It appears in a letter dated 22 March 1919 to his close friend and fellow physicist at Leyden University in Holland, Paul Ehrenfest. Discouraged by the actions of the Entente countries in the aftermath of World War I and by "the excessively dishonest local domestic politics," which he characterizes as "reaction with all its vile deeds under abhorrent revolutionary disguise!," he looks to his fellow Jews to find some "pleasure in human affairs": "I get most joy from the realization of the Jewish state in Palestine. It does seem to me that our ethnic comrades really are more congenial (at least less brutal) than these horrid Europeans."[23] Einstein thus makes an important distinction between the aspirations of Jewish nationalism, which he views positively, and the postwar actions of the Entente Powers and the suppression of the revolutionary forces in Germany, both of which he views negatively.

At this stage, Einstein's commitment to Zionism was still quite minimal. He expressed his sympathy for Jewish nationalist aspirations and his joy that these aspirations may be progressing towards realization. It was an emotional reaction to current events—more significant as it was through these aspirations that his "pleasure in human affairs" was reestablished. From its timing, Einstein's statement was most likely a reaction to Woodrow Wilson's announcement issued earlier that month in which Wilson expressed his support for the Balfour Declaration and "that the allied nations . . . are agreed that in Palestine shall be laid the foundations of a Jewish commonwealth."[24] In his own statement, Einstein also implicitly defines the Jews as non-Europeans. We have seen that two years earlier Einstein had classified himself as an "indolent Oriental." Indeed, a day earlier, he had referred to the Jews as "Asia's sons and daughters."[25] This identification of the Jews as non-European is highly significant. It is another case (if we are to believe Blumenfeld's rendition of the one arm/two arm parable) in which Einstein expressed his belief that in practice (and in essence), Jewish nationalism was essentially different from European nationalism, more humane and less brutal. This provides us with an explanation as to why Einstein was willing to embrace Jewish nationalism as a noble cause and view it in contrast to other forms of nationalism at the time. The question arises, was Einstein being influenced by contemporary Zionist thought in distinguishing Jewish nationalism as non-European?[26] There is no way we can determine this with any certainty, as there are no specific references by Einstein to this view in Zionist thought. In his statement, it is also intriguing that Einstein writes about a "Jewish *state* in Palestine" (my emphasis), and not a homeland or a commonwealth (even though Wilson's statement refers specifically to a "Jewish commonwealth"). Einstein's statement is also significant for what it omits: it doesn't mention the Eastern European Jews or the Hebrew University project, which are often cited as the main motives for Einstein's support of Jewish nationalism. The use of the phrase "*Jewish* state in Palestine" (my emphasis) is significant in that it clearly ignores the local Arab population in the region. In this, Einstein was not far behind the German Zionists: the issues of the Arab population in Palestine and Arab nationalism had only begun to be discussed by the German Zionist Federation at the end of 1918.[27]

Following this statement in his letter to Ehrenfest in March 1919, there occurs a hiatus of six months during which we do not have any correspondence to or from Einstein on Jewish or Zionist-related issues. One possible reason for this lacuna may be that there is much less material extant in Einstein's papers from the period prior to September 1919 than from that month on. Alternatively, it may have been a period in which Einstein's interest in these matters was dormant, particularly as he was in Switzerland over the summer, coping with his terminally ill mother, the difficult situation of his first family, and lecturing intensely in Zurich.[28]

In September 1919, after his return to Berlin, we find the first documented evidence of Einstein's interest in establishing a university in Jerusalem. In a letter to Einstein dated 11 September, Einstein's colleague Paul Epstein, a fellow physicist at Zurich University, refers to Einstein's interest in the "Jerusalem university." Epstein had been informed that Einstein would even consider taking up a position at the university were it not for his belief that he could not master a sufficient command of the Hebrew language.[29]

Viewed in the context of the evolving plans for the establishment of the Hebrew University, the timing of Einstein's burgeoning interest in this institution was far from coincidental. A conference had just been held in London by the Zionist Organization in late August and early September 1919 at which a formal plan for the establishment of the university had been presented by Hugo Bergmann, executive secretary of the Zionist Organization's Education Department. In contrast to previous plans, the new plan envisaged the establishment of a wide variety of research institutes to fulfill the educational needs of the local Jewish population in Palestine. At the conference, a university committee was selected, which immediately started to organize a conference of Jewish scholars to assist in planning the university.[30]

In his reply to Epstein in early October 1919, Einstein revealed that his relation to Zionism was an emotional one: "I was very pleased about your letter because the Zionist cause is very close to my heart." And his reply also reveals that his commitment to Jewish nationalism had grown since his last extant statement on this matter six months previously. At this point, he took a further step toward commitment by urging others to become active on behalf of the Hebrew University: in his letter, he urged Epstein to join the planned university and informed him that he had mentioned him the previous day to "a few gentlemen involved in the organization." This proves that Einstein continued to meet with Zionist functionaries. In his letter, Einstein reveals that he was aware of the larger picture regarding Zionism's aspirations and that he was not merely focused on the university project. He also predicts an influential role for himself in the plans for the university. And he again describes the Zionist project in emotional terms. Furthermore, he reveals another motive for his support, and this may have been the most crucial motivation of all: Zionism as a means to overcome national alienation:

At present, there are more urgent concerns that the founding of the university, of course. But I shall, no doubt, when the time is ripe, gain influence in the shaping of things and will then certainly think of you. I have great confidence in the enjoyable development of the Jewish colony and am glad that there should be a little patch of earth on which our ethnic comrades will not be considered aliens.[31]

Similar to other German Jewish intellectuals for whom Zionism provided a solution to the vexing issue of social and personal alienation,[32] Einstein revealed that he viewed Jewish nationalism as a means to overcome his own growing sense of strangeness. Earlier in this chapter we saw to what extent Einstein viewed himself as an "alien" in Germany and as being "rootless." And we saw in the previous chapter how deep-seated that sense of strangeness had been since his childhood. It is also highly significant that in this passage, Einstein employs the term "ethnic brethren" (*Stammesbrüder*), emphasizing once again the ethnic character of his affinity to his fellow Jews.

In this letter, Einstein also expresses his belief that both the funding for the university and its studentship would have to be derived from Diaspora Jewry and could not merely rely on local Jewish students in Palestine. This is the main motivation for his interest in the Hebrew University project at this stage—to provide a refuge for Jewish academics from Eastern Europe. Einstein clarifies he is aware the university would need to be supported financially by Diaspora Jewry. The student body would also come mainly from outside Palestine. As for the anticipated academic level of the university, it was "up to us to ensure that this University is on a par with its better-quality European sister institutions. There is certainly no lack of fine minds."[33] Most important, this letter reveals that Einstein's attention in Zionist matters was not focused on events in Palestine itself but rather on the interaction between the Jewish Diaspora and Palestine.

A week later, we find evidence of Einstein publicly lending his support for a Zionist cause for the first time: the Zionist functionary Julius Berger acknowledged that the prominent graphic artist and Zionist Hermann Struck had informed him "that you have agreed to sign the appeal for the campaign on behalf of the Palestine Foundation Fund." The appeal "For the Development of Jewish Palestine" called on German Jewry to participate in the planned public funding for creating an infrastructure in Palestine. The printed version of the appeal contained a long list of committee members, including many prominent German Jewish figures, among them Einstein.[34] His willingness to sign this appeal demonstrates that his commitment to certain aspects of the Zionist cause had moved from vague ideological support to a concrete action.

Two days later, a significant correspondence began between the Berlin and London bureaus of the Zionist Organization. Julius Berger informed Hugo Bergmann of the recent contacts between Einstein and the German Zionists. Intriguingly, there seemed to be some confusion about Einstein's exact whereabouts: "During his visit to Switzerland, Prof. Weizmann attempted in vain to contact Prof. Einstein personally. Dr. Berthold Feiwel is informed of this." This confusion was caused either by Einstein's two-month sojourn over the summer in Switzerland or by the London Bureau's erroneous belief that Einstein still resided in Zurich. In his letter, Berger went on to inform Bergmann that the Zionists in Berlin had met with Einstein personally on many occasions. According to Berger, Einstein was "very closely affiliated" with the Zionist movement and "thrilled about the idea of a Jewish state in Palestine and even more about the possibility of the Hebrew University in Jerusalem." The word "thrilled" is very telling in Berger's letter. It is a highly charged word and reflective of Einstein's increasingly emotional commitment to Zionism. This description of Einstein's enthusiasm for the Zionist cause is all the more interesting as it reveals a ranking within Einstein's interest in Jewish nationalism: the project of establishing the Hebrew University was obviously more significant to him than the establishment of a Jewish state. Berger went on to write that Einstein "would consider himself fortunate to be allowed to lecture at the University and is very concerned about whether this would be possible, as he is not conversant in Hebrew." Intriguingly, Einstein's foreseen commitment was described as "lecturing" at the university, yet not as becoming a lecturer there. Also, in light of Einstein's important position within the scientific community, the phrase "allowed to" is either indicative of Einstein's humility or (if one views this more cynically) of his coyness. Einstein's greater degree of commitment to the cause is also reflected in Berger's relating that Einstein had informed them that he knew of "several top-ranking Jewish scientists . . . who would be very pleased to follow a call to the Jerusalem University." Intriguingly, concerning himself, merely lecturing at the university was on the cards, yet in regard to others, the talk was of taking up full academic positions. Then, in a passage that introduces a new phase in Einstein's involvement in the Zionist movement, Berger urged the London Bureau, the up-and-coming central headquarters of the Zionist Organization, to contact Einstein directly. He added, "There is no need to draw special attention to the extraordinary importance of [this] scholar."[35] Within a few days, this new stage in Einstein's involvement would lead to a concrete initiative.

In the meantime, Epstein and Einstein continued their exchange on Zionist issues. On 15 October, Epstein referred to the motives for establishing the Hebrew University: in Palestine, there were high school students who required tertiary education, and in Poland, Jewish students were subjected to anti-Semitism and discriminatory quotas. This would have been highly meaningful for Einstein, reconfirming his view that Palestine in general, and the Hebrew University in particular, could function as a safe haven for young Eastern European Jewish academics. As we saw in the previous chapter, Einstein had been interested in finding just such a refuge for intellectuals, a "secular monastery" for like-minded individuals.

In his letter, Epstein also mentioned the next stage of Einstein's contacts with the leaders of the Zionist movement: "As I hear, Dr. Ruppin will be in Berlin in the next few weeks and will contact you in regard to university matters."[36] This was important, as Arthur Ruppin was director of the Palestine Bureau in Jaffa, and (as far as I can tell) this constituted Einstein's first meeting with a high-ranking functionary of the international Zionist Organization since his burgeoning interest in Zionism, as opposed to his meetings with local representatives of the German Zionist Federation.

A week later, Hugo Bergmann, secretary of the Education Depart-

Figure 3. Hugo Bergmann, 1926

ment of the Zionist Organization in London, informed Einstein about the concrete initiative which the Zionist Organization wished him to be actively involved in. Einstein had originally met Bergmann in Prague when he was a member of the Zionist circles that Einstein had maligned privately merely three years previously. Bergmann informed Einstein that a conference of Jewish scholars on plans to establish the Hebrew University was to be held at a neutral place, probably in Switzerland. The purpose of the conference was twofold: to obtain the scholars' expert advice and to attach their academic reputation to the project. Bergmann stated that this was only a private letter, asking Einstein whether he, "whom the world rightly calls the greatest Jewish scientist," would be prepared to accept an invitation to the planned conference. This initiative demonstrates that the conference of Jewish scholars (many—if not most—of whom would *not* be Zionists) was obviously a continuation of the Zionist Organization's policy to reach out to non-Zionist elements in the Jewish community. The aim was to get these Jewish non-Zionists (of whom Einstein was arguably the most prominent) to lend their support to projects that would not only interest the most committed adherents of Jewish nationalism. Moreover, it is remarkable that Bergmann alluded to Einstein's international renown two weeks prior to the public announcements regarding the astronomical verification of Einstein's general theory of relativity, which led to his worldwide fame. As for Einstein's personal role in the conference, Bergmann saw it as twofold: to participate himself and to catalyze other Jewish scholars to offer their support.[37]

In a letter to the prominent Jewish philosopher Martin Buber, Bergmann revealed another motive for convening the conference: it was hoped that it would play a role in raising funds for the university.[38] As for its timing, the scholars' conference was to complement and precede the annual Zionist conference (the first since 1913 to have representation from both previously Allied and Axis countries) to be held 18–23 January 1920. The general annual conference was to deal primarily with a peace treaty with Turkey, the relations between Palestine and England, and the involvement of the League of Nations.[39]

In his reply to Bergmann of 5 November 1919, Einstein first expressed his warm interest in the new colony in Palestine and "especially" in the planned university. He stated that he would "gladly do everything in my power for its cause." This demonstrates a far greater level of commitment than hitherto on Einstein's part. However, he was obviously still conflicted at this stage as to what the degree of his commitment should be, as he immediately qualified this statement by agreeing to participate in the conference "provided that circumstances permit it." Therefore, the process of Einstein becoming increasingly committed to certain aspects of the Zionist cause was clearly a gradual one, with advances and setbacks. As for Bergmann's request for recommendations of other scholars who might be interested in taking a role in establishing the university, Einstein named four: Paul Epstein, with whom, as we have seen, he had corresponded intensively on this subject; Paul Ehrenfest, with whom he had exchanged some correspondence on Zionist plans; and the Göttingen mathematicians Edmund Landau and Richard Courant.[40]

Einstein's reply was apparently deemed so significant by the Zionist Organization that a transcription was forwarded by Bergmann to Chaim Weizmann eleven days later.[41] This represented a further intensification of Einstein's involvement in the Zionist movement, as his correspondence was now being sent to the highest-level functionary in the organization.

Two days after Einstein's reply to Bergmann, on 7 November, sensationalist headlines appeared in the *Times* of London on the verification of the general theory of relativity.[42] Intriguingly, it was Hugo Bergmann who sent clippings of these reports to Einstein from the Central Zionist Bureau in London.[43] Thus, it seems that it was the Zionist movement that informed Einstein of the very beginnings of his worldwide fame.

In a letter to Paul Ehrenfest sent the very next day, Einstein revealed his main motivation for supporting the cause of the Hebrew University: "I believe a great future lies in store for this affair. This university will contribute toward less Jewish talent, particularly in Poland and Russia, having to go wretchedly to waste."[44] We have already seen how the plight of young Jewish Eastern European academics reverberated in Einstein, and this is reflected here in the intense language he employed to describe the crucial salivatory role the university could potentially fulfill.

Another important issue in this context is how Einstein viewed the Ostjuden, especially in contrast to Western Jews in general and German Jews in particular. At this stage it is not entirely clear that he would have agreed with representatives of the first and second generations of German Zionists, who viewed the Ostjuden as "better Jews" who were "in closer touch with their tradition and culture" than themselves. He may also not have concurred with members of the younger generation of German Zionists who idealized the Ostjuden.[45] But it is clear that Einstein felt a great deal of sympathy and even empathy for the young Jewish academics from Eastern Europe.

One month after his burst to fame in the Anglo-Saxon countries, Einstein became a household name in Germany as well. The caption accompanying a full-size portrait of Einstein on the cover of the *Berliner Illustrirte Zeitung* read, "A new great figure of world history."[46]

Einstein's new celebrity status did not go unnoticed among the Zionists in London. On 17 November 1919 Hugo Bergmann requested that the Berlin Zionist Bureau urgently send a picture of Einstein to the Zionist Organization in London as he was "*the* hero of the day" and "all newspapers are assailing us for it."[47] Thus, intriguingly, the Zionist Organization was functioning as a publicity agent for Einstein.

That same day, Arthur Ruppin met with Einstein in Berlin to discuss Einstein's recommendations regarding which Jewish intellectuals would be interested in the planned conference on the Hebrew University. The location of the conference was also discussed.[48]

Four days later, Hugo Bergmann informed Einstein that the conference would take place in Basel on 14–16 January 1920. More important, he acknowledged just how significant Einstein's participation in the conference would be for the university by stating that his "warm interest" in the project was "a guarantee for us that our plan will succeed." Furthermore, they were convinced that the university would not merely be a minor institution, but rather that it "would be worthy of constituting an intellectual center for Jewry throughout the world."[49] It seems, therefore, that the Zionists believed that Einstein's involvement in the university project actually legitimized their own endeavors: if a scholar of Einstein's caliber supported their efforts, then they must be heading in the right direction. Einstein and the Zionist movement very much needed each other at this juncture in time. Indeed, his personal needs and their organizational needs coalesced to form a highly advantageous constellation for both parties.

However, this newfound cause in his life gradually began to cause a certain amount of tension between Einstein and his close friend, Paul Ehrenfest. In a letter dated 24 November 1919, Ehrenfest subtly accused Einstein of not being sufficiently critical toward Jewish nationalism and of supporting a mass political movement. Ehrenfest declared that, in contrast to Einstein, he himself lacked "[an] unquestioning faith" toward Zionism. He was particularly "alienated" and "demoralized" by instances in which the movement organized "mass actions." In his letter, he also mentioned that his colleague, the Dutch physicist and prominent Zionist Leonard Ornstein, had heard of Einstein's interest in Zionism and wanted to discuss this with him.[50] We should not underestimate the importance of Einstein's scientific colleagues, most notably Paul Epstein, Ehrenfest, and Ornstein, to the formation of his views on Zionism. Besides his exchanges with prominent German Zionists such as Kurt Blumenfeld, these physicists provided Einstein with the opportunity to air his views on Zionism informally and receive feedback on these issues from fellow scientists, who formed his most important reference group.

In late November 1919 Einstein received a circular letter from Shmarya Levin, head of the Zionist Organization's Education Department in London, which constituted the official invitation to the planned scholars' conference on the Hebrew University. In his letter, Levin declared that the goal of the conference was to discuss the university's general plan and concrete steps for its establishment. The conference would hold both plenary sessions and committee discussions dedicated to the specific institutes that were to be established. Intriguingly, a Hebrew version of Levin's letter was also sent to Einstein.[51]

The organizers envisaged a concrete role for Einstein at the conference: British Zionist Selig Brodetsky sent a draft agenda for the conference to Solomon Ginzberg, secretary of the Zionist Organization's University Committee, and recommended that Einstein "act as chairman and then take an active part in the organization." The Berlin Zionist Bureau decided that Einstein would head the "mathematical-philosophical section" at the conference. However, interest in the conference among the invitees was only moderate: of the seventy-five invitations that were issued (and ten to fifteen more would follow), only thirty to fifty recipients actually agreed to attend the conference.[52]

In reply to Ehrenfest's letter of 24 November, Einstein stated that he thought the planned technical institute (i.e., the Technion) and the Hebrew University should not be two separate institutions. In reaching this conclusion, Einstein had apparently been influenced by another scientist, the German Jewish mathematician Constantin Carathéodory, "who is also occupied with the problem of an exotic university (Saloniki)."[53] It is intriguing (and revealing) that Einstein thereby classifies the Hebrew University as an "exotic university." This is another indication that Einstein viewed Zionist aspirations as a non-European endeavor and, in light of his anti-European senti-ments at the time, as therefore worthy of his support. As for Einstein's motives for attending the conference, he wrote to his close friend Michele Besso that he would participate "not because I think I am qualified but because my name, in high favor since the English solar eclipse expeditions, can be of benefit to the cause by having an en-couraging effect on lukewarm ethnic comrades." This tongue-in-cheek remark constitutes Einstein's first allusion to leveraging his sudden worldwide fame for ideological causes. It shows that Einstein was well aware of the use that was being made of his reputation for political purposes. This was obviously an arrangement that suited the needs of both Einstein and the Zionist movement. However, Einstein was apparently quite conflicted about attending the conference. He appended a note to the end of the letter that stated, "I have promised [to attend], but doubt that I will keep the promise. Too much blather-ing and unnecessary fatigue is attached to such occasions."[54] Einstein was possibly influenced in his hesitance by Ehrenfest's criticism of his involvement in Zionist activities and by his wife Elsa: in her post-script to Einstein's letter to Ehrenfest of 10 December, Elsa Einstein alluded to her husband's poor health as a reason for her wanting Ein-stein to refrain from attending the planned conference.[55]

In his reply to Einstein of 9 December 1919, Ehrenfest expressed his ever-deepening reservations about Zionism:

> I hear, for example, that your accomplishments are being used to make propa-ganda with the "Jewish Newton, who is simultaneously a fervent Zionist" (I haven't yet personally *read* this, but have only *heard* it mentioned). Or that one wants to establish here an association to combat anti-Semitism. All this is ter-ribly *damaging* for the future of Judaism . . . I cannot go along with the propa-gandistic fuss with its *inevitable* untruths, precisely *because* Judaism is at stake and *because* I feel myself so thoroughly as a Jew.

Here again, Ehrenfest voices his criticism of Einstein's involvement in Zionism, this time quite scathingly. Simultaneously, he characterizes

Einstein's endorsement of Zionism as "propagandistic fuss." He also indirectly criticizes Einstein for not having a sufficient amount of Jewish sentiment to avoid being manipulated by the Zionist movement. Possibly assuming that Einstein would not be amused by his remarks, Ehrenfest appended a note at the end of this passage that precedes the letter's postscript: "A reply to this letter is completely unnecessary!"[56] Unfortunately, it seems that Einstein complied, as no reply is extant.

In an interview published a week later in a Viennese newspaper, Einstein publicly expressed his view that theoretical physics should be the last subject to be considered in the new university's curriculum: "One does not start constructing a house with the roof." Thus, Einstein had begun to publicly voice his opinions on concrete institutional issues that were being discussed by the planners of the Hebrew University.[57]

By mid-December, Einstein had more or less decided not to attend the planned conference. In a letter to his colleague Felix Ehrenhaft, who had invited him to lecture in Vienna, he informed him that "with my unstable health I shall not be able to undertake the trip to Basle or to Vienna."[58] And in a postscript to a letter to his close Zurich friend and confidant Heinrich Zangger, the full scope of Einstein's internal conflict regarding the issue of whether he should attend or not becomes apparent: First he wrote, "I wanted to travel to Basel in January for consultations about the organization of the yet to be founded Jerusalem university. But it is too time-consuming and exhausting for me."[59] However, then he changed "wanted to" to "will" and crossed out the second sentence. This also makes it clear that ill health was not the real reason Einstein was hesitant about attending the conference. It therefore seems safe to conclude that it may well have been Ehrenfest's recent criticisms of his Zionist involvement that had the greatest impact on Einstein's decision whether to attend or not.

Merely a few days later, Einstein no longer had to torture himself about his attendance at the conference. Just three weeks before it was due to take place, the conference was suddenly postponed: "The conference in Basle will not be taking place, so I do not need to travel now."[60]

Various factors led to the postponement. On the political level, members of the Zionist Executive had to remain in Paris due to a planned peace conference with Turkey, and therefore they would not have been able to attend the scheduled annual conference or the scholars' gathering. On the organizational level, there were many negative responses to the invitations owing to the conference not taking place in the academic winter break. Poor communication between the Berlin and London Zionist bureaus also seems to have been a contributing factor.[61] Einstein was apparently unaware of the organizational factors. In a letter to Besso, he explained that the conference "has been postponed indefinitely—as it seems, due to serious political reasons."[62]

In the interim (probably to compensate for the postponement of the conference), a working committee was established by German Zionist leader Otto Warburg in December 1919 to generate interest on behalf of the university in the German-speaking countries and prepare initial plans for the research institutes. Einstein was named a member of the committee by the Berlin Zionist Bureau to generate interest on behalf of the university in the German-speaking countries. The additional committee members were other prominent Zionist and non-Zionist German Jewish academics, namely Leopold Landau, Eugen Mittwoch, Franz Oppenheimer, and Warburg himself.[63]

At the very end of December 1919, Einstein utilized his new fame to publicly voice his support for his fellow Jews from Eastern Europe in the daily newspaper *Berliner Tageblatt*. In reaction to calls by right-wing and anti-Semitic politicians for legal measures against the Jewish refugees in Germany who had recently fled Poland, Einstein protested these calls on humanitarian, political, and economic grounds. Most important for us, he argued that the 15,000 Eastern European Jews wanted to find refuge in Germany until they could emigrate to other countries. He expressed his hope that "many of them would find a genuine homeland in the newly-established Jewish Palestine as free sons of the Jewish people."[64] This was Einstein's first public statement in support of both Jewish emigration to Palestine and of the establishment of a Jewish entity in Palestine. Of note, however, it constituted a call for the emigration of Ostjuden to the Jewish homeland, not of German Jews like himself.

In mid-January 1920 the issue of whether Einstein was actually a Zionist or not (especially in light of such public statements) was becoming of increasing interest to an ever-widening audience: Charlotte Weigert, a family acquaintance, sent Einstein a Danish newspaper clipping that characterized him as a "zealous Zionist." In her accompanying letter, she asked, "Are you, incidentally, really a Zionist? I never heard that before and would be interested to know!" In response, Einstein wrote,

> My Zionism is not such a terribly serious matter. But I am pleased that our people will once more have their own home and I am particularly interested in the yet to be established university in Palestine. It is even alleged that I myself want to emigrate to Jerusalem; but this is *also* a legend, as are many other things that are published about me. I am also too old to feel comfortable with the Hebrew language, which is almost entirely foreign to me.[65]

Einstein simultaneously reveals his involvement with and distance from Zionism in his reply to Weigert: he claims his affinity with the movement by writing "*My* Zionism" (my emphasis), yet also seems to weaken his affiliation with the words "is not such a terribly serious matter." Furthermore, he implicitly denies that he is a "zealous Zionist" by referring to the many "legends" that have published about him in the press. As we will see elsewhere, Einstein also distances himself from the Hebrew cultural renaissance advocated by the Zionist movement by describing the language as "almost entirely foreign to me."

During this period, rising anti-Semitism in Germany was increasingly becoming an issue Einstein could not ignore. This was especially true in academic circles. Following his signing a petition in January 1920 of a few Berlin professors in support of the German Jewish academic and pacifist Georg Nicolai and against "skirmishes" at his lectures, Einstein had to contend with an anti-Semitic incident in his own lecture hall the very next month.[66] In an attempt to find an impromptu (if limited) solution to the influx of young Eastern European Jewish students in Berlin, Einstein threw open his lecture hall to unregistered students, some of whom were from Eastern Europe. This antagonized the more anti-Semitically inclined students among his audience and led to an "uproar" in his lecture hall on 12 February 1920 during which Einstein's lecture was disrupted and he was forced to abandon the auditorium. In a statement published the next day in the *8-Uhr Abendblatt* newspaper, Einstein declared that he would continue his lectures "in a different format," yet threatened that he would cease lecturing altogether if such scenes reoccurred. To his mind, no anti-Semitic remarks were expressed by the protesting students, but there may have been such an "undertone" in their interjections. However, the anti-Semitic nature of the "uproar" was acknowledged by both the right-wing and left-wing press.[67] It is interesting that Einstein apparently decided to underplay the anti-Semitism of the incident. It is not clear whether this was a form of denial or whether he did so for tactical reasons.

In any case, Einstein decided to move on this issue with uncharacteristic swiftness: one week later, he and Leopold Landau petitioned the Prussian Ministry of Education to allow the courses of Berlin University professors to be accredited by the state. Three weeks later, the Social Democratic minister Konrad Haenisch agreed to such a state accreditation.[68] For our purposes, Einstein seeking non-Zionist (and institutional) alternatives to ease the plight of Eastern European Jewish students is a clear sign that despite his recent enthusiasm for Zionism, his commitment to the movement was definitely still limited at this point in time and by no means exclusive.

Even with the postponement of the Jewish scholars' conference, plans for the Hebrew University proceeded: in mid-January 1920, Bergmann informed Einstein of recent developments toward establishing the university. Most significantly, funds for the purchase of a first building on Mt. Scopus had been approved by the Zionist Organization's Action Committee. A brochure for "propaganda" purposes was also to be prepared. The goal of the propaganda was twofold: to disseminate the idea of the university among "broad strata of the people" and to raise the necessary funds required for further purchases of property and for books and laboratory equipment. Bergmann also asked Einstein "to write some words for this brochure and to express your views on the necessity ... of a Hebrew University in Jerusalem."[69]

Einstein prepared a draft for the propaganda brochure, yet it does not seem to have been published. The statement reconfirmed Einstein's emotional attachment to the university project by expressing his "warm joy"; moreover, it demonstrated his ideological commitment by opining that the university would provide both an institutional solution for the Jewish students of Eastern Europe and contribute to the education of young academics in Palestine.[70] The statement also reveals that Einstein believed the university would become the intellectual center of Palestine and would contribute significantly to its cultural development. In fact, he uses the same term ("intellectual center") which Bergmann had employed two months previously. Einstein voices his concern that the university not contribute to isolationist tendencies within Zionism, thus staking a claim within Zionist ideology for a specific universalistic brand of Jewish nationalism, one that opposes narrow-mindedness and closed thinking. Einstein thereby clearly objects to nationalist interests having priority over universalistic research interests. Perhaps most important, Einstein sees the university as potentially constituting "a new holy sanctuary" for the Jewish people. With this Einstein's commitment to the university project shifted to a higher level. We saw in the previous chapter how science (*Wissenschaft*) and research (*Forschung*) became imbued with "holiness" as part of Einstein's secularized worldview and his forsaking of traditional religion. By this stage, Einstein has embraced the university project fully as his own, as part of what constituted a holy endeavor for himself and the Jewish people. It is also noteworthy that Einstein uses the phrase "our people" in his statement. It is the first time since his newfound support for specific aspects of Zionism that he refers to the Jews as "our people," thus exhibiting a greater degree of attachment to his ethnic background.

In reaction to Ehrenfest's complaint that he had not heard any more news about the university, Einstein informed him in early March 1920 that he was impressed by the obvious progress made and expressed his belief that one could rely on those involved in the planning of the university at the London central bureau of the Zionist Organization. He also revealed that at this stage of his involvement, his thinking about the university was quite abstract: "In the end, the people matter more than the formal structure."[71]

At the end of March 1920, possibly in reaction to the "uproar" at Einstein's lecture the previous month, Julius Brodnitz, chairman of the pro-assimilation Central Association of German Citizens of the *assim group* Jewish Faith (the Centralverein), invited Einstein to attend a discussion of Jewish lecturers "for the purpose of discussing the tactically most appropriate way to educate Germany's university community."[72] Thus, Einstein was being invited to another gathering of Jewish scholars, this time by the assimilationist adversaries of the German Zionist Federation. The wider background to the discussion was the rising wave of anti-Semitism in Germany, which reached a climax in the failed Kapp Putsch of 13–18 March 1920 in Berlin, during which Jewish passersby were subjected to anti-Semitic attacks and physical assaults.[73] Intriguingly, prior to contacting him the Centralverein had apparently requested background information on Einstein and the extent of his interest in Jewish matters from a distant relative by the name of I. Moos.[74]

By early April 1920 Einstein was ready to share his newly developed positions on Jewish nationalism, anti-Semitism, and assimilation with a wider public. He published an article on anti-Semitism in the daily newspaper the *Berliner Tageblatt*. In one of the two drafts of the article, Einstein provides his definition of Jewish nationality: for him it is a "community based on race and tradition." Einstein clarifies that by race he means both "lineage" (*Abstammung*) and "a sense of being different," which leads to a "feeling of alienation" already in the young Jewish child. This is clearly an ethnic definition of Jewish nationhood, not a religious one. However, it is not certain to what extent Einstein was influenced by his contacts with German Zionists in postulating this ethnic definition; as we have seen, he had already conceived of his Jewish identity in ethnic terms on previous occasions. Moreover, according to Einstein, the traditions that segregated the Jews from the Gentiles were mostly not religious in essence. Yet he does not characterize these traditions apart from stating they are predominantly not religious. Therefore, he does not present a *positive* definition of Jewish culture. The article also reveals a very positive view of the Ostjuden: "Eastern European Jewry contains a rich potential of the greatest human talents and productive forces that can well stand the comparison to the higher civilization of Western European Jews."[75] Thus, even if at this stage, Einstein does not subscribe to the romantic idealization of the Ostjuden that was prevalent among many German Zionists, he certainly sees them as at least coequal partners with the Jews of Western Europe.[76]

Early April also saw Einstein's most scathing rejection of the assimilationist worldview to date. In his reply to the Centralverein, Einstein rejected their invitation to discuss ways to combat academic anti-Semitism, as he did not believe in the success of such an undertaking. He, and the ideological movement he was now associated with, had other priorities. He dismissed the Centralverein's position that it was the Jews' responsibility to convince the Gentiles of the errors of their ways. Instead, Einstein believed it was the Jews' own "servile disposition" that needed to be combated. He therefore advocated "[m]ore dignity and independence in our ranks!" To his mind, only when Jews respect themselves and "dare to view themselves as a nation" would they earn the respect of others. He went on to claim that anti-Semitism as "a psychological phenomenon" would exist "as long as Jews come into contact with non-Jews." He saw no harm in this and even believed it was this exposure to animosity that had contributed to the survival of the Jews "as a race." Thus, by now Einstein subscribed to the perception advocated by the proponents of the younger generation of German Zionists that the Jews should focus on their own national dignity and self-respect rather than on anti-Semitic attacks by non-Jews. In his response to the Centralverein, Einstein also clarified his definition of Jewish identity. Similar to his recent article in the

Berliner Tageblatt, he rejected a religious definition of Jewishness and embraced an ethnic one that was clearly influenced by Zionist ideology. He also accused the Centralverein members of denying their solidarity with their Eastern European fellow Jews:

> I cannot suppress a pained smile whenever I read "German citizens of the Jewish faith." What is at the bottom of this nice characterization? What, then, is Jewish *faith*? Is there a kind of non-faith that makes one stop being a Jew? No. But there are two confessions of noble souls hidden in this characterization, namely (1) I don't want to have anything to do with my poor Eastern European Jewish brethren; and (2) I don't want to be seen as a child of my people, only as a member of a religious community.

Einstein goes on to claim that the Jews should not be concerned with what non-Jews think about them. He indirectly attacks such a position as being one of "pussyfooters." In the letter, Einstein also reconfirms his Jewish identity in a positive manner and once more rejects his German identity:

> I am neither a German citizen nor is there anything in me that can be described as "Jewish faith." But I am a Jew, and I am glad that I belong to the Jewish people, even though in no way do I consider them to be the chosen ones. Let us leave anti-Semitism to the goy and let us keep the love of our brethren.[77]

These more pronounced statements by Einstein on Jewish identity, nationality, and nationalism raise the question to what extent Einstein was influenced by the ideological movement he was becoming associated with. To examine this issue, we need to differentiate between the ideologies of the two generations Einstein encountered within the German Zionist Federation. Whereas the members of the first generation were born in the late 1860s and early 1870s, the members of the second generation were born in the mid- to late 1880s. Einstein was born in between the years in which the members of the first and second generations were born, and his ideological leanings seem to have been located somewhere in between those of the two groups as well. Einstein's definition of nationality seems to be more

*an
inter-
national-
ism*

closely related to that of the first generation of German Zionists, for whom the main focus of their Zionism was the "common bond of Volk and nationality" they felt with Jews throughout the world. He would have subscribed to the first half of this generation's belief that the Jews "had the right to preserve their own ethnicity while remaining loyal Germans," yet he would have rejected the second half of that belief, or rather he would have belittled it. On the other hand, he seems to have been indifferent to the second generation's support for the creation of a separate Jewish and Hebrew culture. In his focus on alleviating the suffering of the Ostjuden, his Jewish nationalist sentiment was, in part, a "Zionism out of pity," as critics of the first generation's ideology called it. However, that was only part of the story. Einstein was definitely one of "those assimilated German Jews who had discarded Judaism with its traditions but continued to sense a lack in their personal lives." He would have agreed with the second generation's belief that "the individual Jew was rootless in Germany."[78] We have already seen how Einstein viewed himself as a rootless individual within the context of German society and as lacking a sense of belonging to a physical location or national entity. Like members of the second generation, Einstein had encountered anti-Semitism as a child and teenager, yet unlike them, he had not encountered academic anti-Semitism in Germany during his student years. At that time, he was attending the Zurich Polytechnic. Even though he did (at least in part) ascribe his difficulties in securing an academic position after completing his studies to academic anti-Semitism at German-speaking universities, his first personal encounter with virulent anti-Semitism in academia did not occur until early 1920. Nevertheless, he did become convinced during this period, as did the members of the second generation of German Zionists, that anti-Semitism was inevitable, and therefore there was "no need to fight the anti-Semites." Einstein also subscribed to the second generation's view that Germany was "merely . . . a temporary stopover on the way to Zion," but in contrast to that generation, he did not regard it as a stopover for himself but rather only for the Ostjuden. What did appeal to Einstein in Zionism was that it "offered the psychologically rewarding affirma-

tion of Jewish identity." He would definitely have concurred with Blumenfeld's concept of *Distanz* (distance), whereby "the ethnically and
nationally conscious Jew deliberately had to maintain a certain distance between himself and the German world."[79] However, because of
his reluctance to adopt ideological causes outright, Einstein maintained a distance even from the Zionists and would not commit himself entirely to their ideology, at least during this period. And in light
of his position midway between the ideologies of the first and second
generations of German Zionists, a slightly facetious way of describing
Einstein's support for Jewish nationalism and his position vis-à-vis
these two generations would be to see him as tending toward the ideology of the first and a half generation of German Zionists.

The central question at this stage in Einstein's involvement in Jewish
nationalism is to what extent Einstein's mobilization on behalf of
Zionism was a "conversion," and what the degree of his commitment
toward that ideological movement was. In other words, how much
of a Zionist did Einstein become during this initial period of his
involvement?

Conversions are seen by sociologists as changes of worldview.
They used to be seen mainly as "sudden crisis events," but newer
scholarship has discovered that the process can also be a gradual one.
In any case, a conversion "represents a transformation in a person's
self-image." One major theory on conversions is Rosabeth Kanter's
commitment theory. This theory differentiates between three levels of
commitment: commitment to the organization (instrumental commitment), commitment to other persons in the group (affective commitment), and commitment to "the rules, regulations, ideas, and mores
of the group" (moral commitment). These are forms of commitment
"to the institutional system, the belonging system, and the meaning
system, respectively."

What was the degree of Einstein's commitment to Zionism on
each of these levels? Instrumental commitment demands both sacrifices by the individual on behalf of the organization and investment in
the organization. During the initial period of his involvement, the

level of Einstein's sacrifices and investment went from minimal—lending his name to an official invitation, joining a provisional committee—to moderate: urging others to become involved, being prepared to attend a conference (although we have seen that he was clearly conflicted as to his participation in the planned gathering). The level of his instrumental commitment during this initial stage was by no means strong: he did not join the German Zionist Federation officially, thereby maintaining his distance, and he definitely did not take on any formal roles within the movement. The organization per se was not what his involvement was about at this stage.

As far as Einstein's affective commitment is concerned, his emotional dependence on the Zionists oscillated between minimal and moderate. When affective commitment is strong, the members of an organization become the individual's "primary set of relations." Yet this was by no means the role the German Zionists played for Einstein. Affective commitment involves both a detachment from former ties and a change of reference group, as well as an "emotional solidarity with others."[80] The Zionists did not become Einstein's "surrogate family"; yet there was actually a more complex process at work here. Einstein saw the German Zionists as an instrument he could use to further his own goals, namely, his engagement on behalf of young Eastern European Jewish academics. It was these young academics who to some extent became Einstein's surrogate offspring. In the autumn of 1920, Einstein voiced the rather shocking statement that he did not see his own sons as his "temporal continuation."[81] By this stage Einstein believed that his own role on the frontier of scientific discoveries was practically over and that the next generation of academics would take over. Thus, Einstein's primary affective commitment to Zionism was not to Kurt Blumenfeld and his associates but rather to those to whom he and the Zionists could extend their assistance—the young Jewish academics from the East. However, there was yet another process at work here: because Einstein saw Zionism as a means to overcome his own personal alienation, his personal ties with the Zionist functionaries, especially with Blumenfeld, were definitely of some significance to him.

Einstein's moral commitment to Zionism also went from minimal to moderate during this initial period. Einstein was certainly not a full-fledged ideologue on behalf of Jewish nationalism. He carefully chose which tenets of Jewish nationalism he wanted to subscribe to. Moral commitment gives the adherents of a belief system a "sense of ultimate purpose and meaning."[82] Jewish nationalism certainly gave Einstein a certain sense of purpose and meaning, but not an ultimate sense. That decisive aspiration was reserved for his scientific pursuits. Even when he described the university project as being "a new holy sanctuary," it was endowed with such holiness because of its potential scholarly role, and not because of its ideological or national one.

But was Einstein's turn toward Zionism a one-time event or a gradual development? In the late 1970s, sociologist John Lofland developed the process model of conversion. In contrast to previous models, this model does not view conversion as a single event that brings radical change but rather as a gradual process. From the available sources, it is clear that even Blumenfeld saw Einstein's "transformation" as a gradual process. In his model, Lofland postulates four situational contingencies that influence the social interaction between potential converts and the group into which they are converted. The first is a turning point in life. There were certainly many of these in Einstein's life during this period: the fall of the Kaiser and the proclamation of the new republic, the finalization of his divorce from his first wife and his marriage to his second wife, his mother's death, and his sudden access to worldwide fame overnight. The second contingency is close intragroup affective bonds: recruits are usually "gained through pre-existing friendship networks." Whether this was the case or not in regard to Einstein's induction is unclear. If the initial contacts between Einstein and the Berlin Zionist Bureau came about owing to Einstein's acquaintance with Hans Mühsam, then that would be the case. However, if Blumenfeld's account of his fellow Zionist Felix Rosenblüth suggesting Einstein's name is accurate, then preexisting friendship networks did not play a role in Einstein's induction. The third contingency is the "weakening of extra-group affective bonds." As we have seen, Einstein's affective commitment to the Zionist functionaries and the subjects of their activities that is, the Ostjuden—was quite a complex one. The final contingency is an intensive interaction with other group members and a sense of unity with its goals.[83] This was rather weak in Einstein's commitment to Zionism and does not seem to have played a major role during his initial involvement.

Einstein was not the only prominent figure to turn to Zionism during this period. Other famous individuals about whom scholars have had animated discussions on their path to Jewish nationalism are the founder of political Zionism Theodor Herzl, U.S. Supreme Court justice Louis D. Brandeis, and the Czech Jewish author Franz Kafka.

Unlike Einstein, Herzl experienced blatant anti-Semitism at the University of Vienna during his student years, leading him to resign from the pan-German student association to which he had belonged. And in contrast to Einstein, who was at the apex of his career when he became active on behalf of Zionism (yet who had begun to feel that his productive scientific output was waning), Herzl has been perceived as turning to politics after having failed as a playwright. Similar to Einstein, Herzl was very critical of the Jewish bourgeoisie and viewed Zionism as a way of overcoming its "corruption and decadence." Furthermore, they both came to reject assimilation as a viable solution for the Jews. As Herzl put it, "We are still decried as strangers." We have seen how a similar sense of alienation played a major role in Einstein's turn toward Jewish nationalism. Standard accounts ascribe great weight to the treason trial of the French officer Alfred Dreyfus in Herzl's conversion to Zionism. However, recent research confers more importance on the impact of anti-Semitic politics in Vienna during the mid-1890s, especially the election of Karl Lueger as mayor. Moreover, Herzl is seen as having passed through two stages on his path to Zionism—first a growing awareness of the "Jewish problem" and then later, after Dreyfus's second trial, the adoption of a Zionist program. When he wrote his seminal work *The Jewish State*, he was approximately the same age as Einstein was when he forged his initial contacts with the German Zionists—both were in their late thirties.

In general, their aims for Jewish nationalism were similar: both viewed it as a means to modernize Jewish culture and society. Interestingly, the situation of the Jews in Prague played an important role for both Herzl and Einstein. Their being scorned by both German and Czech nationalists led Herzl to the perception that the Jews had nowhere to escape to. We have seen that Einstein's sojourn in Prague awakened in him a greater realization of the political situation in which the Jews found themselves. For Herzl, finding a political solution to the "Jewish question" became an increasingly urgent necessity, as he saw in intellectual anti-Semitism "the seeds of the destruction of European culture."[84] Even though his turn toward Jewish nationalism occurred more than twenty years after Herzl's (and under entirely different circumstances), discouragement with European politics was also a highly significant factor for Einstein. However, the most important difference between Einstein and Herzl vis-à-vis Jewish nationalism was that while Herzl developed a fully fledged ideological program, Einstein's commitment was far more tentative and restricted.

Justice Louis D. Brandeis's conversion to Zionism came far later in his life than Herzl's and Einstein's turn toward Jewish nationalism: he was roughly twenty years older. Similar to the change in Einstein's views, Brandeis's transformation has puzzled historians. American scholar Sarah Schmidt has stated that he changed from "a typical American, assimilated Jew and an outspoken foe of the dual loyalties assumed to exist among so-called hyphenated Americans, into an active partisan of a Jewish state." Like Einstein, Brandeis originated in a nonobservant family, yet the future justice's social status was far more elevated than Einstein's. Another similarity between them is that they both came to Zionism at the height of their careers. Israeli scholar Allon Gal believes that (like Herzl) Brandeis actually underwent two transformations: first "from a marginal Jew to a positive and committed member of the Jewish community" in 1905 and then to an ardent Zionist in 1913. Another interesting parallel between Einstein and Brandeis was the impact their contact with Eastern European Jews had on their affinity with Zionism. For Brandeis, it was his encounter with the New York garment workers' unions in 1910 that brought about his first contact with a Jewish collective, which he came to see as "a living model of democracy." His first public appearance on behalf of Zionism occurred in 1913. Dramatically, he took on the leadership of the Zionist Organization of America merely one year later. Schmidt believes it was Brandeis's perception that even America "had failed to eliminate the anti-Jewish prejudice" that was a decisive factor in bringing about his support for Zionism. Like Einstein, the return of all Jews to Palestine was not the primary purpose of Zionism for Brandeis. It was more "the effect which the return to Palestine would have on the lives of Jews who chose to live outside the national homeland." And like Einstein, Brandeis saw Palestine as potentially an exemplary society for the other nations of the world. Schmidt attributes great importance to Brandeis's adoption of the American Jewish philosopher Horace Kallen's vision of Palestine as a model utopian democracy that would embody "the same concepts of liberty and equality that symbolized America for him." She claims this resolved the problem of dual loyalty for Brandeis—he could advocate that being a better Jew was a means to becoming a better American.[85] In this, Brandeis's case was quite dissimilar to Einstein's, who was not at all interested in a solution to the issue of dual loyalty. He certainly did not view being a better Jew as a means to being a better German. As he did not define himself as a German during this period, he did not believe he had to solve the issue of a divided loyalty. Moreover, Einstein does not seem to have perceived any tension between his ethnic identity as a Jew and his national identity as a Swiss citizen.

The major difference between Kafka's relationship to Zionism and the ties of Einstein, Herzl, and Brandeis to the movement is that, in his case, there was no public dimension to his involvement. Indeed, as his close friend and confidant Max Brod described it, Kafka's Zionism was not practical or political but rather "private." Many scholars have noted both Kafka's yearning to belong to a community and "his fear of dissolution as a self in a group." However, in both Einstein's and Kafka's cases, there has been much discussion about the extent of their commitment to Zionism. Some of Kafka's closest friends (who were themselves Zionists) claimed he became a full-fledged Zionist. British scholar Ritchie Robertson has claimed "there are both proofs of Kafka's anti-Zionism as well as his pro-Zionism." Kafka biographer Ernst Pawel has argued that "Kafka never allows himself to be encapsulated in a simple category." Literary critic Robert Alter has said that Kafka exhibited a "growing interest in Zionism, which like all the important interests in his life was oscillating and ambivalent." And German scholar Hans-Richard Eyl has argued that "Kafka was just as incapable of making a choice with regard to Zionism, as he was with regard to many other realities of life, for example, marriage, profession, or living independently."[86]

Like Einstein (and Brandeis, for that matter), Kafka viewed Zionism as a means to something else: "Zionism . . . is merely the entrance into that which is more important."[87] And like Einstein and Herzl, Kafka was sharply critical of assimilated Western Jewry. He famously described his father as a "four-day Jew,"[88] and saw Western Jewry in a state of decay. As with Einstein and Brandeis, his encounter with Eastern European Jewry (in his case during the years 1910–12) brought about a significant transformation in Kafka's relationship to his Jewish identity. He was greatly intrigued by a Yiddish theater group from Lemberg that toured Prague. The troupe revealed a new kind of Jewishness to Kafka. Yet the resulting heightened interest in Jewish matters did not lead him to outright Zionism, of which he said, "I admire Zionism and am disgusted by it." In line with his deep feeling of solitude, Kafka saw himself as standing outside tradition, history, and society. Therefore, like Einstein's close friend and colleague Paul Ehrenfest, he felt he could not belong to such a collective endeavor as Zionism. Like Einstein and Brandeis, Kafka did not believe that Palestine offered a practical solution for *all* Jews; it was "not the only and obligatory place to be for a Jew." Dutch scholar Niels Borhave has described Kafka's attitude toward Zionism as "alternat[ing] between irony and scepticism to antagonism." In contrast to Einstein, Kafka took an active interest in the Hebrew cultural renaissance: in 1922 he took lessons in Hebrew with Puah Ben-Tovim, a nineteen-year-old student from Palestine. The following year he moved to Berlin and began living with Dora Diamant, with whom he possibly planned to emigrate to Palestine. However, Borhave has claimed that the pull of Eastern European Jewry played a stronger role for Kafka and that he actually wanted to go east to Galicia from where Dora originated.[89] His death the following year made all these plans irrelevant in any case.

We have seen how Einstein affiliated himself with the Zionist movement in the immediate aftermath of World War I. By all accounts, he had to overcome a considerable amount of initial resistance before he could allow himself to become associated with a political ideology he had originally scorned. Merely two years earlier he had derided Zionism and viewed it as being no different than other forms of nationalism. This sea change in Einstein's views necessitated his conceiving of Jewish nationalism as qualitatively different from other nationalistic ideologies. This required a fair amount of special pleading on Einstein's part: as the Jews had been stateless and persecuted for so long, he was willing to be more tolerant of their nationalist aspirations. Thus he could have his national cake and eat it too.

In line with his skepticism toward political ideologies and mass movements in general, Einstein did not embrace Zionism unreservedly, and maintained a considerable amount of distance from its central tenets and its charismatic leaders during this initial phase of his involvement. Yet Zionism provided Einstein with a renewed sense of purpose at a time when he felt great disenchantment with Germany and with Europe in general. Jewish nationalism clearly appealed to Einstein on an emotional level. He could give expression to the deep-seated affinity he felt with his "ethnic comrades." Yet there was also quite a degree of expediency in his alignment with the Zionist movement. He sensed that this was a movement that could further some of the goals of utmost importance to him: the creation of a refuge for Jews from Eastern Europe and the elevation of the self-esteem of his fellow Jews in the West so that they could overcome their "servile disposition" toward their host countries. By the end of the gradual process in which Einstein adopted some of the main tenets of Jewish nationalism, his commitment to Zionism had progressed to the point that he was ready to become actively involved in a major educational project that was of prime importance to both himself and the movement with which he was now aligned. This more intense activity on behalf of the Zionist movement would soon bring him into contact with some of its most fervent supporters in the New World.

CHAPTER 3

THE "PRIZE-WINNING OX" IN "DOLLARIA"

Einstein's Fundraising Trip to the United States in 1921

"Harvard absolutely declines Einstein."[1] This dire statement was sent by American Zionist leader Julian Mack to the president of the international Zionist Organization, Chaim Weizmann, at the end of March 1921. The telegram came at precisely the time Einstein was crossing the Atlantic on his first voyage to the New World. But how did the planned visit of such an illustrious scientist elicit such an emphatic response from one of America's most prestigious universities? And what did all this have to do with Zionism?

Following his intense involvement in Jewish affairs in late 1919 and early 1920, which led to his mobilization on behalf of the Zionist movement, Einstein's preoccupation with Jewish nationalism seems to have receded for a few months. However, this does not mean he did not have to deal with Jewish issues at all during this period. Following the uproar in his lecture hall in February 1920, a major clash between Einstein and the emerging antirelativists, some of whom were also fervent anti-Semites, occurred at the Berlin Philharmonic in late August 1920.[2] Persistent rumors of Einstein's intention to leave Germany ensued, only to be denied. The following month he had to defend himself in a major confrontation with the antirelativists at the annual conference of the Society of German Natural Scientists and Physicians in Bad Nauheim in southern Germany.[3]

Einstein renewed his activities on behalf of Zionism with increased intensity in the spring of 1921. Invited by Weizmann to join him as part of his delegation to the United States, Einstein ventured beyond European soil for the first time. This intense and hectic tour would last two months.

Einstein's first trip to the United States has been viewed by historians of Zionism as a small if interesting footnote to two major events in Zionist history that occurred in the spring of 1921: the visit to North America by Zionist leader Chaim Weizmann and other high-ranking members of the European Zionist Executive and the imbroglio over the Keren Hayesod (the Palestine Foundation Fund) between the rival factions of Weizmann and American Zionist leader Louis D. Brandeis. This fund was to be the Zionists Organization's central means to raise funds for immigration and settlement in Palestine. Weizmann's plan to establish an office of the fund in the United States during his visit was seen as a provocation by the Brandeis faction. Control of the funds raised in the United States for Zionist colonization in Palestine was at the core of the conflict. Historians have pointed to the clash between the personalities of the major protagonists (most notably Weizmann and Brandeis), their differing mentalities, and differing organizational cultures as causing the rift between the factions.[4]

Einstein scholars tend to view his U.S. trip as a direct result of his interest in the Hebrew University and as playing a major role in deepening Einstein's ties with his American scientific colleagues.[5] Yet neither historians of Zionism nor Einstein scholars have explored the intricacies of the Zionist aspects of Einstein's trip to the United States. Furthermore, these studies have not examined the plethora of material on the trip available in the various relevant archival repositories.

Often overlooked by both Einstein scholars and historians of Zionism is the intriguing fact that the trip, which eventually took place under Zionist auspices, was preceded by Einstein's plans for a personal trip in the form of a comprehensive lecture tour. In October 1920, both Princeton University and the University of Wisconsin invited Einstein to hold extensive lecture series at their universities.[6] Encouraged by his close friend Paul Ehrenfest to demand exorbitantly high fees for the proposed tour, Einstein seems to have hoped that by requesting $15,000 from each institution, he would succeed in "frightening them off," as he apparently preferred to remain at home.[7] Yet there were two major reasons for him to demand such high honoraria. The first was a personal reason: to secure adequate funding for the future needs of his first family in Zurich by achieving "economic freedom." The second pertained to the status of science in Germany: in Ehrenfest's opinion, the prestige of "German science" would only be advanced if Einstein were to be invited by the "2–3 *most prestigious* universities in America."[8]

Even though both Princeton and Wisconsin were very enthusiastic about Einstein's proposed lecture series, his pecuniary demands turned out to be too extravagant for both universities. In late December, Princeton's president informed Einstein that his university could not afford the honorarium of $15,000 Einstein had requested. Dean Charles Slichter of the University of Wisconsin deemed the proposed fee "so preposterous" that he informed the president of the University of California, Berkeley, the third university at which Einstein was to lecture, that he would "not tak[e] the trouble to send [him] a copy of the documents in the case."[9] Slichter then informed prominent German American banker Paul M. Warburg, who was representing Einstein in his financial negotiations with the U.S. universities, that "the finances involved are quite beyond the ability of our American institutions to meet." Paul Warburg's brother, Max M. Warburg, subsequently informed Einstein that the U.S. universities could not afford the honoraria Einstein had sought.[10] However, that was not the end of the matter. As we will see, these negotiations for a personal lecture tour in the United States were to have a considerable impact on the arrangements for Einstein's lectures at American universities, which were now to be planned under Zionist auspices.

In mid-February, Einstein expressed his relief to Paul Ehrenfest at not having to embark on the planned voyage: "I am glad not to have to travel [to America]; it really isn't a pretty way to make money, and otherwise truly not a pleasure."[11] Yet within three days, all this would change quite dramatically.

Einstein first learned of Chaim Weizmann's invitation to join him on a tour of the United States around 16 February 1921, when the German Zionist leader Kurt Blumenfeld brought him a telegram he had received from Weizmann.[12] However, the invitation to Einstein had been preceded by intricate planning behind the scenes for a first tour by Weizmann of the United States, to which Einstein was not privy.

Weizmann had first planned to visit the United States in February 1920 as a consequence of the growing discord between European and American Zionists over the planned formation of the Keren Hayesod. However, Brandeis and his supporters feared that a visit by Weizmann or his followers would strengthen the local American opposition to their leadership of the Zionist Organization of America. A few months later, the conflict between Weizmann and Brandeis and their followers over the Keren Hayesod erupted at the Zionist Conference in London in July 1920.[13] As tensions mounted and it became clear that Weizmann intended to proceed with his trip in early 1921, the president of the Zionist Organization of America, Julian W. Mack, even asked him to refrain from visiting the United States, as it "would be 'unseemly' . . . for a British Zionist to appear to be directing the Zionist movement in the United States." However, Weizmann decided to proceed with the planned trip nevertheless, arguing that its "sole purpose . . . was to achieve a compromise with the American Zionists."[14] Whether Weizmann sincerely wanted to achieve a compromise or whether he wanted to impose his will on the Brandeis faction is a matter for debate. However, it is apparent that Weizmann believed it was time to "enlist the help of the whole Jewish people in the task of building up Palestine."[15] This provides us with at least part of the motivation for Weizmann inviting Einstein to join him on his tour of the United States, as he apparently believed that Einstein, the prominent Jewish non-Zionist, would have great appeal to American non-Zionist Jews.

After receiving Weizmann's telegram, Blumenfeld immediately requested an audience with Einstein. In his telegram, Weizmann (who had recently returned from a trip to Palestine) informed Blumenfeld that he was planning to depart for the United States in mid-March to organize a large-scale fundraising campaign for the development of

Palestine. He asked Blumenfeld to urgently forward his telegram to Einstein, as he could not cable him directly, lacking Einstein's private address. And he stressed that Einstein's involvement "will contribute toward guaranteeing the success of the general campaign and in particular [for the] University."[16] Weizmann's emphasis on Einstein's beneficial role for the Hebrew University is similar to previous statements (by Einstein and others) in which his interest in the university project clearly outranked more general colonization efforts in Palestine. However, it is notable that Weizmann envisioned Einstein playing a substantial role in the planned general campaign—thereby belying the claim made in most previous studies that the sole purpose of Einstein's participation in the Zionist delegation was to raise funds for the Hebrew University.

Three days later, Blumenfeld informed Weizmann that Einstein was prepared to travel with him to the United States.[17] A day later, Blumenfeld sent Weizmann a lengthy report on his recent meeting with Einstein. This document is a fascinating account of Blumenfeld's subtle attempts at manipulating Einstein and thus constitutes a continuation of the tactics he employed during the physicist's mobilization on behalf of Zionism. Yet, perhaps surprisingly, the report indirectly reveals that Einstein also manipulated Blumenfeld.

Blumenfeld apparently first met with strong opposition from Einstein to the invitation. He cited two major reasons for his initial refusal: first of all, he had rejected an offer to lecture at six U.S. universities "under . . . brilliant material conditions" just a few days prior to their meeting.[18] Thus, Einstein apparently portrayed the outcome of his negotiations with the American universities to Blumenfeld as the opposite of what had actually occurred: as we have seen, *they* had rejected *his* financial demands. Furthermore, Einstein had been preparing for the planned Solvay Congress "for months" (the conference was to be held in early April in Brussels) and was the only German scientist invited. The Solvay Congress was the most prestigious international meeting of physicists and was to be the first such conference since World War I. In addition, Blumenfeld's work was made even more difficult by the fact that Einstein did not see why the establishment of the Hebrew University was an immediate urgent requirement. He confessed to Einstein that the German Zionists were not aware of the reasons why Weizmann had "insisted" that Einstein join him on the trip at this point in time, yet he argued that the Zionist leader's invitation should be seen as a sufficient reason for Einstein to agree to his request. According to Blumenfeld, this is what made Einstein decide to embark on the trip. As Einstein had yet to meet Weizmann, it seems unlikely that this was the deciding factor in influencing Einstein to change his mind about his participation. Yet it is possible that Blumenfeld chose to present this version of events to flatter Weizmann.

Blumenfeld requested that Weizmann send a telegram to Einstein immediately—his participation in the trip was of particular importance, as he had "close friends" among the wealthiest U.S. Jews, notably Paul Warburg and another of Paul's brothers, Felix. Blumenfeld informed Weizmann that Einstein preferred not to travel to the United States via London, as he had received an invitation from Manchester University and was "constantly" being asked by English scholars to visit the UK. Blumenfeld clarified to Einstein that, nevertheless, he would have to travel via London but that the arrangements would be such that he need stay there only one day. Blumenfeld admitted to Weizmann that he had done so as he assumed that Weizmann might need Einstein in London. Thus, we can see how Blumenfeld tried to manipulate Einstein and convince him to take actions that were not necessarily in Einstein's best interests. At this point, however, Elsa Einstein clarified to Blumenfeld that a trip via London was "out of the question."[19]

Despite Blumenfeld attributing Einstein's willingness to embark on the mission to Weizmann's influence, Weizmann himself claimed that it was the German Zionist who had convinced the prominent physicist to embark on the tour. He announced to the Zionist Organization's Action Committee several months after the trip that "it was a tremendous feat that Blumenfeld succeeded in winning over Prof. Einstein for the journey to America."[20]

Einstein hastily did his best to reconstruct the canceled lecture tour to the United States, which was now to proceed under Zionist auspices. He contacted the president of Princeton and informed him that he was "compelled to travel to America at the invitation of the Zionist Organisation in the middle of March, because of the founding of the University to be established in Jerusalem." The word "compelled" reveals that perhaps he was not entirely happy with his decision to embark on such an extensive mission. In his letter to President Hibben, he clarified another main motivation for his willingness to travel to the United States: "I would not like to neglect anything that could contribute toward an improvement in international relations, which is close to my heart."[21]

Yet the U.S. universities were not the only institutions and individuals Einstein had now to quickly inform of his Zionist trip and convince of its legitimacy. He immediately wrote to his prominent Dutch colleague, Hendrik A. Lorentz, who had orchestrated Einstein's invitation to the Solvay Congress.[22] This must have been particularly difficult for Einstein in light of the degree to which he felt intimidated by his elder Dutch colleague.[23] Though he regretted having to withdraw from the Congress, he was "convinced" Lorentz would approve.

Einstein had been invited to the Solvay Conference despite some apparent confusion over the issue of his citizenship. According to Emile Tassel, the head of the secretariat of the Solvay Institute, even though the institute planned to invite to the conference "only scholars from Allied countries or from countries which remained neutral," an exception was made for Einstein and Ehrenfest. In Einstein's case, his nationality was seen as being "uncertain," and particular note was taken that he had been "the subject of quite a bit of agitation in Berlin during the War because of his pacifist sympathies which remained the same at all times."[24] According to the British scientist Ernest Rutherford, "The only German invited is Einstein, who is considered for this purpose to be international."[25]

Einstein's letter to Lorentz revealed how Einstein perceived of the purpose of his trip to the United States. He believed that the "decisive constituent sessions" would take place during his tour. Yet this was not how the Zionist Organization viewed the goal of the trip. So we can conclude that they presented the planned events to Einstein in a manner they believed would be palatable to him. He does not seem to have had any awareness of what was going on behind the scenes in regard to the Keren Hayesod. However, he seems to have been more aware of the Zionist Organization's plans for the Hebrew University project in the United States:

> The people have now entreated me to come to these meetings because they believe that through my personal involvement the rich Jews of America will be made more eager to donate. Strange as this may sound, I do believe that these people are not mistaken. Since this enterprise is very close to my heart, and I as a Jew also feel obliged to contribute toward its success, to the best of my abilities, I accepted.[26]

The issue of funding for the Hebrew University was not explicitly mentioned in Blumenfeld's report to Weizmann on his meeting with Einstein. However, Einstein must have learned about this goal for the tour from the Zionist propagandist, as his letter to Lorentz followed only three days after his meeting with Blumenfeld. Einstein asked Lorentz to inform Ernest Solvay and "the colleagues" that he had decided, "with a heavy heart," not to participate in the congress.[27] As this was to be the first congress to take place after World War I and as Germany would have no representatives at this most prestigious of scientific meetings, it is highly significant that Einstein chose a Zionist cause as a reason to be absent from the conference.

A day later, Weizmann sent Einstein the telegram Blumenfeld had requested, which constituted the first direct contact between Einstein and the president of the Zionist Organization. The cable's first sentence is perhaps more revealing of Weizmann's real intentions than he intended: "Extremely happy about your consent, our friends and I wholeheartedly appreciate your readiness at such a decisive hour for the Jewish people."[28] It is a matter of debate whether this was "a decisive hour" for the Jewish people in their entirety, yet it certainly was a critical moment for Weizmann in his power struggles within the Zionist Organization.

Inviting Einstein to tour the United States, however, was not a simple or straightforward matter. Four days after Weizmann's cable, the president of the American Jewish Physicians Committee, Nathan Ratnoff, sent Einstein a parallel invitation to visit the United States, yet made no mention of the Zionist Organization or Weizmann's invitation: "american jewish physicians committee cordially invites [you to] visit america [at] our expense in interest [of] medical school and university hospital[.] visit would assure complete success [of] plans for completion [of the] university project."[29] The committee had just been founded in New York by Ratnoff and a group of fellow physicians, apparently in expectation of Einstein's pending visit. The Jewish physicians were subsequently described as having been "inspired to the effort by the plea of the great Jewish scholar, Albert Einstein."[30] It would seem that this subsequent description of the founding of the committee was slightly hyperbolic: there is no evidence that Einstein issued a plea for such a committee to be established. In any case, the committee was not an integral part of the Zionist Organization; it was founded as "an independent body but in harmony and co-operation with the Zionist Organization." The committee's first immediate objective was to fund the establishment of a research institute in microbiology.[31]

Hearing of Einstein's plan from Lorentz, Paul Ehrenfest quickly wrote to Einstein, expressing his delight at the proposed trip, which he termed "your Jerusalem procession to Dollaria." He was very pleased that Einstein was prepared to make this sacrifice on behalf of the university, aware of "all its ugly hullabaloo and tantara around it."[32] Thus, Einstein could travel to the United States knowing he had the blessing of his close friend, who had previously expressed such harsh criticism of Einstein's involvement in Zionist affairs.

In his reply to Ehrenfest's letter, Einstein revealed his great relief that he approved of his planned trip and "for not scolding [him] about [his] Zionist escapade." More important, he expressed how he viewed the opposition to his trip, which was beginning to gather momentum: "There is quite some outrage here, but it's all the same to me. Even the assimilation Jews are wailing or scolding."[33] Whether Einstein had already learned that his friend and colleague Fritz Haber was displeased with his upcoming trip is unclear, but Haber wrote to him the very next day. In his plaintive letter, Haber made it clear that it was particularly the fact that Einstein would be traveling to the United States "with English supporters" of Zionism and that he would be subsequently visiting England "at the invitation of the government over there" that deeply alarmed him. Only one of these assumptions was correct: Weizmann had become a naturalized British citizen in 1910. However, Einstein was invited to the UK not by the British government but by individual universities.[34] Haber was also deeply disturbed that Einstein had chosen to forgo the Solvay Congress because of the planned tour, especially as he was the only German to be invited. For Haber, Einstein's decision to travel to the United States at this point in time, when President Warren Harding had just delayed discussions on ratifying the Versailles Peace Treaty and the British government was tightening its sanctions against Germany, would have fundamental consequences for Einstein himself and for Germany as a whole.[35] Haber believed that Einstein would be declaring "publically before the whole world that you do not want to be anything but a Swiss, who by coincidence resides in Germany." In his melodramatic view, now was the time "when belonging to Germany is an act of martyrdom." Einstein's decision would also have disadvantageous consequences for German Jewry—according to Haber, it would prove the Jews' "disloyalty." Haber pleaded with Einstein to delay his trip by a year and not to place in jeopardy "the narrow ground on which rest the livelihoods of academic teachers and students of the Jewish faith at German universities" by participating in his planned trip on behalf of the Hebrew University.[36]

Einstein replied to Haber the same day. He clarified that he had received Weizmann's invitation a few weeks earlier, prior to the current "political imbroglio." He tried to downplay his own role in the upcoming U.S. tour: "I am not needed for my abilities, of course, but only for my name. Its promotional power is expected to generate considerable success among our rich ethnic comrades in Dollaria." Yet he clearly needed to legitimize his participation in the Zionist delegation: "Despite my declared international disposition, I do still always feel obliged to speak up for my persecuted and morally oppressed ethnic comrades, as far as it is within my powers." And this led (at least according to this description for Haber's benefit) to his rapid assent to the tour. He claimed to accept the invitation "gladly, without more than five minutes' consideration, even though [he] had just declined all offers from American universities." Furthermore, he rejected Haber's accusation of disloyalty, stating that it was "far more an act of loyalty than of disloyalty." As we saw in Blumenfeld's description of Einstein's acceptance of the invitation, it certainly took longer than five minutes to convince him of the merits of the tour. Yet Einstein clarified that his motivation for embarking on the tour was tied not only to his enthusiasm for the planned establishment of the Hebrew University but also to his recently seeing "from countless examples how perfidiously and unkindly fine young Jews are being treated here in an attempt to deprive them of educational opportunities."[37] This was obviously an issue Einstein felt passionately about, as he clearly identified with the plight of young Jewish academics. Other instances of anti-Semitism Einstein had experienced personally over the past year had clearly also increased his Jewish self-awareness: he specifically mentioned the refusal of the secretary of the Prussian Academy, Gustav Roethe, to express solidarity with him against the antirelativists, the aversion of fellow academy member Ulrich von Wilamowitz-Moellendorff to add his name to a petition because Einstein had already signed it, and his clash with the antirelativists at Bad Nauheim.[38] These and other recent instances "would compel any self-respecting Jew to take Jewish solidarity more seriously than had been necessary or seemed natural in former times." Einstein thus revealed the more immediate catalyst that led him to embrace a more nationally minded attitude toward Jewish affairs.

Haber's accusation of "disloyalty vis-à-vis [his] German friends" clearly unsettled Einstein, and he swiftly defended himself by enumerating the various offers to take up a position abroad he had rejected: "This I did, by the way, not out of attachment to Germany but to my dear German friends. . . . Any affinity for the political structure of Germany would be unnatural for me as a pacifist." Einstein's primary allegiance in Germany was clearly to his colleagues and not to the German state itself.

Einstein informed Haber he would not change his plans regarding the tour, but he did agree to cancel his lecture in Manchester "[i]f the dreary political situation were to continue." Regarding the Solvay Congress, Einstein claimed he had forgone participation "with a heavy heart. [The prominent chemist Walther] Nernst was furious, by the way, when he heard that I had been invited over there and was intending to go."[39]

In the meantime, practical arrangements began to be made in anticipation of the trip. In a letter that swiftly followed his cable of 23 February, Weizmann acknowledged Einstein's "great sacrifice" in undertaking the voyage and expressed how honored he was to "be allowed" to collaborate with Einstein on this "historic project." He proposed that the Zionist Organization's secretary for university affairs, Solomon Ginzberg, serve as Einstein's secretary during the voyage and U.S. tour. Officially, this was to facilitate Einstein's work and provide him with access to "all the necessary information connected with any questions that might arise."[40] It seems safe to assume that this was also a means to keep a watchful eye on Einstein. However, we will see that Ginzberg had his own agenda, which was not necessarily identical to Weizmann's.

On 8 March, Einstein expressed his very restrained enthusiasm about the upcoming trip to his longtime friend, Maurice Solovine. He stated that he was not "keen on going to America, but am doing it on behalf of the Zionists, who have to beg for dollars for the educational institutions in Jerusalem, for which I must serve as famed bigwig and decoy." He explained that his main motivation in undertaking the trip was his willingness to do whatever he could for his ethnic brethren "who are being treated so vilely everywhere."[41] Yet by defining his participation as being "on behalf of the Zionists," he actually distanced himself from their goal for the tour and refrained from defining himself as a Zionist. At least outwardly, he was undertaking the tour in their interests, not his own.

Apart from his personal agenda of inviting Einstein to the United States to assist him in his power struggles with the Brandeis faction, Weizmann's invitation came in the context of fundraising efforts for the Hebrew University. The Zionist Organization had decided to establish a separate University Fund because its central fundraising instrument, the Keren Hayesod, was "overburdened with other tasks."[42] In January 1921, a plan was devised to establish university committees "composed, if possible, of scholars of standing, of prominent Jews, (including financiers and bankers) and representatives of any academic or other bodies and organisations which may be induced to take part in that work."[43] Just prior to Einstein's departure for the States, Ginzberg termed these proposed committees university aid committees, which the Zionist Organization hoped to establish in the UK and the United States, the two destinations on Einstein's trip. Regarding the establishment of such an aid committee in the United States, Ginzberg hoped that Einstein's "prestige may win for that Committee the personal and material support for many important new members who might stand aloof otherwise." He also surmised that the visit would "probably prove the acid test of the possibility or otherwise to collect important funds for the University."[44] During the planning stages of the trip, Ginzberg believed that the Hebrew University was a "nonpolitical institution of great spiritual appeal" and that Einstein's participation in Weizmann's delegation to the States would motivate "wealthier non-Zionist Jews" in the United States who did not support Zionist projects to contribute to its funding.[45] Thus, Ginzberg had very high hopes for the planned fundraising tour. However, powerful factors would subsequently challenge his (and Einstein's) agenda.

In early March, Einstein received his third official invitation to visit the United States—a reflection of the complex internal politics within the Zionist Organization. Erich Marx of the German Zionist Federation clarified to Einstein that the London Zionist Organization Bureau, the Zionist Organization of America, and the German Zionist Federation "greatly value[d]" Einstein's participation in the deliberations for the establishment of the Hebrew University, which were to take place at the end of March in New York. Marx stressed that "[p]articularly the gentlemen from New York" had informed the Berlin Bureau repeatedly that it was primarily Einstein's participation which would advance the project "in the most appropriate manner." He also claimed that the American Zionists preferred the deliberations to be concluded in the United States, as they viewed their participation as "a matter of honor." Marx specifically mentioned Paul and Felix Warburg as being particularly interested in Einstein's presence in the United States. He also informed Einstein that "a group of members from our central leadership" would accompany Einstein and Weizmann on the voyage. A day later, Einstein confirmed his acceptance of the official invitation.[46]

The composition of the delegation to the United States took some time to be decided on. Originally, besides Weizmann and Einstein, the other delegation members were to be Menachem Ussishkin, member of the Zionist Executive, and Isaac A. Naiditch and Hillel Zlatopolsky, both founding directors of the Keren Hayesod. Eventually, Ben-Zion Mossinson, a member of the General Zionist Council and one of the directors of the Keren Hayesod, replaced both Naiditch and Zlatopolsky.[47]

Meanwhile, preparations for the arrival of the Zionist delegation in the United States got under way. Almost immediately after receiving Einstein's consent to travel with him to the States, Weizmann sent a cable to Julian Mack in which he informed him that Einstein would be "coming with me [to] work for [the] University. Please prepare ground. Recommend appointment [of] special University commission." In response, the Zionist Organization of America set up a "Weizmann Committee" to plan for the visit. The committee was informed by Mack that "Dr. Einstein's coming is specifically in the interest of the Hebrew University,"[48] thereby perhaps hoping that Einstein would not become involved in other outstanding issues between the Europeans and the Americans.

However, obviously fearing that Einstein *would* get involved in the anticipated conflict on a wider scale than merely the Hebrew University, the Brandeis faction seemed to be less than enthusiastic about the planned visit. In a letter to his mother-in-law, Brandeis himself wrote with ironic humor: "You have doubtless heard that the Great Einstein is coming to America soon with Dr. Weizmann, our Zionist chief. Palestine may need something more than a new conception of the Universe or of several additional dimensions."[49]

In the meantime, back in Europe, the Zionist "preparation" of Einstein for the tour got under way. In mid-March, Blumenfeld reported to Weizmann on several meetings he had had with Einstein to prepare him for the trip. As Weizmann had not yet met Einstein personally, he had to remind him that "Einstein is, as you know, not a Zionist," and asked him not to undertake any attempts to persuade him to join the Zionist Organization. Blumenfeld thus defined the relationship between the prominent physicist and the Zionist movement as one that was quite complex. Similar to the period in which the Zionist propagandist mobilized Einstein on behalf of Zionism, its leaders were aware they had to respect their famous supporter's aversion to formal links to political or organizational entities. In Blumenfeld's opinion, Einstein "will always be at our disposal, when we need him for specific goals." To some extent, this comment reveals the rather cynical and manipulative use of Einstein by the Zionist movement, yet we must also remember that, to a large extent, Einstein was a willing participant in this manipulation. Furthermore, it cannot be ignored that he was also using the Zionist movement for his own purposes and interests. Blumenfeld obviously made some attempts to get Einstein to toe the party line. He stressed to Einstein that Weizmann was the ultimate authority on Zionist issues and that "it was naturally completely out of the question that two differing opinions be presented in America." Thus, Blumenfeld apparently tried to make it clear to Einstein that only Weizmann's views were legitimate in the Zionist leader's conflict with the Brandeis faction. Blumenfeld specifically discussed the issue of the Keren Hayesod with Einstein and reported to Weizmann that he believed he had convinced Einstein that only a unified, centrally organized fundraising effort could enable the kind of development required by Jewish "national needs." The views of Weizmann and his supporters were thus presented to Einstein as the only legitimate option. Furthermore, their needs were equated with Jewish "national needs" in general. Blumenfeld believed that Einstein would be immune to the views of "our American opponents," as the main motivation for Einstein's interest in Zionism lay in his antipathy toward assimilated Jews. Thus, it is clear that Weizmann and his supporters did not hesitate to demonize the Brandeis faction and had little or no understanding of their complex concept of Jewish nationalism.[50]

In his report to Weizmann, Blumenfeld referred to the criticism Einstein had received at home for his decision to embark on the voyage: "His trip has earned him outraged letters from German assimilationists, which he has dismissed with a smile." Yet we have seen from his defensive response to Haber that Einstein did not relate to all critics with such nonchalance. In his letter to Weizmann, Blumenfeld revealed a further motivation for Einstein's participation in the tour: he believed that Einstein's trip would give the Zionist Organization the opportunity to recruit new members for a European Keren Hayesod committee, and asked Weizmann to arrange for Max Warburg and prominent German Jewish politician Walther Rathenau to be recruited for such a committee by using Einstein's contacts with American Jewish leader Louis Marshall and Paul and Felix Warburg. Ironically, Weizmann's faction aimed at recruiting just those same assimilated American Jews, whom they otherwise derided.

Blumenfeld clearly did not trust Einstein fully and was wary of what Einstein might say during the tour: he advised Weizmann to carefully prepare speeches to be given by Einstein together with him, as Einstein "is a poor orator and sometimes says things out of naivety which are unwelcome to us."[51]

Meanwhile preparations for Einstein's arrival in New York were proceeding haltingly—and they seem to have been made at the last minute. Just prior to Einstein's departure, Solomon Ginzberg reported that although "no detailed arrangements have been made with Einstein concerning his activities in America," he presumed that he would be prepared "to address small but influential private gatherings, or to approach important individuals who are likely at all to sympathise with the cause of the University, provided a list of such individuals is available in time."[52] This lack of detailed arrangements in quite remarkable and seems to have had a decisive impact on the outcome of the tour.

Before leaving for the United States, Einstein gave an interview to the German Zionist *Jüdische Rundschau* in which he expressed his opinions on the planned Hebrew University and the Keren Hayesod. In addition to some of the views we have already encountered, he also elaborated on his previously expressed vision of the university as an "intellectual center" by articulating his hope that it would become "a focus of Jewish intellectual life."[53] Thus, in this interview at least, Einstein seems to have decided that the university would play a role as a center of *Jewish* scholarship and not merely as a universal intellectual center. With plans for the university's Institute for Jewish Studies gathering momentum, this was an issue Einstein could not ignore.

In the meantime, the German Zionist preparation of Einstein for the trip even extended to their controlling the flow of information he received. Blumenfeld apparently channeled Einstein's Zionist-related correspondence via the German Zionist Federation's Berlin office. In mid-March he informed Weizmann that the Office for Workers' Welfare of the Jewish Organizations of Germany had asked Einstein to raise $20,000 in the United States for the welfare of Eastern European Jews. He stated that he had blocked similar requests to Einstein in the past, yet did not believe that this specific campaign would harm the efforts on behalf of the Keren Hayesod.[54] However, it is unclear whether Einstein actually undertook any action to raise the requested funds, as there is no material extant relating to this issue. A day before his departure, the primary proponent of the Jewish National Library in Germany, Heinrich Löwe, asked Einstein to wield his influence to raise two million marks on behalf of the planned national and university library.[55] As we will see, Einstein did raise funds for the library on his U.S. tour.

On 21 March, Einstein left Berlin together with Elsa to travel to Rotterdam to embark on their trip to the United States.[56] Two days later, Einstein finally met the leader of the Zionist Organization, Chaim Weizmann, and his wife, Vera. Accompanied by the other del-

Figure 4. Einstein and members of the Zionist delegation to the United States, Ben-Zion Mossinson, Chaim Weizmann, and Menachem Ussishkin, on board the SS *Rotterdam*, March 1921

egation members, the Einsteins and the Weizmanns set sail from Plymouth on board the *Rotterdam* en route to New York.

After their departure from Europe, Weizmann apparently did his best to follow Blumenfeld's advice on how to relate to Einstein. Later that year he reported to the Zionist Action Committee that he had traveled with Einstein for eleven days and "took care not to attempt to make him into a Zionist. We were quite natural in our contacts. I never grasped the initiative to rope him in [to the Zionist movement]."[57]

Meanwhile, the hesitant arrangements for Einstein's visit continued. However, the apparently patchy communication between the Zionist Organization and the American Zionists regarding Einstein's trip resulted initially in grave misunderstandings about his planned lecture tour. The increasingly widening differences between the Eu-

Figure 5. Einstein with Ussishkin, Weizmann, Vera Weizmann, Elsa Einstein, Mossinson on board the SS *Rotterdam*, March 1921

ropean and American Zionists swiftly led to a conflict over Einstein's visit, with the preeminent physicist becoming a political football between the two sides. On 2 March, Zionist Organization of America leaders Mack and Felix Frankfurter sent a cable to Weizmann in which they inquired how long Einstein planned to stay in the United States and voiced their concern that "it will materially injure his Palestine propaganda if he does not accept invitation from American university groups as [a] scientist." They added that "[s]uch invitations will not be extended unless assurance is given that they will be accepted."[58] Without first conferring with Einstein or the European Zionists, the Zionist Organization of America had obviously made inquiries at some of the U.S. universities about his planned lectures.[59] Yet it is intriguing (and somewhat mystifying) that the leaders of the organization believed that Einstein's giving scientific lectures in the United States was vital for the success of his actions on behalf of the Hebrew University.

It seems fair to speculate that it was their own status vis-à-vis the elite American universities that they were ultimately concerned about. Weizmann informed Einstein of the cable, assuring him that he "naturally" did not want to influence Einstein in his decisions, but stressed that it would assist "our work ... in a most forceful way" if he would accept a few such invitations. To this end, Weizmann asked Einstein to inform him of what arrangements he had already made with universities in the United States.[60] Einstein's reply is not extant, but in mid-March, Weizmann informed Mack by cable of the status of Einstein's negotiations with the American universities: "Following from Einstein begins negotiating with Wisconsin Washington Princeton Yale for month without definite result prepared [to] negotiate further personally." Two days later, Weizmann sent Mack another cable in which he clarified that he understood that Einstein's negotiations with the American universities had "not yet broken off," and he was prepared to undertake further negotiations. Furthermore, it was "impossible" to cancel the journey now, as it "would cause grave prejudice."[61]

The next day (which was only five days prior to the delegation's scheduled departure), Mack, Frankfurter, and Brandeis supporter Bernard Flexner informed Weizmann that owing to opposition from "scientific bodies and universities," Einstein's visit was "bound to be [a] failure and bitter disappointment because his negotiations sought to commercialize science and thereby raised the indignation [of] scientists." They "solemnly" urged Weizmann "to avoid the tragedy of his visit and the irreparable waste of so great an asset. If invitations come later, with a view to visiting [in the] fall under scientific circumstances he would be invaluable to [our] cause."[62] The leaders of the Zionist Organization of America had obviously learned of Einstein's negotiations with some of the U.S. universities and of the exorbitant fees he had requested. Yet it is possible they were desperately seeking for a reason for Einstein not to accompany Weizmann on his U.S. tour. It seems they were extremely fearful of the impact Einstein's support for Weizmann would have on the local Zionists and the Jewish community. Otherwise, why would they suggest that Einstein visit in the autumn rather than in the spring, particularly as his lecture tour under "commercial" auspices had been called off? Weizmann refused to cancel Einstein's participation, claiming that he had already left Berlin. The exchange of cables reveals just how tenuous Weizmann believed Einstein's backing to be: "Such insult equivalent [to] losing finally his support. Very harmful [to] our position [in] Europe. Shall endeavour [to] persuade Einstein [to] offer universities [a] couple [of] lectures without payment, thereby smoothing opposition." He reassured them that "Einstein's personal character [was] above commercialism. Whole trouble certainly due [to] misunderstandings."[63] As we have seen, Einstein apparently did not inform Blumenfeld that it was the U.S. universities that had refused his request for exorbitant fees. Therefore, Weizmann was under the erroneous impression that Einstein was "above commercialism."

The harsh criticism of Einstein actually extended beyond the universities directly involved in the negotiations. Dean William F. Magie of Princeton apparently sent Washington University physicist Arthur H. Compton a letter in which Einstein's financial demands were severely criticized. Compton forwarded the letter to the chancellor of Washington University, Frederic A. Hall, who then replied to Compton (with a hint of anti-Semitism): "Einstein intends to lead the world in more directions than one. I am sorry to find a man so gifted and yet so grasping."[64]

In the meantime, the confusion over the plans for Einstein's visit continued right up until his arrival. In a cable sent to the Zionist Organization of America from on board the *Rotterdam* just five days prior to their planned disembarkation in New York, Chaim Weizmann asked for information on the "arrangements [for the] first few days" and asked, "What are plans for Einstein?" Two days later he sent a cable to Julian Mack stating that he "[c]annot take any steps [regarding] Einstein before [I] meet [to] discuss [this] with you."[65] It seems the mere fact of Einstein's participation was a huge coup for Weizmann and that concrete plans were not deemed necessary or feasible. Another possible explanation is that Weizmann did not want to inform his American colleagues of his plans for Einstein's visit, so as to keep them guessing as to his true intentions in light of his anticipated clash with the Brandeis faction.

Even a day prior to the arrival of the delegation, communication between the Zionist Organization of America and Weizmann was still confused and was quickly becoming more dramatic: "Einstein situation extremely difficult[.] expedient [that] you explain [to] us fully his exact negotiations[,] who represents him here[.] am also awaiting you[r] promised cable whether he accept[s] your suggestion [to hold a] couple [of] university lectures [for] free."[66]

On a parallel track, preparations for the fundraising for the planned medical faculty at the Hebrew University had begun prior to the delegation's arrival in the United States. A day before they disembarked in New York, the Zionist Organization of America informed Weizmann that "we have [a] group [of] physicians preparing [for] Einstein banquet [in the] interest [of the] University."[67]

In a statement on board the *Rotterdam*, just prior to their arrival in New York, Weizmann outlined the Zionist delegation's hopes for Einstein's participation. He stated that "Einstein attaches the utmost importance to the early inauguration of the Jerusalem University and is prepared when the time arrives personally to associate himself with its activities—a course in which he will there is reason to hope be followed by other Jewish scholars and scientists of world wide reputation."[68]

On 2 April the delegation arrived in New York, where they received an official welcome by Mayor Hylan's office. As their motorcade proceeded through the Lower East Side to the Hotel Commodore in uptown New York, they were greeted by enthusiastic Jewish crowds.[69] In a cable to the Yiddish newspaper *Der Tog*, which Einstein cosigned with Weizmann, Ussishkin, and Mossinson, the two main goals of the delegation were conflated: "spiritual regeneration symbolized by Hebrew University, of which realization guaranteed by whole-hearted participation [of the] Keren Hayesod."[70]

By the time he reached the United States, Einstein had apparently succeeded in reconstructing part of his originally planned personal

Figure 6. Einstein in a motorcade, New York, April 1921

lecture tour. In early April he informed the prominent Jewish surgeon Carl Beck from Chicago that he had "already arranged with the Princeton University to deliver five lectures there in the second week of May. I may also lecture at one or two other Universities." He was also busy preparing for the future stops on his tour. Beck, with whom he had previously corresponded, would apparently assist in organizing the fundraising efforts for the Hebrew University in Chicago. But the main efforts were to concentrate on New York, where "an important University Committee" was to be established "on a big scale." This was obviously the proposed university aid committee that Ginzberg had mentioned prior to the visit. The purpose of the committee was to mobilize non-Zionists such as retired Jewish American politician Oscar S. Straus and prominent scientists such as the biologist Jacques Loeb on behalf of the Hebrew University.

In a letter to Beck, Einstein gave the establishment of contacts with American scientists equal weight with his fundraising efforts: "I mentioned already my journey has also a second object namely, to come into contact in so far as possible under the circumstances with the American World of Science."[71] However, he did not give such equal weight to both goals in every case. When he wanted to reject an invitation, he cited the primacy of his fundraising efforts.[72] His letter to Beck also provides us with an insight into the hectic atmosphere of Einstein's visit: he declined Beck's invitation to lodge with him during his stay in Chicago, as "my work for the Jerusalem University makes me a center of so much noise and coming and going that I would be really a most troublesome guest."[73]

In contrast to the popular welcome by the crowds on their immediate arrival, an official reception to greet the delegation to the United States was held by the Zionist Organization of America at the Metropolitan Opera House in New York on 10 April.[74] Weizmann held a grandiose speech, yet Einstein did not speak at all, thus proving that the Zionist leader's supporters were keeping a tight lid on Einstein's statements to avoid the "unwelcome things" he might say.

Two days later, a popular reception was held at the 69th Regiment Armory in New York to welcome the delegation. Twenty thousand people "turned the . . . Armory into the scene of a near riot." The crowds shouted, waved "Jewish and American flags" and "stormed the police lines." The "demonstration" was organized by "[m]ore than eight hundred Jewish organizations. . . . It was the people's welcome to their leaders." The constraints on Einstein's freedom of speech continued at the mass rally. Following Weizmann's speech, Einstein "made the briefest speech of the evening, when he said: 'Your leader, Dr. Weizmann has spoken, and he has spoken very well for us all. Follow him and you will do well. That is all I have to say.'"[75] Thus, Blumenfeld's preparation of Einstein to ensure that he would toe the party line seemed to be working, at least at this stage of the tour.

In the meantime, the delegation's visit caused a serious political dispute within the City Council of New York. Initially, the freedom of the city was refused to Weizmann and Einstein because of the sole opposition of Republican alderman Bruce M. Falconer, as the vote was supposed to be unanimous. However, a few days later the council ignored Falconer's opposition, reversed its decision, and conferred the freedom of the city on Weizmann and Einstein.[76]

After two weeks of "promising" negotiations between Weizmann and the Brandeis-Mack faction over the Keren Hayesod, the talks broke down and a deep rift between the factions ensued.[77] Much has been written about the clash between Weizmann and Brandeis and their supporters, which came to a head during the delegation's visit. According to two historians of the conflict, Urofsky and Levy, the opponents of Brandeis and Mack in the Zionist Organization of America presented Weizmann with an ultimatum: either he broke off talks with Mack or "they would desert him, and make sure that he no longer led the movement." Weizmann then "suddenly broke off all further meetings on 17 April, and without consulting Mack, proclaimed the establishment of the *Keren Hayesod* in the United States."[78]

The breakdown in the negotiations did not have an immediate impact on Einstein's efforts. In mid-April, Brandeis supporter Judah L. Magnes proposed to organize a gathering of intellectuals interested in the university. However, Einstein stressed that he was pressed for time and not interested in a mere "exchange of views." He would participate only if it would "contribute to the success of my Mission, to secure support for the Jerusalem University—then I shall be glad to come to such a gathering." Einstein made it clear to Magnes that it was "essential" that the meeting "include as many influential people as possible. If any of your friends, outside the intellectual circle, but of possible use for the University, are interested, I will be glad if those too are invited." It is clear that, intriguingly, the fundraising efforts for the Hebrew University were of primary importance for Einstein during his tour, even over intellectual discussions about the essence of the planned institution.

However, Magnes was not interested in a fundraising gathering and decided to skip the meeting.[79] It is ironic that Einstein, whom one views primarily as an intellectual, would prefer a fundraising meeting with bigwigs rather than a gathering of scholars to discuss plans for the Hebrew University.[80]

Meanwhile, the European Zionist leaders were forming their own opinions about Einstein, whom they had not met previously. Shmarya Levin expressed his view of Einstein's role as a member of the Zionist

Figure 7. Louis D. Brandeis, ca. 1921

delegation to the United States as follows: "Einstein is as helpless as an innocent baby, and to look after him is no small matter. But he is a darling all the same."[81] We can see how patronizing at least some of the Zionist leaders were in their perception of Einstein, whom they were utilizing for their own purposes.

Perhaps the Einsteins sensed some of the condescension of the fellow members of the delegation. In any case, when they visited Washington, D.C., where they met with President Harding and Einstein addressed the National Academy of Sciences, they were not accompanied by anyone from Weizmann's entourage.[82]

Whatever the reason for this lapse in the Weizmann camp's shadowing of Einstein, it led to a dramatic development. On 26 April, Einstein and Elsa had an important meeting with Louis D. Brandeis, the leader of the "enemy" camp. Brandeis wrote two accounts of the meeting. To his wife, he described the Einsteins as "simple lovely folk." On the Einsteins' role in the dispute between himself and Weizmann, Brandeis wrote, "It proved impossible to avoid some discussion of the 'break,' though they are not in [it]. They specialize on the University." To his colleagues at the Zionist Organization of America, he reported that he had met with the Einsteins for an hour, "in which I struggled in German to make them understand the situation as they were eager to talk on it and discussion was unavoidable." Brandeis understood that as a consequence of the meeting, "the only subject they will concern themselves with is the University and they are ready to do whatever we deem necessary to ensure an honest efficient administration of University affairs and I told them—in view of that—I felt sure that a way could be found for working out the University project alone." He recommended that his New York allies talk with the Einsteins "not in the presence of W[eizmann]."[83]

This was a watershed meeting for Einstein. Suddenly he was confronted with the viewpoints from the other side in the conflict, views from which Weizmann and Ginzberg had managed to shield him for almost a month (actually longer, if one includes the voyage and the preparatory buildup to the trip). It is intriguing that Ginzberg apparently did not accompany Einstein on his trip to Washington, thus paving the way for Einstein's meeting with Brandeis. Obviously, Brandeis hoped that Einstein would not become involved in the wider contentious issues.

A few days later, Brandeis supporter Bernard Flexner sent Frankfurter a report on the meeting that contained dramatic references to all the intricate and complex issues which arose during Einstein's visit to the United States. First, Flexner revealed that the meeting had been arranged by the prominent biologist Jacques Loeb. According to Flexner, Loeb, "of whom Einstein has seen more than anyone else in this country," informed Flexner that he "is not only heartbroken on account of Einstein's having come to this country, but so is Einstein." After the meeting of the National Research Council in Washington on 26 April, Einstein asked Loeb his opinion of Brandeis and "was utterly astounded at what Loeb had to say of L.D.B. and told Loeb that Weizman [*sic*] had given him an utterly different opinion

THE "PRIZE-WINNING OX" IN "DOLLARIA"   113

and had urged him not to go to see L.D.B. when they were in Washington." According to Flexner, Einstein returned from the meeting with Brandeis "amazed and delighted, he said unequivocably to Loeb that he would stand to the uttermost with L.D.B. and the Americans as against Weizmann; that he was utterly sick of what Weizman [*sic*] had done and was trying to do."

Regarding the Hebrew University, Flexner asked Brandeis "how he Einstein, could be of help and reported that L.D.B. said he could help in the university matter upon insisting on the right sort of committee." As far as what Einstein knew about the preparations for the trip was concerned, "He had no idea of what he was coming into and he meant to get away at the earliest opportunity." Flexner added at this point in his letter to Frankfurter that "we have always believed as you know, that he never saw any of our cables to Weizmann." Regarding the circumstances of their lodgings in the United States, Flexner reported that "Mrs. Einstein expressed to Loeb her horror at the luxury in which they were living and said she meant to get out of it all with her husband, at the earliest moment." As for the issue of Einstein's request for payment for his lectures at U.S. universities, Flexner reported that Einstein told Loeb that

> his one thought in doing so was to provide enough money to be held in trust in this country with which to educate his two children; that he was not a well man and while he could without the slightest question take care of his own needs and those of his family during his lifetime in Germany, the thought of dying without having made provisions for the education of his children was a nightmare to him.

Flexner informed Frankfurter that he therefore intended to "get together" a fund "so as to relieve Einstein of the necessity of lecturing and leave him free for research work." Loeb thought that $10,000 would be needed for such a fund, but was not certain.[84]

Obviously, one needs to be very cautious with the claims made by Flexner. He was a staunch Brandeis supporter and clearly very hostile toward Weizmann. Some of his claims seem quite exaggerated, to say the least. Despite their alleged "horror," the Einsteins did not pack their bags immediately and leave the United States; they did not even leave the luxurious Hotel Commodore. Yet the letter does give us a clear sense of the Einsteins' apparent outrage at the claims being made by Brandeis and at the fact that Weizmann and his supporters had evidently not informed Einstein sufficiently about the basic details regarding the University Fund. In light of Flexner's report, one can also not escape the conclusion that the Einsteins were venting their frustration at the way they were being treated by the Weizmann camp, as well as making statements which the Brandeis camp would have liked to hear.

Following up on their meeting, Einstein sent Brandeis a letter, which is unfortunately not extant. In response, Brandeis sent Einstein an innocuous-looking reply, which would swiftly turn into a major political issue in the clash between Brandeis and Weizmann and their respective supporters: "I am asking one of my associates to send you the data for which you ask—so far as they are available here. It was indeed a great pleasure to see you and Mrs Einstein."[85] However, Brandeis's innocuous reference to "data" was actually a profound understatement that masked his dramatic disclosures to Einstein at their recent meeting. In the correspondence that ensued, it became apparent that, at the meeting, Brandeis had accused Weizmann and the Zionist Organization of the misappropriation of funds that had been earmarked for the Hebrew University and had allegedly been used for various Zionist infrastructure projects in the Haifa area.[86]

Adding to the drama, Brandeis's reply was apparently lost temporarily, as it was sent to Chicago c/o a "Mr. Lubin," in accordance with advice Mack allegedly received from Elsa Einstein. To add to the confusion over the letter, Mack apparently believed that Brandeis had "referred the subject matter" to him "for answer," as the letter did not name the individual who was to send the requested information to Einstein.[87] When he forwarded the letter to his staunch supporter Jacob de Haas to be transmitted to Einstein, Brandeis informed him that he had told Einstein "of the misappropriation of which we learned in London" and had "mentioned the diversion also of a University Fund & our apprehension as to further diversion. I had in mind a sum—mentioned to me in London by W[eizmann], as well as by [Weizmann supporter Julius] Simon—of at least £1000."[88]

In mid-May, the lost letter from Brandeis to Einstein became one of the hotly contested issues between the Mack-Brandeis faction and Weizmann's supporters. Weizmann's faction apparently believed that Brandeis had intentionally failed to reply to Einstein.[89] As for the allegation of the misappropriation of funds, according to Mack, the Zionist Commission in Palestine "had not only used up the J.D.C. [Joint Distribution Committee] relief money, but they had used up the University Fund, the Jewish National Fund, and other special funds." Mack concluded that the University Fund had "disappeared."[90] In response, the Zionist Organization's political secretary, Leonard Stein, a staunch Weizmann supporter, sent Mack a detailed and lengthy reply in which he defended the Zionist Organization against what he saw as "a charge of downright malversation" and rejected the accusation that the University Fund had "disappeared." Stein found that the charge that the fund had "vanished" was

> indeed amazing, most amazing of all when not only made to a distinguished
> non-Zionist in private conversation, but formally justified in writing by alle-
> gations, which, even if they were in themselves as accurate as they are in fact
> misleading, would, on the face of them, be wholly insufficient to establish its
> truth.[91]

Most important, Stein's letter reconfirms that the Zionist Organization continued to regard Einstein as a non-Zionist, and this fueled their indignation at the accusations even more. In his reply to Stein, Mack claimed that Einstein "desired and had the right to know from Mr. Brandeis why it was that he and his associates would not cooperate with Dr. Weizmann in raising funds for the University." According to Mack, "it was part of simple candor and plain honesty" for Brandeis to explain to Einstein why they had requested "adequate safeguards to insure the maintenance of certain standards in the financial administration of the funds raised. In this connection, Brandeis had occasion to repeat what he had learned from Dr. Weizmann last summer, that university funds had disappeared and had been used for general Zionist purposes."[92]

By this stage, fundraising for the Hebrew University and indeed Einstein himself had become full-blown political issues in the clash between the two factions. Whether the Brandeis-Mack faction's declaration that they would not participate in the fundraising efforts for the university because of the alleged misappropriation of funds was sincere or not, the disclosure of the accusations to Einstein was clearly intended to cause maximum political damage to Weizmann and was an attempt to neutralize the positive outcome of Einstein's fundraising efforts.

In July 1921, Weizmann gave his own account of Einstein's meeting with Brandeis in Washington and the ensuing correspondence between the various factions. According to Weizmann, the Brandeis faction believed that the matter of the Hebrew University should be eliminated "from the contentious issues" between their two factions. According to Weizmann, he had refrained from imparting information on the controversial issues to Einstein: "I didn't let him in on the whole issue, when he asked something, I replied in brief. He heard what we were talking about, we did not keep any secrets from him.

When Brandeis informed Einstein that a fund for the university had been established a year earlier but that "it had disappeared," Einstein confronted Weizmann with this information. According to Weizmann, he "returned from Washington completely distraught by this news and informed me of these allegations. He reproached me for not letting him in on this issue, thereby winning him over for a cause which compromised his name." In reaction, Weizmann "came clean with him and demanded that he write to Brandeis and have these allegations repeated." Subsequently, a letter arrived from Brandeis stating that he had "handed the matter over to Mack and it would be dealt with by his office; however, it has not been dealt with to this day." Thus, Weizmann had clearly paid the price for not informing Einstein in advance of the confusion concerning the "missing" funds. It is possible that he did not even inform Einstein of the existence of the University Fund in general (his text is ambiguous in this regard), thereby possibly leaving Einstein patently uninformed about what was happening behind the scenes. As before, the Zionist Organization was certainly not treating Einstein as an insider, this time to their detriment. However, Weizmann's explanations of the financial arrangements regarding the University Fund "made an enormous impression on Einstein."[93] Yet it is not clear from this description how long it took Weizmann to rebuild the trust between himself and Einstein, which had been shattered by Brandeis's revelations.

On his return to New York, Einstein had an important exchange with the philanthropist and Zionist Organization of America Executive Committee member Solomon Rosenbloom, who had apparently written to Einstein, outlining his ideas for the planned Department of Jewish Studies at the university. In his response, Einstein rejected the idea that "the *teaching* and the *research* of our future professors at Jerusalem are to be bound by the orthodox jewish [*sic*] laws or conceptions." He added that "any such restriction of the freedom of thought would be intolerable (except, perhaps, in a frankly theological institute or department) and would defeat your own purpose—to further a *free and creative* synthesis of faith and reason." Yet Einstein was prepared to concede to Rosenbloom that there should be special arrangements for the appointment of faculty members and for the funding of the proposed department. Einstein suggested "a special committee of 3 or 4 scholars who may be trusted to preserve the proper *spirit* at that department without insisting too much on any purely *external* conditions." To counterbalance the power of this committee, "the Governing Body of the University should also have a certain amount of control over those appointments." He thus implicitly agreed that only scholars who were religious would have the decisive say in determining the faculty of the proposed department. In light of Einstein's deep-seated secularism, this statement to Rosenbloom is quite astonishing. In his view, the special circumstances of the institute also had implications for its funding: it was "imperative that the latter be established not by the General University Fund (which could hardly impose any restrictions) but by a special Foundation for which the funds should be collected by yourself and your friends who share your feelings in this matter."[94]

Meanwhile, once the immediate crisis in the Weizmann camp over Brandeis's disclosures had passed, Einstein attempted to make his decisive push to establish a framework for mobilizing the Jewish non-Zionists on behalf of the university project. He did so by sending a circular letter on 9 May, inviting a number of prominent Jewish personalities to a discussion on the establishment of the Hebrew University on 19 May at the Hotel Commodore. In his invitation, Einstein stated that the objective of the gathering was "to discuss with a number of gentlemen of standing in the community, the proposed Hebrew University in Jerusalem, and to consider ways and means of organizing in its support the sympathy which is certainly felt toward that scheme in wide circles of American Jewry, both Zionist and non-Zionist."[95] According to Ginzberg, Einstein invited "about 50 non-Zionists" to the planned meeting. The purpose of the gathering was not to raise funds but rather to establish a university aid committee and elect its principal officers.[96] Thus, this was supposed to be the pivotal meeting at which the principal goal of Einstein's mission was to be realized.

A number of prominent figures rejected the invitation, some of them declaring their ideological opposition to the whole project. The rejection that must have been the most difficult for Einstein to deal with was that of Paul M. Warburg, who had acted as a conduit for the negotiations regarding his planned personal lecture tour and who allegedly had been one of the prominent Jews who had insisted on Einstein's participation in the Zionist mission.[97] Warburg stated that his "presence would be of no use; on the contrary, I fear that, if at all, its effect would rather be to cool things down." He had "the greatest doubts relating to the Zionist plans" and anticipated "their consequences with genuine consternation."[98] According to Ginzberg, the breakdown in the negotiations between the Weizmann and Brandeis factions over the latter's cooperation with the University Fund had a deterrent effect on the non-Zionists' willingness to participate in these efforts: "the non-Zionists were naturally even less inclined to cooperate than they would normally have been." These negotiations

broke down at the last moment "because of the Fund's official desig-
nation ('of the Keren Hayesod')." Thus, "when [Einstein] invited several dozen notables to a drawing-room meeting in the interests of the
university, very few responded."[99] In fact, it is not clear whether the
meeting ever took place, as there is no documentation which would
indicate that it did. In any case, the primary goal of Einstein's mission
to the United States, the establishment of an American university aid
committee, could not be realized, for political reasons.

Solomon Rosenbloom also realized that the Weizmann-Brandeis
clash had a detrimental impact on Einstein's mission: "It is very unfor-
tunate that the break between the European leaders and the American
leaders should have occurred at this time. Aside from the fact that the
general Zionist work is bound to suffer on account of this misunderstanding, I feel that even the work for the University is suffering on
account of this break" To his mind, Einstein's presence in the United
States "could have been better utilized in behalf of the University had
there been no division in the ranks."[100]

In additional to the detrimental effect of the Weizmann-Brandeis
clash on Einstein's mission, there was another crucial organizational
issue that further hampered Einstein's work, the divergent agendas of
Weizmann as president of the Zionist Organization and of Ginzberg
as secretary of the organization's University Committee. In his remi-
niscences on his time at the university, Ginzberg quoted one of the
pro-Weizmann New York Zionists, who, when asked "why adequate
preparations had not been made for Einstein's visit to his city," responded, "we did not want Einstein to take away the money that was
waiting for Weizmann."[101] Thus, at least some of Weizmann's supporters saw this as a competition for funds, even though the monies
for the university were to be raised from new, non-Zionist, sources.

In mid-May the controversy over the alleged misappropriation of
funds continued. Julian Mack made detailed allegations against the
Zionist Commission in Palestine concerning its alleged use of the
University Fund (and other funds), its "utter disregard . . . of budget
limitations," and its using "the £1,000 monthly sent for University
purposes and devoted to other Zionist work." He apparently planned

to "lay the documents in support of these statements before" Einstein in person. However, Einstein was due to depart for Boston, and the meeting had to be canceled. Therefore, Mack planned to send Einstein the details in writing.[102]

Yet it seems that Mack *did* meet with Einstein (together with Bernhard Flexner) upon Einstein's return to New York from Boston. On 22 May, Brandeis shared with Frankfurter that "It must have been a painful interview for Mack & Ben F. I am glad that W. admitted on the 19th his statement to me about the use of U. of University [*sic*] Funds, whatever W's view of the proprieties."[103]

As a result of their meeting, Mack sent Einstein thirteen cables between Weizmann and the Brandeis faction pertaining to the preparations for Einstein's visit. He probably did so, as Mack and Flexner's accusations do not seem to have changed Einstein's mind: "I deeply regret that Mr. Bernard Flexner's and my efforts, to explain our thought and purpose in this whole matter, namely, to protect you against unjust attacks and to protect the organization against the results of such unjust attacks, have not appeared to you to be convincing."[104]

In mid-May the next crisis struck. On 18 May Einstein visited Harvard University, where he participated in a reception at President Lowell's residence. In the evening he attended a dinner held by the New Century Club in Boston, at which $25,000 was raised for the university library. However, there had apparently been an "incident" in regard to Einstein's brief visit at Harvard earlier that day.[105]

Felix Frankfurter, who besides being a prominent leader of the Zionist Organization of America was a professor of law at Harvard, believed that Einstein had been informed he had played a role in the cancelation of a lecture by the physicist at Harvard. Apparently, Weizmann's supporter Shmarya Levin had accused Frankfurter of "preventing [Einstein's] appearance at Harvard." Frankfurter strongly denied these allegations as "an absolute untruth" and as "absolute slander."[106]

As we saw at the beginning of this chapter, the controversy over a possible visit by Einstein to Harvard had actually been brewing for six weeks: at the end of March, Julian Mack had sent Weizmann a cable stating that Harvard refused to welcome Einstein. Whether this was

Figure 8. Felix Frankfurter, ca. 1921

due to rising anti-Semitism at Ivy League universities or only to Einstein's exorbitant demands regarding fees to other prestigious universities is unclear.[107]

There is also uncertainly as to what constituted the "incident." The only event we know of that had been canceled was his planned talk at the Harvard Liberal Club. Although Einstein had been invited to lecture at the club on international conciliation, there is no indication that the lecture actually took place.[108] It is possible that the Brandeis-Mack faction intervened with the heads of Harvard to cancel Einstein's lecture because of the views of the club, which may well have been deemed too progressive, and that this is what constituted the incident. In his reply to Frankfurter's letter regarding the Harvard incident, Einstein expressed his "fear" that he had behaved "unjustly" toward him. He reassured Frankfurter that "nothing was withheld from me, nor falsely represented to me, and no one said anything unfavorable about you, not even Levin, whom, you were suspecting." Einstein was clearly not willing to admit that he had not been informed at all times by Weizmann and his supporters to the extent that would have avoided the ensuing misunderstandings. Possibly out of unease at having "unjustly" accused Frankfurter, Einstein felt the need to explain the background of the tour and the hectic events of the past two months from his own perspective. He explained to Frankfurter that "good friends in Holland" had advised him to "set such high demands" so that he could achieve "economic independence." When his financial requests generated "greatest indignation," he was actually pleased that the tour was canceled. This account is accurate—it corresponds to the known documents at our disposal. It is intriguing to note that Einstein provided Frankfurter with a truthful account of his negotiations with the U.S. universities, yet he apparently did not do the same with Blumenfeld. He seems to have placed more trust in Frankfurter, whom he had met only once, than in Blumenfeld, whom he had met frequently. He next gave an account of receiving Weizmann's cable. Similar to his account to Fritz Haber prior to his trip, he claimed "not to have hesitated for a moment." As he had previously been in negotiations with Princeton, he renewed his contacts with them and informed them "that [he] was willing to deliver lectures there without setting any kind of financial conditions." This account is less accurate than that on his negotiations. We have seen that he *did* hesitate quite considerably about his participation in the Zionist mission. Furthermore, one could conclude from Einstein's letter that he had not requested any remuneration from Princeton. However, as we have seen, he had actually left the issue of the fee up to the university. Einstein went on to describe to Frankfurter his reception in the United States and his interpretation of the cables that had been shown to him: "Having now encountered a much friendlier welcome everywhere than I could have expected (even disregarding the incident), I could not interpret your information by telegram as sincere." In Einstein's opinion, it would not have been "terrible" if he had not received any invitations from the American universities, even though he did know that "it is a Jewish weakness to be forever anxious about keeping the Gentiles in a good mood." From his description it is clear that Einstein interpreted the Zionist Organization of America's cables as "insincere." He seems to have lacked an understanding of the American Zionists' motivations. For the Brandeis-Mack faction, there was a direct connection between efforts on behalf of the Hebrew University and the Ivy League's perception of Einstein's "grasping" "commercialism." It seems the missing link in this was that for the elite of U.S. Jewry to embrace the cause of the university, the project (and its main protagonist, Einstein) had to be deemed respectable by American academia.

For his part, Einstein had concluded that Weizmann and his faction held the moral upper hand. In his view, the American Zionist leadership had "grave sins on its conscience" and that "in the interest of the holy cause a purge ought to take place, without bitterness, but also irrespective of the persons involved." He then qualified this statement by declaring that this was merely his personal opinion and that he had "no right to any official say." Indicative of the internal struggle regarding his attitude toward Zionism and his own role within it, he then wrestled with the wording of a sentence in which he first stated, "It can't be my task to have a *say* in these political matters, as I am a novice in *Zionist matters*" (my emphasis). But he then amended the sentence to read, "It can't be my task to *participate* in these political matters, as I am a novice in *Zionism*" (my emphasis). The correction is intriguing, possibly revealing that Einstein did believe after all that he had a right to have a *say* in a conflict he perceived as being highly political. He then distinguished between Frankfurter, whom he held blameless for the current situation, and the rest of the American Zionist leadership group, which he viewed as being morally compromised. He concluded the letter by declaring that a "splendid success has been achieved."[109]

A day before his departure, Einstein sent Frankfurter another letter, practically begging him to forgive him for his "unsavory behavior." It was very unlike Einstein to describe his own actions in such a manner—he would often expect others to forgive him for his conduct. And indeed, he swiftly justified his behavior, insisting that Frankfurter must realize that Einstein *had* to reach the conclusion that he had been the victim of "an unseemly political maneuver."

And he continued to express his loyalty to Weizmann, voicing his regret that Frankfurter had participated in the "campaign against Weizmann." Einstein was convinced that were he to "[g]et to the bottom of the matter on the basis of solid documents," he would change his mind.[110] The two months of being in the Weizmann camp had obviously had their effect on Einstein. Even if he had amended his views on Weizmann and the Zionist Organization temporarily, he was firmly back in their camp by the time he left for England.

In his response to Einstein's accusations against the Brandeis-Mack faction and his defense of Weizmann, Frankfurter pointed out that Einstein did "*not* know the facts upon which alone a judgment can be based. I wish the facts did not exist. But they do. And perhaps there will be occasion, one of these days, for you to know them. Disillusionment is hard, I know."[111] Whether there was additional information about Weizmann that Brandeis and Mack had not disclosed to Einstein or whether Frankfurter did not know the details of what they had revealed to him is unclear.

Toward the end of his visit, Einstein's interest refocused on the planned medical faculty. On 21 May he attended a reception and banquet "tendered by the American Jewish Physicians Committee" to Weizmann, Einstein, Levin, and Mossinson at the Waldorf-Astoria in New York. At the banquet, which was attended by eight hundred Jewish physicians, the fundraising goal for the establishment of a medical faculty at the Hebrew University was set at $1,000,000, and $250,000 was subscribed.[112] On the same day, a formal agreement had been drawn up between Weizmann, Einstein, Isaac Naiditch, Levin, and Ussishkin, on the one hand, and the president and officers of the American Jewish Physicians Committee on the other that laid out the preliminary statutes of the American Jewish Physicians Foundation to be established for raising funds for the planned medical faculty in Jerusalem.[113]

Even with this obvious success in raising funds for the planned medical faculty, from the point of view of Solomon Ginzberg, who represented the university's interests most closely, Einstein's campaign on behalf of the university had failed dramatically. In a memorandum written half a year later, he concluded that the efforts had not been crowned with success. As the interests of the fund were seen as competing with those of the Keren Hayesod, "the University Fund did not receive from the Zionist bodies anything like the support which might be expected in the face of the repeated Zionist demands for a beginning of a University." Regarding efforts in the United States, Ginzberg was disappointed that the endeavors surrounding Einstein's mission centered on utilizing "Prof. Einstein's prestige for the work of the Keren Hayesod and to safeguard the authority of the latter by insisting on donations to the University Fund being paid in through the Keren Hayesod treasury, even when that condition led to difficulties (as was the case with the New Century Club of Boston)." Furthermore, Weizmann's supporters were "dominated by the fear lest 'Prof. Einstein should take away money available for Weizmann' as some Chicago Zionist put it." Therefore, Einstein "has never been put in contact with well-to-do Jews of the semi-Zionist baalei-batim type, and practically no preparations were made by local Zionists for Prof. Einstein's activities." This was a reference to the lay Jewish leaders who supported religious institutions. The situation was different only in Boston, "where a special function for the University Fund was arranged (as that was insisted upon by Prof. Einstein as a condition of his visit), and as a result about 17,000 dollars were subscribed." In Ginzberg's opinion, "This example shows what might have been done with the active support of the Zionist machinery." Ginzberg concluded that the University Fund had also not succeeded in motivating Jewish millionaires in the various communities to make donations to the fund, even if their interest in Zionist causes in general was lukewarm":

> Experience has shown . . . that the well known generosity of Jews towards the Universities of their countries must be as a rule rooted in motives other than interest in the cause of science; motives absent in our case. The fact is that both the effort of Prof. Einstein in America (made it is true at an unfavorable time, but backed by his unique prestige in those very circles of the Jewish "Society") and an attempt to form an influential University Aid Committee in England, were not rewarded with any appreciable success.[114]

It seems highly ironic that the Zionist establishments of both Europe and the United States contributed substantially to the relative failure of Einstein's mission in the United States and the UK. It is also clear that the Brandeis-Mack faction warned Weizmann that Einstein should not visit the United States at the planned time, and then subsequently did its share to undermine his mission.

Just before his departure for England, Einstein penned his first written summary of the trip. His claims were not modest. He informed his close friend Michele Besso that he had "two immensely strenuous months behind me but have the great satisfaction of having been very useful to the Zionist cause and of having assured the founding of the university." Einstein obviously believed that his tour had constituted a crucial step in securing the establishment of the Hebrew University. However, this was far from the case. Einstein must have been aware that the primary goal of his mission, the establishment of a university aid committee, which was to have been facilitated by the proposed drawing-room meeting in mid-May, had not been realized. It is intriguing that Einstein could not admit this failure, even to such an intimate friend as Michele Besso.

In a typical tongue-in-cheek remark, Einstein described his personal role during the tour in the following manner:

> I had to let myself be shown around like a prize-winning ox, speaking countless times in large and small assemblies, deliver countless scientific lectures. It's a miracle I endured it. But now it's finished and what remains is the fine feeling of having done something truly good and of having worked for the Jewish cause despite all the protests by Jews and non-Jews—most of our ethnic comrades are smarter than they are courageous, I could definitely see that.[115]

Thus, Einstein was well aware of the opposition his visit had raised in various quarters. In characteristic fashion, he portrayed himself as struggling heroically against all forms of antagonism—once again, the "valiant Swabian" had prevailed. Yet his European view of the conflict between the various Zionist factions he had just witnessed had apparently not enhanced his understanding of the motivations of the elite of American Jewry. He interpreted their unwillingness to contribute to the cause he had been advocating for the past two months as lack of courage, thus arguably misunderstanding their goals and interests.

The failure of the efforts in the United States to establish a university aid committee seems to have had a decisive impact on similar plans in the UK. On the occasion of Einstein's planned visit to London in early June, the Zionist Organization planned to hold a banquet on behalf of the Hebrew University under the chairmanship of Lord Rothschild or Sir Alfred Mond. The main aim of the planned event was to "invite a lot of money-bags to canvass them for pledges beforehand, so as to be able to announce at least some donations during the banquet, and so to induce others." By early May the planned banquet had been canceled, as "Einstein refuses banquet [in] London because tired[.] consider inadvisable [to] insist[,] success being doubtful."[116] On 30 May, after two exhausting months, Einstein and Elsa left New York for Liverpool on board the *Celtic*.[117]

Upon his arrival in England, an article on Zionism attributed to Einstein was published in the *Jewish Chronicle*. Four days later a revised version of this article was published in the *Jüdische Rundschau* under the title, "How I Became a Zionist." However, Einstein was displeased with the article being attributed to him rather than being defined as an interview with him.[118] Thus, we find further evidence of Einstein distancing himself from identification as a Zionist.

After his return to Berlin, Einstein wrote to Ehrenfest of his impressions of his tour to the United States and the UK. As with Besso, he first described the personal toll of the voyage: "The trip to America and England was so taxing that I'm only good for vegetating." And as he had told Besso, he described his mission on behalf of the university as "very successful." More specifically, he informed Ehrenfest that the establishment of the medical faculty had been secured by the Jewish middle classes, "not by the rich."

As Ehrenfest had doubted the usefulness of "mass actions" on behalf of Zionism when his friend initially became involved in the movement, Einstein apparently needed to reassure him of the value of Zionism for world Jewry. In his view, "Zionism really offers a new Jewish ideal that can give the Jewish people joy in its own existence again." Einstein thus ascribed to Zionism a crucial role in improving the psychological condition of the Jewish people. Yet he was also writing about himself—Einstein's involvement in the Zionist movement and the trip had obviously contributed to his own emotional reinvigoration (despite the strenuous tour). This could provide us with a partial explanation for why he obviously could not accept that his mission was, in part, a failure. As with Besso, he could also not admit this failure to Ehrenfest.

Einstein voiced his general satisfaction with having agreed to Weizmann's invitation to embark on the tour, yet also revealed a certain degree of apprehension about the future direction of Jewish nationalism itself: "There are ... in some places signs of high-strung Jewish nationalism that threatens to generate into intolerance and small-mindedness; but this is hopefully just a childhood disease."[119] This constitutes the first expression of a concern that would come to preoccupy Einstein more and more over time.

The contrasting success of the efforts to raise funds among the various sectors of the Jewish community in the United States during Einstein's visit on behalf of the Hebrew University was confirmed by Ginzberg. Whereas the efforts among the wealthier members of the community were not fruitful, those among the Jewish physicians met a great deal of success. Originally, the main goal of the University Fund was to "try and reach such wealthy Jews as may be interested in a Jewish University although indifferent towards Zionism generally." However, not even Einstein's visit could help the fund attain this goal: "It seems that Jewish millionaires are on the whole viewing a Hebrew University in Jerusalem in the same light as Zionist work generally; only those who are in sympathy with the latter are also in sympathy with the former, so that in these circles no special source of revenue for the University has until now been found." In contrast, the efforts of the American Jewish Physicians Committee were seen as "a successful experiment ... with regard to organisation of Jewish professions for the work of the University. ... In these circles there really seems to be a very strong feeling in favour of the University, even amongst non-Zionists."[120]

Apparently as a consequence of the perceived success of his tour and the enthusiasm it had generated among the Jewish masses in the United States, Einstein fever finally hit Berlin. At a Jewish "mass rally" on 27 June at the Blüthner Hall, Einstein was welcomed "with rapturous rejoicing." According to press reports, the lecture led to such a large crowd outside the lecture hall that the police had to regulate the street traffic.[121]

In early July, the German Zionist Federation held an event at the Hotel Kaiserhof in Berlin in honor of Einstein's return from his trip to the United States. According to the press, Einstein declared in his speech that "in supporting Zionism, he believes that he is above all advancing the cause of peace, as the regeneration of the Jews being carried out under the leadership of the Zionists ought to provide the friends of world peace with renewed optimism."[122] This ties in with a tendency we have previously seen in Einstein, namely, that he resolved the apparent contradictions between his general disdain for nationalism and his enthusiasm for Zionism by interpreting his support for Jewish nationalism as a contribution to international reconciliation.

In early July, an alleged interview with Einstein was published in the Dutch daily *Nieuwe Rotterdamsche Courant*, recounting his impressions of the United States.[123] Regarding the Zionist aspects of his trip, Einstein was quoted as expressing the "great emotion" he felt at his first encounter with "Jews *en masse.*" The interview also quoted Einstein as criticizing the wealthy Jews, who were allegedly more interested in their public image than in contributing towards establishing the Hebrew University:

> [T]he American millionaire prefers of course to buy power with his wealth, and this he accomplishes mainly by getting public opinion on his side, but what is it to public opinion whether he also donates a large sum to the Zionist fund, even though perhaps a chair will be named after him?[124]

After a partial translation of the interview, which contained many other provocative statements on the Unites States and its inhabitants, was published in the *New York Times*,[125] Ginzberg sent Einstein two cables in which he conveyed the Zionist Organization of America's outrage at the alleged reports and pressed Einstein for immediate denial. In the more relevant one, Ginzberg conveyed the organization's obvious concern that Einstein's alleged criticism would undermine support for Weizmann. The *Times* report ascribed Einstein's "derogatory" views toward America to his "spleen to failure of [the] Weizmann mission." The Zionist Organization of America requested a correction to this report that would emphasize the success of Weizmann's campaign in the United States. The next day Ginzberg also sent a letter, explaining that the alleged misrepresentations of Einstein's views would be detrimental not only to his own prestige in the United States but also "to the Zionist cause in this country."[126]

The very same day, Einstein replied to both communications from Ginzberg. He denied "any responsibility for unauthorized notices about me. Specifically re Dutch article." He also claimed that his private remarks had been "completely distorted" and that he "approved without reservation Weizmann's leadership." Furthermore, he viewed the U.S. trip as a "huge success."[127]

With these denials, the immediate aftermath of Einstein's tour of the United States was over. Yet not long afterward, preliminary discussions regarding what would eventually become Einstein's next journey on behalf of the Zionist movement were already being held between the prominent scientist and the president of the Zionist Organization. Einstein had apparently asked Weizmann whether arrangements could be made for him to visit Palestine. In reply, Weizmann wrote, "I am sorry I cannot yet give you a definite answer to your question. However, it seems to me there is as yet no great urgency to travel to Palestine." In the meantime, fundraising was deemed to be more of a priority. Therefore, Weizmann asked whether Einstein would be prepared to "make a tour of two months in America." In light of the short time that had elapsed since Einstein's completed visit, the exhaustion it had caused him, and the limited success of the mission, it is amazing that Weizmann had the audacity to ask Einstein to embark on yet another tour of the United States. Weizmann told Einstein that "action in America would perhaps be more important, and that a later journey to Palestine would be more opportune."[128]

However, Einstein would not follow Weizmann's advice; he would actually not visit the United States again until 1930, and this would not be under Zionist auspices.

This detailed analysis of Einstein's two-month tour of the United States at the height of a crisis among the leaders of international Zionism has been based on a plethora of previously unused sources. It has enabled us to completely reassess Einstein's first visit to American soil. As a consequence, Einstein's participation in the Zionist delegation can no longer be seen as a mere adornment to the "clash of heroes" between Weizmann and Brandeis. The vast array of new sources has also made it possible to present the broad spectrum of the major players' perspectives on Einstein's participation in the Zionist delegation.

An analysis of these sources has revealed that two crucial factors formed the background for Einstein's trip. From Einstein's own perspective, it was his failed negotiations for a personal lecture tour at elite American universities. From the perspective of the Zionist movement, it was the evolving conflict between Chaim Weizmann and the leadership of the Zionist Organization of America over the issue of the establishment of the Keren Hayesod.

The internal Zionist correspondence also shows us to what extent German Zionist leader Kurt Blumenfeld had to utilize his extensive persuasive powers to convince Einstein to participate in the tour. The invitation itself and Blumenfeld's preparation of Einstein for the tour were undoubtedly further instances of the Zionist movement's subtle manipulation of Einstein. Yet we have also seen how Einstein manipulated Blumenfeld in regard to certain aspects of the planned trip, most notably in his account of his canceled personal lecture tour.

The invitation also resulted in a critical exchange of correspondence between Einstein and his close friend Fritz Haber that revealed fundamental differences between the assimilationist and anti-assimilationist stances among German Jewry represented by these two prominent scientists on the role of the Jews in German society.[129] Moreover, despite Einstein's alleged bravado in reaction to the criticism of assimilated Jews to his planned trip, Fritz Haber's criticism of Einstein's acceptance evoked a detailed and (at times) agonized response.

Einstein had various motivations for accepting the invitation: the primary incentive was his continued passion for the Hebrew University project, which he viewed as an instrument for improving the plight of his less fortunate "ethnic comrades." His own experiences of heightened anti-Semitic attacks in the year following his mobilization on behalf of Zionism had further deepened his sense of identification with young Jewish academics from Eastern Europe. The tour also afforded Einstein the opportunity to advance another of his cherished goals, reconciliation among the international scientific community. Yet there were also personal motivations for his acceptance. The visit enabled him to promote his scientific theories at the most prestigious American universities, even though his hopes for such a tour had just been dashed by his failed negotiations with those institutions. The tour also provided Einstein with an opportunity to become more prominent on the international stage, even though publicly (and privately) he abhorred the media circus this entailed. It is intriguing that these goals were obviously of greater importance to Einstein than his participation in the most important annual European conference in physics, the Solvay Congress, which he was invited to attend as the only German scientist. We must also consider the possibility that Weizmann's charisma was one factor swaying Einstein to accept the invitation, thus providing us with another example of his susceptibility to the influence of charismatic figures.

A question that emerges from Einstein's participation in the delegation is how well informed he was in regard to what was happening behind the scenes in the Zionist movement. From the accounts of Blumenfeld, Weizmann, and Ginzberg, we can conclude that Einstein was very poorly informed about what was going on, particularly in regard to the Keren Hayesod controversy. Indeed, it is even unclear how well informed Einstein was about the Zionist Organization's plans in regard to the Hebrew University in the United States. He initially thought he would participate in "constitutive discussions" on the university. Yet it is not clear how much he knew about the proposed university aid committee prior to his arrival in the United States. Arguably, Einstein could have tried to be more informed. Yet he seems to have been disinclined to do so, perhaps out of a lack of interest or his natural tendency to try to remain "above politics." In any case, Einstein's lack of information seriously backfired for Weizmann and his supporters, at least temporarily.

Once they joined the Weizmanns and the other members of the Zionist delegation on board the *Rotterdam*, the Einsteins began to belong to the "Weizmann camp." Yet Weizmann later reported that he had made no attempts to "rope Einstein in" as a Zionist. As with Blumenfeld's efforts to persuade Einstein of the merits of Zionism, Weizmann preferred to use an indirect approach. However, there was an important contradiction here: Einstein clearly belonged to the Weizmann camp, yet he was not treated by the members of the delegation as an insider, as they obviously did not fully trust him. This situation eventually damaged the relationship between Einstein and Weizmann.

The multiple invitations Einstein received for the trip are indicative of the manifold agendas of the individuals and organizations involved in the planning and execution of the tour. Einstein's mission to the United States came at a time when widely divergent agendas were jockeying for power within the Zionist movement.

Publicly, Einstein's participation in the Zionist delegation was linked to the Zionist Organization's fundraising efforts in the United States on behalf of the Hebrew University. Weizmann's overt agenda was for Einstein to bolster the fundraising efforts for the Hebrew University and to a lesser extent for the Keren Hayesod. As secretary of the Zionist Organization's University Committee, Solomon Ginzberg's agenda was most closely aligned with the direct objectives of the Hebrew University. He saw the primary goal of Einstein's participation in the Zionist delegation as forming the catalyst for major funding from influential Jews in the United States and the UK. This campaign was to be aimed primarily at wealthy non-Zionist Jews, because of the perceived nonideological nature of the university project (which, of course, was also its appeal for Einstein). The instrument for the funding was to be the university aid committees that were to be established in both countries. The public agenda of the American Jewish Physicians Committee was to utilize the prestige of Einstein and (to a lesser extent) of the other members of the Zionist delegation to raise funds for the planned medical faculty of the Hebrew University.

However, the hidden and semihidden agendas of the dramatis personae were of far greater consequence. The most important semihidden agenda was the appeal the preeminent physicist was deemed to have as a leading international non-Zionist to American non-Zionists in the planned establishment of a university aid committee in the United States. It was also hoped that this would have a positive impact on plans for such committees in the UK and Germany.

Weizmann's hidden agenda in soliciting Einstein's participation was his need for the support of a prestigious international Jewish individual to offset the influence of the Brandeis-Mack faction within the Zionist Organization of America. He hoped to thereby strengthen the faction of his supporters within the American Zionist movement and push forward with the establishment of an office for the Keren Hayesod in the United States.

The faction of Weizmann's supporters in the Zionist Organization of America had two semi-hidden agendas: to raise money for the Keren Hayesod (as opposed to raising money for the Hebrew University, which they saw as a rival for funds) and to gain the upper hand in their internal struggle within the Zionist Organization of America.

The Brandeis-Mack faction within the organization had a number of semihidden agendas: to contain Weizmann's growing domination of the Zionist Organization and to gain the upper hand in the internal Zionist Organization of America struggle. Their hidden agenda was to limit Einstein's involvement to issues related to the Hebrew University and to avoid his becoming involved in the broader issues concerning the Zionist movement, most notably the Keren Hayesod. They also aimed (and succeeded) at neutralizing Einstein's appeal to wealthy non-Zionist Jews, thereby effectively torpedoing the initiative to establish a university aid committee in the United States.

This study has shown that Einstein's mission to the United States on behalf of the Hebrew University was a relative success: the "prize-winning ox" certainly succeeded in unlocking some of the dollars the Zionist movement had hoped to raise in the United States. We have seen that the effective outcome of the fundraising efforts for the planned medical faculty and the library were undoubtedly major accomplishments. However, the varying and conflicting agendas had a decidedly negative impact on the success of the tour. The inadequate communication between Weizmann and the Brandeis-Mack faction regarding Einstein's arrangements seriously impeded both his planned lecture tour in the United States and his fundraising mission on behalf of the Hebrew University. The negative perception of Einstein among some of the elite American universities clearly had a detrimental impact on the Zionist Organization of America's leadership. And, in turn, their restrained enthusiasm about Einstein's tour and his mission had a damaging effect on the willingness of wealthy non-Zionists to get involved in the Hebrew University. Along with its chief proponent (the "commercialized" scientist), the university was deemed to be lacking in respectability.

It is important to note that as far as the Hebrew University was concerned, Einstein's priorities during the tour lay in fundraising and not in intellectual discussions. In contrast, this was not the case in regard to his meetings with his American scientific colleagues.

It is intriguing that Einstein agreed to the constraints on his freedom of speech during his U.S. tour. Blumenfeld had advised Weizmann to prepare Einstein's speeches for him. However, as he was not naturally inclined toward compliance or submission, Einstein's toeing the party line only worked for a while. Eventually he broke out of the Weizmann camp and met with Brandeis without a "shadow" from the Zionist leader's court. As we have seen, it was possibly the Einsteins' frustration at how they were being treated by some of the members of the Weizmann camp that led to their meeting with Brandeis. Viewed symbolically, Einstein's breaking out on his own to travel to the U.S. capital constituted a trip from "Pinsk" (as Weizmann's faction is referred to in U.S. Jewish historiography) to "Washington" (as the Brandeis faction is referred to).[130] Following the crucial meeting in

Washington, fundraising for the Hebrew University and Einstein himself became full-blown political issues in the clash between Weizmann and Brandeis. In disclosing their accusations to Einstein, Brandeis and Mack clearly wanted to inflict major political damage on Weizmann. However, to date, none of this seems to have been noticed by historians of the conflict.

Weizmann and his supporters made a grave miscalculation in employing Einstein in their attempts to mobilize the wealthy Jewish non-Zionists for the cause of the Hebrew University. They overestimated the impact that a German-Jewish non-Zionist would have on his American counterparts. In general, there was a fundamental mutual lack of understanding between the two sides. As with the conflict over the Keren Hayesod, there was a clash of cultures and mentalities between the European and American Zionists.

It is intriguing that Einstein was temporarily swayed by Brandeis following their meeting in Washington at which the justice disclosed the alleged misappropriation of funds by Weizmann. However, soon after Einstein returned to the Zionist leader's camp, Weizmann succeeded in controlling the damage that Brandeis had temporarily inflicted on Einstein's relations with the European Zionists. Thus, Einstein's trip to "Washington" was a brief excursion. Soon after his return to New York, he was firmly back in the environs of "Pinsk."

As Weizmann's supporters were fearful of available funds going to the university and not to the Keren Hayesod, they did not facilitate contacts between Einstein and the "semi-Zionist" wealthy Jews who backed them financially. Therefore, it is highly ironic that the Zionist establishments of both Europe and the United States (and, within the United States, both the pro-Weizmann and the pro-Brandeis factions) contributed fundamentally to the relative failure of Einstein's mission in the United States and, by means of a knock-on effect, in the UK. All these players undermined Einstein's position and endangered, at least temporarily, the success of the university.

The incident over the planned lecture (or lectures) at Harvard revealed the divergent political positions of the various players and further exacerbated the tensions between the two factions. The Weizmann faction apparently supported Einstein's willingness to address a politically progressive group, whereas the Brandeis-Mack faction may have been fearful of the effect this would have on their relations with the Ivy League universities.

Despite his aversion to discussing intellectual matters regarding the university during his tour, there was one exchange in which such issues were deliberated. Einstein's correspondence with Solomon Rosenbloom on the planned Jewish studies institute at the university amazingly revealed the limits of the physicist's concern about the freedom of scholarship. His willingness to concede to Rosenbloom that only religious scholars would decide the composition of the faculty highlighted the unease of such a secular academic in dealing with the possible impact of religion on scientific research.

Einstein's accounts to his closest friends about the trip were quite astounding: despite failing to achieve the main objective of his mission, the creation of the proposed university aid committees, he informed them that he had "secured" the establishment of the university. Apparently, the "valiant Swabian" could not admit failure to his closest friends—or even to himself, for that matter.

What impact did this extensive tour have on Einstein's attitude toward Zionism in general? The trip brought Einstein into close contact with the highest echelons of the European and American Zionist leadership, and he became embroiled in the intense conflict between them. This experience does not always seem to have been a pleasant one. Einstein returned from the trip somewhat disenchanted with his "ethnic comrades," whom he found "smarter than they are courageous." Yet he remained convinced that he had made a substantial contribution to the success of the mission on behalf of the university. Ultimately, this was the most important factor for Einstein: to advance a cause he viewed as being high above petty Zionist political infighting. The visit also brought Einstein face-to-face for the first time with "Jews *en masse*" (most of whom were of Eastern European origin).[131] Furthermore, he was very favorably impressed by the efforts of the American Jewish Physicians Committee (many members of which were also originally Eastern European Jews). In this context, one could almost say (in a slightly exaggerated manner) that Einstein, who had been mentored by an Eastern European student of medicine, could repay the favor by contributing to the establishment of a medical faculty for a planned refuge for Ostjuden—the Hebrew University. All in all, the tour does not seem to have brought about a substantial change in Einstein's basic position on Zionism. It continued to be a cause he supported in part and could utilize for his own interests. And despite his intense interaction with the heads of the Zionist Organization (or perhaps because of it), he also continued to maintain his distance from the movement's leadership.

Einstein fever finally hit Berlin (at least among certain sectors of the Jewish community) following his return home. Neither sensationalist reports on the verification of his theories nor controversy surrounding those theories in the German scientific community had led to public enthusiasm about Albert Einstein in his native country. It was Einstein's tour to the United States on behalf of the Hebrew University that finally brought traffic to a standstill in the German capital.

In the next chapter I examine how Einstein fever played out in another important venue for the preeminent scientist, Palestine under the British Mandate, and what impact this encounter with the land of his forefathers would have on his views on Zionism.

CHAPTER 4

SECULAR PILGRIM OR ZIONIST TOURIST?

Einstein's Tour of Palestine in 1923

"Great Brother! Step up onto the stage which has been waiting for you for 2,000 years!" This is how prominent Zionist leader Menachem Ussishkin welcomed Einstein to give a talk on relativity at the future site of the Hebrew University in Jerusalem in the winter of 1923.[1] The lecture was held on Mt. Scopus overlooking the Old City of Jerusalem in a large hall that had been decorated festively for the occasion with Zionist and British insignia. Ussishkin's words hint at the messianic fervor and anticipation among the Jewish community evoked by their preeminent guest's brief visit to Palestine to tour the country. As the local press proclaimed, wisdom was once more "going forth from Zion." But how did this tour come about? What was going on behind the scenes? What was Einstein's own reaction to the land of his forefathers, and how did it affect his views on Zionism?

In biographical studies of Einstein, his tour of Palestine in February 1923 has usually been treated cursorily, as a mere by-product of his extensive trip to the Far East in the autumn and winter of 1922–23. A few biographies have attributed some importance to the visit. Philipp Frank stresses the exceptional nature of the visit: he claims that Einstein was "unable to be simply an unparticipating observer" as he "had carried on propaganda for the development of a Jewish national home in Palestine." Frank also emphasizes Einstein's reception as a public figure by both the Mandate authorities and the Jewish community. In contrast to Frank, Ronald Clark has a far less sympathetic view of

Einstein's visit: in his opinion, Einstein's and Palestine's enthusiasm for each other was "almost embarrassing." He also claims that Palestine "strengthened [Einstein's] Zionist sinews." Albrecht Fölsing tries to explain Einstein's enthusiasm for Palestine in the wake of his visit; he believes that "[i]t was th[e] practical work [of the Jewish construction workers], more than anything else, that made him believe in the future of a Jewish Palestine."[2] It seems that Frank based his study on personal interviews with Einstein; Clark utilized English-speaking newspapers and British memoirs; and Fölsing made use of Einstein's travel diary. However, as with his trip to the United States, none of the biographical studies of Einstein to date have utilized the archival material from behind the scenes or the vast coverage of the tour in the local Hebrew press.

In the aftermath of his extensive tour of the United States, Einstein's intense involvement with Zionist activities continued. When Chaim Weizmann visited Berlin in December 1921, Einstein participated in some of the critical discussions that took place between the Zionist Organization's leader and the German Zionist Federation. Einstein even hosted one of these meetings in his apartment.[3]

Privately, however, his intense interaction with the Zionists during the American tour and its aftermath had taken a toll on Einstein. We have already seen how he expressed his exasperation about the Zionists being "shameless and importunate" in the wake of his U.S. tour, yet at the same time he restated his favorable disposition to the cause.[4]

In the spring of 1922, Einstein began to plan his next trip abroad. In May 1922 he accepted an invitation to undertake a six-week tour of Japan.[5] However, soon after his acceptance, the circumstances under which he would embark on his trip to the Far East changed radically. On 24 June 1922 the Jewish German foreign minister Walther Rathenau was gunned down in Berlin by right-wing extremists. Rathenau's assassination was a traumatic watershed event for Einstein; he had been close to Rathenau and was "deeply shocked" by his violent death. Einstein received death threats in the wake of the assassination and had to go into hiding temporarily. His immediate reaction to the assassination was a desire to flee Berlin and not return. He also contemplated leaving Germany altogether. Believing that in light of the assassination he could not represent German intellectuals, Einstein decided to withdraw from participation in the League of Nations' newly founded International Commission on Intellectual Cooperation. In the ensuing weeks, though, he apparently viewed the situation more calmly. He changed his mind about leaving Germany altogether and decided to remain in Berlin following his planned return from the Far East. Nevertheless, prior to his departure, it seems that he was still planning to give up all his official positions in Germany on his return.[6]

Einstein's prolonged absence from Berlin became an issue of great political controversy in late December 1922 during his tour of Japan. The German Jewish journalist Maximilian Harden claimed that the main reason for Einstein's extended trip was his decision to seek refuge outside Germany following Rathenau's assassination. Harden himself had been the victim of an assassination attempt a few days after Rathenau's death. In response to a cable from Wilhelm Solf, German ambassador to Tokyo, Einstein confirmed that his trip had been motivated both by "a yearning for the Far East" and by the "certain amount of imperilment to [his] life" that had existed after Rathenau's murder.[7]

Einstein's extensive tour of the Far East formed the broader context for his visit to Palestine. The genesis of Einstein's specific plan to tour Palestine and of the initial preparations for the visit is slightly complex. As we saw in the previous chapter, Einstein had apparently asked Chaim Weizmann sometime in the autumn of 1921 whether arrangements could be made for him to visit Palestine. Weizmann had replied that such a trip was not urgent. It is important to note that it seems it was Einstein's own idea to visit Palestine, even though we cannot know for certain whether this decision was influenced by individuals such as Blumenfeld. The first evidence we have of Einstein's renewed interest in touring the region can be found in late September 1922. After he had met with Einstein on the very day of his departure from Berlin for the Far East, Blumenfeld noted that Einstein had agreed to accept an invitation to visit Palestine extended by Arthur

Ruppin, the director of the Zionist Organization's Palestine Office.[8] He planned to travel to Palestine on his return trip from Java back to Berlin.[9] This visit was to be brief, only ten days. According to Blumenfeld, Einstein emphasized that this short sojourn was "not to be confused with his actual trip to Palestine." He quoted Einstein as saying "One should only travel directly to Palestine and not incidentally following a trip to other countries." Einstein was also aware that he could not form an opinion on the issues in which he was truly interested in merely ten days. Therefore, it was clearly important for Einstein to get to know the country and its problems thoroughly and to form his own opinion about developments there. This, it seems, is exactly what the Zionist Organization feared. In his notes, Blumenfeld also mentions that it would be necessary to present to Einstein the Zionist Organization's position on the establishment of the Hebrew University as an "official position." This was necessary as "[h]e would then campaign for this cause and be prevented from adopting random opinions on the university plan as his own."[10] Thus, the Zionists' suspicion of Einstein (which we have observed in previous chapters) reemerged: from their point of view, their "prize-winning ox" could evidently not be trusted to form his own informed opinion on crucial Zionist matters and needed to be presented with the party line on every pertinent occasion.

Reports on the planned truncated trip appeared in the press on 6 October and the following days. It was immediately clarified that this was not to be the originally planned trip and that Einstein would remain in Palestine "for several months" after his return from the Far East. It was also reported that the goal of this brief trip was "to acquaint himself with conditions in the country" and he would "in particular . . . pay a visit to the Hebrew University in Jerusalem where it is believed it will be possible for him to deliver several lectures."[11]

However, correspondence regarding the trip between Einstein and Weizmann and among Zionist functionaries took place in parallel to these initial reports in the press. Apparently unaware of Blumenfeld's recent meeting with Einstein, Weizmann suggested to the latter, "Perhaps you will also pass by Palestine and could on your *return trip* take a detour from Port Said." As Einstein had already departed from Berlin, his stepdaughter (and secretary) Ilse Einstein replied to Weizmann that her parents would travel to Palestine unless they were prevented from doing so by extraordinary circumstances. In her opinion, it would be "a great joy for them" were Weizmann also to be there during their planned visit. She surmised they would arrive there in February. However, Weizmann had to visit Palestine urgently in November 1922 and subsequently embarked on an extended fundraising tour of the United States.[12]

Ten days following Weizmann's letter, Ruppin notified the Zionist Executive of information he had received from Blumenfeld. Einstein had accepted the invitation and planned to arrive in Palestine in late February or early March. Ruppin believed the visit might have "great propaganda value" for the Zionist Organization, and in particular for the university project. As a consequence of the Zionist Organization's experiences with Einstein during the U.S. tour, Blumenfeld thought it "absolutely necessary" not only that Einstein be accompanied by "Sicma [sic] Ginzburg" but also that Elsa Einstein be accompanied by a woman, preferably Rosa Ginzburg. Thus, during this trip both Einsteins were to receive official "shadows." Regarding the university, the Zionist Executive should contact Weizmann and find out from him "which projects and explanations you should give to Prof. Einstein."[13]

Simultaneously, Ruppin asked Weizmann to tell the Zionist Executive "which opinions about the university project are to be viewed as the official stance." This was deemed necessary, as, according to Blumenfeld, if one did not give him any directives "the danger exists that he endorse opinions presented to him randomly and this could diminish the propaganda effect of his tour."[14]

It does not seem that Einstein had the time (or the inclination) to prepare himself for the upcoming trip. The only extant book on Palestine in Einstein's personal library was sent to him by its author, the left-wing travel writer Arthur Holitscher, in August 1922. However, the book was never opened by Einstein.[15]

Even though, according to the Zionist Organization's plans, the

Hebrew University was to be the focus of Einstein's upcoming visit to Palestine, the proposed Jewish university played a crucial role in one of his ports of call well before he set foot in the Holy Land. On arriving in Singapore in early November, Einstein learned that Weizmann had decided "to exploit my trip for Zionist purposes." He was asked to solicit a donation for the university from Menasseh Meyer, "the Jewish Croesus of Singapore." It is intriguing to note that Einstein was informed of this plan at the last minute. The Zionist Organization obviously believed it could ask him to do so without any previous notification or preparation. Indeed, after broaching the subject with the millionaire, Einstein was very uncertain whether the mission had been successful: "Finally, the well-planned ambush of Croesus initiated by Weizmann took place (for a contribution for the Jerusalem university), yet I do not know if, in spite of numerous attempts, any of my projectiles penetrated Croesus' thick epidermis." In fact, the success was quite modest: Meyer donated £500 and the rest of the Singapore Jewish community £250. Moreover, behind the scenes (and unbeknownst to Einstein), this limited accomplishment was attributed more to a letter Weizmann wrote to Meyer than to the physicist's visit.[16]

Perhaps Meyer would have been more impressed with Einstein if he had known that within days of the visit to Singapore, Einstein would become a Nobel Prize laureate. Just prior to his arrival in Shanghai, Einstein received word from the Royal Swedish Academy that he had been awarded the Nobel Prize for Physics for the year 1921.[17]

Amazingly, perhaps, his winning the Nobel Prize was not an event Einstein felt the need to record in the travel diary he kept during his trip to the Far East, Palestine, and Spain. This journal is the first extant journal written by Einstein and, as far as we know, it was the first such diary kept by him. The diary is a fascinating document that provides us with gripping insights into the most immediate levels of Einstein's experience of the journey, in particular of the people he met and the places he visited.

The journal offers a snapshot of the Jewish community in Palestine (the *Yishuv*) at the time of Einstein's visit from the perspective of the document's author. The British Mandate had been established in Palestine only two and a half years before Einstein's visit: Sir Herbert Samuel had assumed office as high commissioner in July 1920. Jewish immigration had intensified after World War I—this "Third *Aliyah*" (i.e., immigration wave) consisted mainly of immigrants from Eastern Europe. At the time of Einstein's visit, the Jewish population numbered 86,000 out of the general (predominantly Arab) population of approximately 600,000. Political tensions were high during this period: in early May 1921 (while Einstein was visiting the United States), attacks on Jewish residents in Jaffa by Arab residents and quelling of the riots by the Mandate authorities led to the deaths of forty-seven Jews and forty-eight Arabs. In June 1922, the Churchill White Paper was issued which reaffirmed the Balfour Declaration, yet also declared that Jewish immigration would be restricted, as a reassurance to the Arab population. The Third Aliyah also led to a considerable increase in the number of agricultural settlements. In addition, foundations were laid for the industrialization and electrification of Palestine. In the three large cities—Jerusalem, Tel Aviv, and Haifa—modern urban patterns began to crystallize. There were also significant achievements in the establishment of autonomous institutions of the Yishuv. The Assembly of Representatives (Assefat Hanivcharim) served as the highest legislative body and the National Council (Va'ad Leumi) served as the executive body. The General Federation of Labor (the Histadrut) had been founded in 1920. The economic conditions in the country at the time of the visit were dire as a consequence of World War I. The Jewish economic section could not provide employment for all the Jewish population or for the new immigrants.[18]

At the end of a hectic six-week tour of Japan, the Einsteins set sail for Egypt on 30 December 1922. Following a month's voyage, they disembarked at Port Said on 1 February 1923. However, this was not Einstein's first arrival in the Middle East. Nearly three months earlier, he had had his first encounter with the Levant en route to Japan. Einstein's ambivalent reaction to the region and its inhabitants was similar to that of other European travelers: he was struck by the "screaming and gesticulating Levantines," who flooded the ship upon its arrival. Yet the "bandit-like, filthy Levantines" were also "handsome and gracious to look at."[19]

Einstein's second arrival in Port Said evoked from him a negative reaction without ambivalent feelings: "City a regular den of foreigners with correlating riff-raff."[20] The Einsteins traveled by train from Port Said to El Cantara, where they stayed overnight. On the morning of 2 February they continued by train to Lydda (Lod), their first stop in Palestine. They were met at the station by Ussishkin and Mossinson (with whom they had traveled to the United States) and various other senior Zionist officials. They then continued by train up the Judean Hills to Jerusalem, where they were greeted by Solomon Ginzberg, who once again was to be their official escort during the trip.[21] In Jerusalem, their first hosts were the British Mandate authorities: they were driven by the high commissioner's aide-de-camp to his official residence at Government House, at the former Augusta Viktoria hospice. There they were welcomed with an official gun salute and met Sir Herbert Samuel and the members of his family.[22]

The following morning, a Saturday, Samuel accompanied Einstein on a walk to the Old City of Jerusalem. Einstein entered the Old City with Ginzberg and toured the Dome of the Rock, the El-Aksa Mosque, the Western Wall, and the city ramparts. Ginzberg hosted Einstein for lunch, at which Ruppin was also present, and there ensued "cheerful and serious conversations" After lunch they visited the Bukharian Jewish quarter in West Jerusalem before calling on Hugo Bergmann at his home.[23]

On Sunday, 4 February, Ginzberg and Samuel's daughter-in-law, Hadassah Samuel-Grazovski, drove the Einsteins down to Jericho to tour its ruins. After lunch they drove along the Jordan River to the Allenby Bridge. They then returned to Government House and had tea with Sir Wyndham Deedes, civil secretary of the Mandate authority, and discussed "religion, nationalism." In the evening they had a "cozy conversation" with Herbert Samuel and his daughter-in-law.[24]

Figure 9. Jaffa gate, Jerusalem in the early 1920s

The following day, the official Zionist agenda of the visit commenced. Einstein was first given a tour of the new Jewish garden
suburb of Beth Hakerem. He then visited the Jerusalem headquarters
of the Zionist Executive and toured the Zionist Organization's Museum of Agriculture. In the afternoon he visited Talpiot, another garden suburb, and was then welcomed by Hugo Bergmann at the Jewish National Library, where he received a standing ovation from the
readers. In the evening a private reception was held for Einstein at
Ussishkin's home to which senior Zionist and British officials were
invited.[25]

On 6 February, Einstein first toured the Bezalel Art Academy and
promised to send its director, Boris Schatz, Emil Orlik's portrait of
him for the planned National Museum.[26] He then proceeded to the
major popular reception held in his honor in Jerusalem, which was
held at the Lämel School. The event was organized by the Zionist
Executive and the National Council. The large crowd cheered upon
Einstein's arrival and tried to rush the school gates, but was pushed
back by the ushers. "All the Jewish pupils in Jerusalem" formed a guard

Figure 10. Einstein and Menachem Ussishkin, Jerusalem, February 1923

of honor. At the reception, attended by an audience of two hundred, Einstein was welcomed by the heads of the two organizing institutions, Ussishkin and David Yellin. In their speeches, Einstein was called on to settle in Palestine, termed a "genius" (*Gaon*), and presented with the Jewish National Fund's "Golden Book" in recognition of his services to the Zionist cause. In his speech (as reported in the press), Einstein said that this was "the greatest day of [his] life" and that "only Zionism can heal the sick Jewish soul." In the evening, the Einsteins were invited for dinner at the home of Norman Bentwich, attorney general for the British Mandate.[27]

The next morning, Frederick Kisch accompanied Einstein on another walk to the Old City to visit the Christian Quarter. Kisch explained to Einstein "the political situation and some of the intricacies of the Arab question." Einstein mentioned that Ussishkin had attempted to persuade him to settle in Jerusalem: "He has no intention of doing so, not because it would sever him from his work and his friends, but because in Europe he is free and here he would always be a prisoner. He is not prepared to be only an ornament in Jerusalem."[28]

In the afternoon, the focal event of the entire visit took place on Mt. Scopus: Einstein's lecture at the future site of the Hebrew University. The event was a particularly festive occasion. The lecture hall at the mandatory police school in Gray Hill House was decorated with blue and white stripes and the Union Jack, symbols of the twelve tribes, the slogan "Light and Wisdom" (ora ve-torah), and portraits of Herzl and Herbert Samuel. Officially, the lecture was organized by the Zionist Executive. The invitees included the high commissioner, the British governor of Jerusalem, and other senior officials of the Mandate, as well as the Arab dignitaries; the heads of the Christian and Muslim communities; the Jewish dignitaries and the heads of the Zionist institutions in Jerusalem; foreign consuls; members of the scientific community from Jerusalem and Tel Aviv; and writers, teachers and journalists. However, the Arab dignitaries did not attend.

When Einstein, Samuel, and Ussishkin entered the hall, the audience burst into applause. Ussishkin greeted Einstein with the words quoted at the beginning of this chapter. Einstein began his lecture with a brief greeting in Hebrew, "which I read cumbersomely."[29] According to the press reports, Einstein expressed his joy at "giving his lecture in the land whence the Torah and light emanated to the entire enlightened world, and in the house, which is ready to become a center of wisdom and science for all the peoples of the east." He also expressed his regret that he "could not lecture to them in the language of [his] people." This was greeted by a "storm of uncontrollable ap-

Figure 11. Albert and Elsa Einstein with Tel Aviv mayor Meir Dizengoff and members of the Tel Aviv City Council, 8 February 1923

plause." He proceeded to lecture in French for an hour and a half on the theory of relativity. At its end, he received another ovation and Herbert Samuel eloquently thanked Einstein for his lecture.[30]

In the evening, an official banquet was held for the Einsteins at Government House that was attended by the Muslim mayor Raghib al-Nashashibi, William F. Albright, director of the American School of Oriental Research, E. T. Richmond (who served as assistant civil secretary [political] with special responsibility for Arab affairs), and a number of Christian Bible scholars and Orientalists.[31]

The next morning the Einsteins were driven to Tel Aviv by Mossinson and Ginzberg. The first public event there was a small-scale reception held at the Herzliyah high school, at which the pupils performed gymnastic exercises. A reception at the town hall followed: crowds lined the streets and the visitors were greeted with cheers. Mayor Meir Dizengoff announced that Einstein was to be named the first honorary citizen of Tel Aviv. In response, Einstein stated that he was "ten times happier to be a citizen of the beautiful Hebrew city" than of New York (whose honorary citizenship he had received in April 1921). Mossinson told the crowd that Einstein was learning Hebrew and that he hoped to soon be able to teach them in Hebrew at the Jerusalem University. In response, the crowd cheered and cried, "Long live Prof. Einstein!" After lunch, Einstein was given a tour of the major achievements of the new city: the electrical power station being built under the directorship of Zionist leader Pinchas Rutenberg, the quarantine camp for immigrants with contagious diseases, and the Silikat brick factory. This was followed by a large public reception back at the high school, which was attended by "thousands." Mossinson told the crowd that "Einstein has come as a Zionist to see the country . . . with the hope that he will afterwards be able to settle in it." This announcement was greeted with cheers. In his speech, Einstein expressed his excitement about the "huge efforts" the residents had made "in very difficult circumstances." The reception was followed by a tour of the agricultural experimental station established by the agronomist Yitzhak Volcani, of the scientific evening courses organized by Aharon Czerniawski, and of the Engineers' Association, where Einstein was

Figure 12. Tel Aviv, 1924

named their first honorary member. A private dinner ensued in the home of the agronomist and city council member Shmuel Tolkowsky. A small number of public functionaries, teachers and writers were invited to an intimate gathering at the Herzliyah high school. Even though no lecture by Einstein was planned in Tel Aviv, he spoke on the relationship of relativity to philosophical issues.[32]

On the morning of 9 February, Einstein attended the opening session of the second semiannual conference of the General Federation of Labor. When Einstein entered the hall, labor leader David Ben-Gurion was giving a speech, which came to a halt as the delegates cheered Einstein's arrival. In a short speech, Einstein declared that he was convinced that the future of Palestine and the Jewish people "rests with the masses of Jewish workers." There followed a tour of the agricultural school at the settlement of Mikve Israel, the experimental agricultural school at Ben Shemen, and the small town of Rishon LeZion. At a large reception in front of the town hall, Einstein promised that "until my last moment, I will work on behalf of our *Yishuv* and the country."[33]

The Einsteins then set out by train to Haifa (accompanied by Hillel Jaffe, a member of the board of the Technion). The train arrived in Haifa after the onset of the Sabbath "in spite of Mr. Struck's prior warning." They stayed overnight at the home of Shmuel Pevzner, one of the founders of the Jewish community of Haifa, and his wife Lea, who was Ginzberg's sister.[34]

Because of the institutions' physical proximity, Einstein seems to have had some difficulty distinguishing between the Technion and the Reali school. In his diary, he wrote that on the morning of 10 February (which was a Saturday), he toured the Reali school. However, according to the press coverage he visited only the Technion, where two receptions were held in his honor. The first, which was more popular in nature, was attended by 1,500 people "from all social strata." Both the British governor of Haifa and the inspector-general of police were in attendance. The chairman of the committee of Jewish residents of Haifa welcomed Einstein and announced that the committee had named Einstein a "resident of the Land of Israel." Hillel Jaffe greeted Einstein on behalf of the Technion's board. The second reception, an invitation-only affair, was organized by the board of the young institution. At this stage the Technion was still in the process of being established. In the afternoon the Einsteins visited Weizmann's mother, Rachel-Lea Weizmann-Tchemerinsky, surrounded by her many children. Later they also met an "Arab writer with a German wife." Subsequent correspondence points toward them as having been Asis Domet and his wife Ruth.[35] In the evening a festive banquet was held in Einstein's honor at the Reali, to which alumni of the school and the upper level were invited. Czerniawski gave a brief lecture on relativity. Einstein's speech "was full of emotion and greeted with wild applause."[36]

On the morning of 11 February the Einsteins toured the Reali School and the workshops of the Technion. They then planted trees in the courtyard between the two buildings. A tour of two industrial plants followed—the Nesher factory for building materials in Yagur and the Shemen oil factory in Haifa. Then they drove via the Jezreel Valley to the cooperative farm (*moshav*) of Nahalal, which was being constructed at the time. They spoke to the members over tea about

Figure 13. Einstein planting a tree at the Migdal farm in the
Galilee, February 1923

their working conditions and the differences between the moshav and
other forms of settlement. They continued on to Nazareth and then
arrived at the farm of Moshe Glickin at Migdal for the night.[37]

On 12 February they drove down to the Sea of Galilee, which re-
minded Einstein of Lake Geneva. They traveled from Tiberias via
Magdala (which Einstein referred to as "Mary's hometown") to the
"Communist settlement" of Degania, a kibbutz in the Galilee. It is not

Figure 14. Einstein at the Midgal farm with Jewish settlers,
Palestine, February 1923

clear from Einstein's diary (or the press) whether Einstein visited Degania Alef, the first kibbutz, or Degania Bet, a neighboring settlement
that was founded in 1920. Einstein found the "colonists extremely
congenial, mostly Russians." Upon their return to Tiberias they were
"introduced to the Mufti Sheikh Taher el Tabari and other notables of
the different communities." Then they drove back up to Nazareth,
where they stayed overnight at the German hostel.[38]

The next morning the Einsteins drove from Nazareth via the Jezreel Valley and Nablus back to Jerusalem. In the evening Einstein
gave his second lecture in the city at the Lämel School. This time he
lectured in German (which, as we will see, was highly significant) in a
"jam-packed auditorium." The topic was the main issues of the theory
of relativity. The event was jointly organized by the associations of
Jewish physicians, teachers, engineers, and architects, as well as the
Hebrew technical association and the association for the research of
the Orient. According to the press, "all the Jerusalem intelligentsia"
attended the lecture. Jews as well as non-Jews were in the audience,

Figure 15. Elsa and Albert Einstein, Midgal, February 1923

"especially Germans." Among the 450 attendees were the British participants Lady Samuel and her daughter-in-law, senior Zionist officials, "and more of the distinguished scholars of the nations who re-
side in Jerusalem." The two chief rabbis were invited but did not
attend. The press reported that Einstein held this second lecture because many intellectuals had not had the chance to attend his lecture
on Mt. Scopus. The German Institute for Foreign Affairs was very
pleased that the lecture was a resounding success, despite its irritation
that the invitation to the lecture had only been printed in Hebrew and
did not contain "one single European (let alone German) letter." According to their report, the audience was "mixed: English, French,
American etc., Catholics, Protestants, Templars and the majority:
Jews." The institute also pointed out that it "was the first time since
the War that Jerusalem saw such a large gathering who had come to
listen to a German professor at his German lecture."[39]

The next morning the Einsteins were bid farewell by Colonel

Kisch at the Jerusalem train station and traveled with Hadassah Samuel to Lod, where they changed trains for El Cantara. On 16 February they set sail from Port Said for Spain, which they toured for three weeks before arriving back in Berlin in mid-March.[40]

Thus ended Einstein's brief but intense tour of Palestine. Let us now take a closer look at the impact the visit had on Einstein himself, on his dual hosts, the Zionist establishment and the British Mandate authorities, and on the local population, especially the Jewish community.

Einstein's diary entries on Palestine immediately reveal his sense of entering a country whose landscapes are very foreign to his central European eyes. Upon first entering Palestine, he noted "a plane with very sparse vegetation . . . olive trees, cactuses, orange trees." Yet his reaction to these landscapes also demonstrates his clear enthusiasm for these new vistas: "Admiration of the landscape. . . . The country is exquisitely beautiful, enchanting."[41]

Somewhat surprisingly, perhaps, there are very few explicit references to biblical places and events in Einstein's diary. This highlights the secular character of Einstein's travel journal. Notable exceptions are Solomon's Temple in Jerusalem and the village of Magdala, which he refers to as "Mary's hometown," thus revealing his erstwhile Jewish and Catholic religious instruction.[42]

His most eloquent description of Palestine was reserved for Jerusalem's Old City:

> Past the city walls to picturesque old gate, walk into town in sunshine. Austere bleak hilly landscape with white, often domed white stone houses and blue sky, enchantingly beautiful, likewise the city crammed inside the square walls. Further on into the city with Ginzberg. Through bazaar alleyways and other narrow streets to the large mosque on a splendid wide raised square, where Solomon's temple stood. Similar to Byzantine church, polygonal with central dome supported by pillars. On the other side of the square, a basilica-like mosque of mediocre taste.[43]

Einstein thus renders his perception of the distinct architecture of the Old City and of its most prominent sites, the Dome of the Rock and the Al-Aqsa Mosque. Even though the mosque is a far holier site than the Dome of the Rock, Einstein does not relate to this aspect of these prominent buildings, focusing instead solely on their architecture. In general, his most vivid impression of the Old City was of both beauty and squalor. It is also apparent that his associations for processing the new visual information were European: he compares the Muslim edifices with similar European ones.

His enthusiasm for the landscapes and edifices of Palestine notwithstanding, in his later accounts Einstein makes it clear that it was his encounter with the people of Palestine for which he felt the greatest enthusiasm.[44]

Of all the individuals and groups Einstein encountered during his visit, he was clearly most interested in two groups: the young Jewish agrarian settlers and the Jewish urban construction workers. This was a consequence of his sympathy for the goals of labor Zionism and his hope that the Jewish people would play a more productive role than they had in the past from a social and economic point of view.

The fact that the communal agrarian efforts were being carried out by young Russian Jewish pioneers (*halutzim*) had a special resonance for Einstein in light of his affection for young Eastern European Jews. Einstein admired their "work which was full of self-denial" and their formation of "very close-knit communities." His regard was all the greater in light of the formidable difficulties they faced, which he identified as "malaria, hunger und debt." Yet despite his enthusiasm for the settlers' efforts and his prediction that their endeavors would have a positive effect on the individual members and "breed complete human beings," he concluded that "This communism will not last for ever."[45] Thus, Einstein confirmed his skepticism of radical politics, which we have seen elsewhere.[46] According to notes recorded by Hermann Struck during Einstein's visit, his guest was also convinced that the agricultural sector of the Jewish community was not viable from an economic point of view: "The Jew cannot compete with the Arab in agriculture and is not competitive here with the products on the world market. So nothing will come of the agriculture."[47]

In one of his later accounts of the trip, Einstein claimed that it was the urban "Jewish workingmen's groups" that made the strongest positive impression on him. He was particularly struck by the efficient use of unskilled labor. Yet his concern about the high interest rates charged by American Jews in lending funds to the workers led to a hasty defense of their practices by the lenders, the Palestine Cooperative Company.[48]

Einstein admired in general the urban development in the modern Jewish community in Palestine (the new Yishuv). He was especially struck by the rapid expansion of the first all-Jewish city, Tel Aviv, which led him to positive generalized conclusions about the Jews: "The accomplishments by the Jews in but a few years in this city elicit the highest admiration. Modern Hebrew city with busy economic and intellectual life shoots up from the bare ground. What an incredibly lively people our Jews are!"[49]

Einstein was favorably impressed with the entrepreneurial spirit he encountered in the Yishuv, of which this brisk urban development in Tel Aviv was the most salient expression. His admiration for the urban development in Palestine is all the more remarkable in light of his anti-urban sentiments in respect to such European cities as Berlin.[50] This is another instance in which Einstein displayed special sympathy for an expression of Jewish nationalism, as opposed to his disaffection with other forms of nationalism. Indeed, in Einstein's opinion, the efforts towards urbanization in Palestine were driven by "the strong sense of a nation,"[51] thus demonstrating his positive view of Jewish nationalism at this stage of his involvement in the Zionist movement.

Intriguingly, in his later accounts of his tour, Einstein does not mention the event that was the focal point of his visit, his lecture on Mt. Scopus at the future site of the Hebrew University. His comments on the event in his diary are quite laconic and do not reveal his emotions about this central event. Yet his opening words preceding his lecture (as reported in the press) give a slight indication that he was aware of the significance of the occasion. As we have seen, in his initial statement, Einstein established a connection between the ancient and the modern contributions of Jewish thought to "the enlightened world."

In stark contrast to his enthusiastic views on the dynamic spirit of the new Yishuv, Einstein's perception of the Old Yishuv, the traditional, pious Jewish community that predated Zionist immigration to

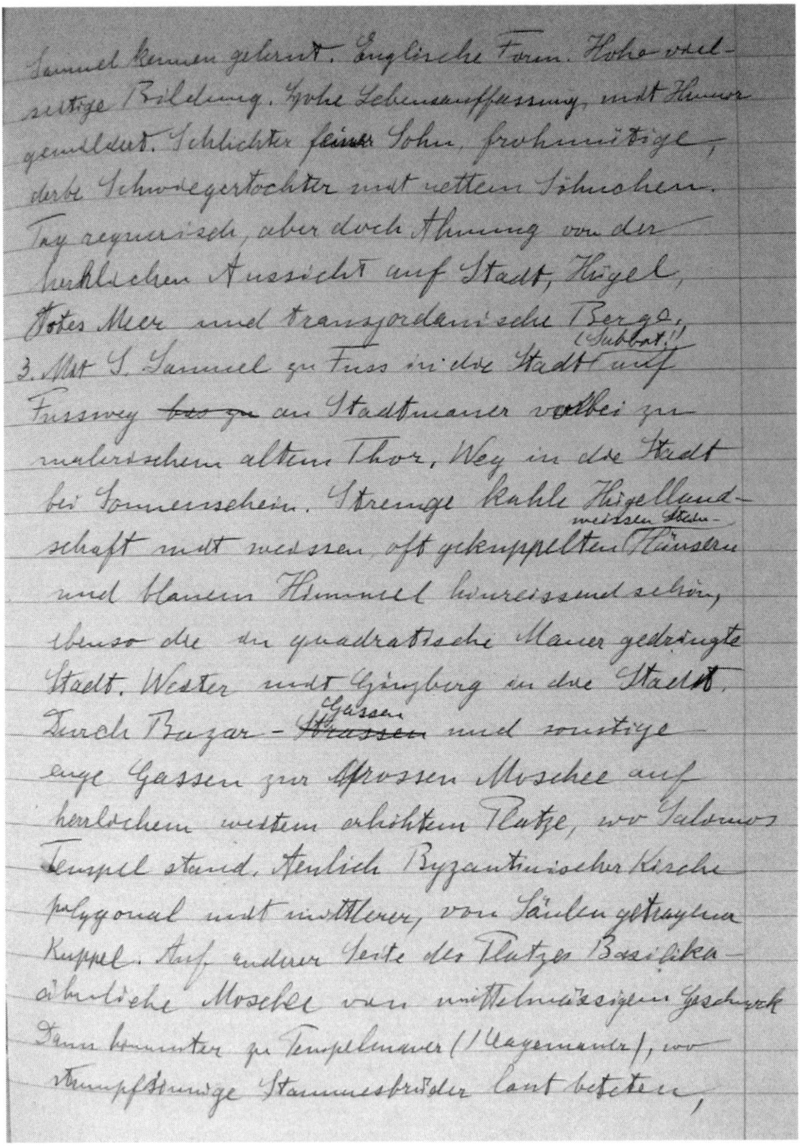

Samuel kennen gelernt. Englische Form. Hohe viel-
seitige Bildung. Hohe Lebensauffassung, mit Humor
gemildert. Schlichter ~~feiner~~ Sohn, frohmütige,
derbe Schwiegertochter mit nettem Söhnchen.
Tag regnerisch, aber doch Ahnung von der
herrlichen Aussicht auf Stadt, Hügel,
Totes Meer und transjordanische Berge.
3. Mit S. Samuel zu Fuss in die Stadt (Sabbat!) auf
Fussweg ~~bis zu~~ an Stadtmauer vorbei zu
malerischem alten Thor, Weg in die Stadt
bei Sonnenschein. Strenge kahle Hügelland-
schaft mit weissen, oft gekuppelten weissen Stein-Häusern
und blauem Himmel hinreissend schön,
ebenso die in quadratische Mauer gedrängte
Stadt. Weiter mit Ginzberg in die Stadt.
Durch Bazar-~~Strassen~~ Gassen und sonstige
enge Gassen zur grossen Moschee auf
herrlichem weitem erhöhtem Platze, wo Salomos
Tempel stand. Ähnlich Byzantinischer Kirche
polygonal mit mittlerer, von Säulen getragener
Kuppel. Auf anderer Seite des Platzes Basilika-
ähnliche Moschee von mittelmässigem Geschmack
Dann hinunter zu Tempelmauer (Klagemauer), wo
stumpfsinnige Stammesbrüder laut beteten,

Figure 16. Einstein, travel diary, description of orthodox Jews praying at the Western Wall in Jerusalem, 3 February 1923

Palestine, was far less favorable. After visiting the Western Wall and the Bukharian quarter in Jerusalem, he noted:

> Then down to the temple wall (Wailing Wall), where obtuse fellow Jews pray loudly, with their faces turned to the wall, bend their bodies to and fro in a swaying motion. Pitiful sight of people with a past without a present. Then

Figure 17. Western Wall during the Sukkot festival, Jerusalem, 1924

diagonally through the (very dirty) city teeming with the most disparate collection of saints and races, noisy, and strangely Oriental. Splendid walk over the accessible part of the city ramparts. . . . Visit in the Bukharian Jewish quarter and in a dark synagogue where pious, dingy Jews await the end of the Sabbath in prayer.[52]

Einstein's perception of the ultra-Orthodox having a past but no present is a view that would have been informed both by his childhood perceptions of unassimilated Jewry and possibly also by the Zionist perception of the Old Yishuv.[53] It is also important to point out that although Einstein perceived the enterprises of the new Yishuv as being rooted in (predominantly Eastern) European Jewry with whom he felt great affinity, the Old City was perceived by him as "strangely Oriental."

Perhaps one reason for Einstein's seemingly unbridled enthusiasm for the new Yishuv was that, as far as I can tell, internal community conflicts were basically hidden from his view. For example, even though he toured private, cooperative, and collective farms, there is no mention in Einstein's diary (or in any of the other available sources) of the contemporary clash between private farmowners in the rural farms (*moshavot*) and the pioneers (*halutzim*) or the *kibbutzim* over the issue of "Hebrew labor," as the private farms were hiring Arab laborers and the kibbutzim wanted them to only hire Jewish workers.[54] Consequently, Einstein's idealistic perception of the Jewish colonization efforts was heightened even further.

In stark contrast to the suppression of internal discord in the Yishuv, Einstein was well aware of the nascent national conflict between Jews and Arabs in Palestine. However, in all his accounts, he downplayed the potential explosiveness of the situation. Even though his visit came less than two years after the violent riots in Jaffa, Einstein's references to the conflict were decidedly low-key. Einstein's first impression of the Arab inhabitants of Palestine reveals his idealistic (and somewhat condescending) perception of them: "Unique charm of this austere monumental nature with its dark, elegant sons in their rags." During his tour, Einstein's direct contact with representatives of the Arab community was limited; he seems to have met with only moderate figures: Jerusalem mayor Raghib al-Nashashibi, a few notables in the Galilee, and the Arab writer Asis Domet, who, as an author who wrote in German about (among other issues) Jewish themes, must have been quite a marginal figure in the local Arab population. As a consequence of his encounter in Haifa with a Jewish pioneer who had an Arab friend, Einstein concluded that "the common folk know no nationalism." Indeed, he later stated that "most of the difficulty comes from the intellectuals—and, at that, not from the Arab intellectuals alone."[55] Thus, in this instance, he was prepared to apportion equal blame for the nationalistic tensions in the region to both Arabs and Jews.

Yet if Hermann Struck's notes of Einstein's remarks while staying in Haifa are to be believed, it was in that city that he also made some less positive remarks on the Arab population: "If there were no Jews here, but rather only Arabs, the country would not need to export [produce], as the Arabs do not have any needs and live off whatever they grow themselves."[56] This statement reveals Einstein's stereotypical perception of the local Arab inhabitants as a unified and undifferentiated economic entity engaged in agriculture. This is curious, as he obviously encountered urban Arabs in Haifa. His disquieting remark on the local Arabs having "no needs" seems to have been informed by the mainstream Zionist image of the Arab populace.[57]

However, it was after his return to Germany that he made his most disturbing comment on the Arab-Jewish conflict, as he had perceived it during his visit. Remarking on the two major difficulties faced by the agrarian settlers, debts and malaria, he stated, "In comparison with these two evils the Arab question becomes as nothing."[58] That Einstein referred to the intercommunal conflict as "the Arab question" and was prepared to associate it so closely with "these two evils" reveals the limits of his ethnic tolerance.

These limits were also apparent in some of his reflections on the future of Palestine, as he saw it. Although he was aware that his ethnocentricity colored his expectations, Einstein still believed the future of the country was Jewish: "[I] believe . . . that the country will be ours. . . . I see it objectively, yet deep down [my] Jewish consciousness may be lurking. If distinctions are to be made between peoples, so I would prefer to be a Jew."[59] This vague statement does not clarify whether what Einstein meant by the country becoming "ours" was that it would eventually be exclusively Jewish or that the Jews would form the majority within the population. It is therefore difficult to judge the degree of Einstein's ethnocentrism in this instance.

Einstein's enthusiasm about his eventful tour, as well as his reserved optimism about the future of the Jewish colonization attempts, is present in what seems to be his first letter to Weizmann following the trip, not sent till October 1923. He wrote, "Palestine was indeed the great experience you predicted it would be for me. I have brought [back] the strong conviction that something will come of this enterprise, even if it will cost a great deal of sweat and sacrifices and cause disappointment." He also concluded that "the difficulties have the advantage of leading to a good selection of our people gathering there."[60] Thus, it seems that Einstein advocated a form of social Darwinism in relation to settlement efforts in Palestine, whereby immigration would lead to a positive natural selection among the Jewish population.

Despite his belief that the Jewish colonization efforts would be successful, Einstein had strong doubts about the economic viability of the Jewish community. Yet owing to his ideological priorities, Palestine's economic independence was "of secondary importance" to Einstein. For him, success would be gauged by the creation of "a unified community which shall be a moral and spiritual center for the Jewries of the world. Here, and not in its economic achievement, lies, in my opinion, the significance of this work."

Yet it was actually Palestine itself that was of secondary importance to Einstein. As we have seen in previous chapters, he viewed the land in instrumental terms, and this does not seem to have changed through his visit to the actual country. He concluded that "Palestine will not solve the Jewish problem, but the revival of Palestine will mean the liberation and the revival of the soul of the Jewish people. I count it among my treasured experiences that I should have been able to see the country during this period of rebirth and reinspiration."[61] In a letter to his close friend Maurice Solovine, he explained what he meant by Palestine not solving the "Jewish problem": "The country on the whole is hardly fertile. It will become a moral center, but not be able to absorb a large section of the Jewish people." And in his article of May 1923, Einstein explained why he believed that Palestine could "heal the sick Jewish soul": "the sincerity and warmth of my reception ... were to me an indication of the *harmony* and *healthiness* which reigns in the Jewish life of Palestine" (my emphasis). Here again we can discern Einstein's lack of awareness of the social tensions in the Yishuv and his conviction that Palestine had both a harmonious and a healthy effect on its Jewish inhabitants. Yet this positive effect was of such crucial importance to Einstein that in Jerusalem he vowed to "bring the message that only Zionism can heal the sick Jewish soul" to the "whole Jewish world."[62]

However, Einstein believed the Jewish colonization efforts in Palestine would have not only an internal positive effect on the Jewish people but external ones as well. Beside the internal benefit of the "social recovery of the Jews," the existence of a Jewish cultural center would strengthen the moral and political status of the Jews worldwide.[63]

Despite his commitment to "spread the gospel" of the benefits of the Zionist efforts, Einstein was not prepared to settle in Palestine himself. As he noted in his diary, this was an issue that came up repeatedly during his visit as he came under considerable pressure from Zionist dignitaries to move to Jerusalem and take up a position at the planned Hebrew University: "One wants me in Jerusalem at all costs and I am being assailed in this regard by a unified front. My heart says yes but my reason says no."[64] This distinction between his emotional and intellectual attitude toward the idea of joining the university faculty is indicative of his attitude toward Zionism in general. As we saw in the chapter on Einstein's mobilization on behalf of the Zionist movement, his emotional commitment to Jewish nationalism was far stronger than his intellectual and ideological commitment.

Despite Einstein's firm decision not to personally settle in Palestine, his trip did lead to intensified efforts on behalf of Zionist enterprises in the region. These efforts began during the visit itself when Einstein wrote to Weizmann from Haifa and appealed to him to assist in enabling the commencement of teaching activities at the Technion.[65] This was an interesting reversal of the usual roles between Einstein and Weizmann—in this case, it was Einstein who reported to Weizmann on the current state of affairs of a Zionist institution and asked him to raise funds to enable its opening. Einstein's commitment to be active on behalf of Palestine was also reconfirmed by Elsa Einstein prior to their return home to Berlin. From Barcelona, she wrote to Arthur Ruppin's wife Hannah, "the country is so immensely beautiful. My husband has also embraced it in his heart and he will campaign and be creative on its behalf."[66]

The Hebrew University continued to be the institution in Palestine on whose behalf he labored most intensely following his return to Berlin. Only two weeks after his return, Einstein hosted a discussion on the university in his Berlin apartment. At the meeting, which was attended by several Jewish scholars and the chairman of the German

Keren Hayesod, Oskar Wassermann, Einstein described "in a most appreciative manner" his impressions of the situation of the future university and Gray Hill House on Mt. Scopus. And when the Society of Friends of the Jerusalem Library was established in late July 1923, Einstein was appointed its deputy chairman.[67]

The most significant event planned in the aftermath of his visit was the initiative that Einstein give the closing speech on the Hebrew University at the Zionist Congress in Karlsbad in August 1923. Yet Einstein declined the invitation, claiming that he "[had] to have some rest at all costs." Plans for the congress also reveal Einstein's new status in the Zionist movement. Special invitations for Einstein, Herbert Samuel, the Anglo Jewish politician Lord Melchett, and Baron Edmond de Rothschild to attend the congress as "guests" were to be issued.[68] Einstein was thus elevated by the Zionist Executive to membership in an illustrious group of prominent Jews who sympathized with some of the goals of the Zionist movement, yet were not fully committed to Zionism in all its ideological aspects. Significantly, Einstein was the only one of these esteemed individuals who was neither wealthy nor the holder of a high office.

Einstein's views on his other hosts during his visit, the British Mandate authorities, can also be gleaned from the sources at our disposal. Of all his brief notes on the individuals he met during his tour, his description of the high commissioner is especially revealing of his perception of upper-class British formality: "Became acquainted with Herbert Samuel. English formality. High-minded, versatile cultivation. Elevated view on life, alleviated by humor." Einstein had much praise for the efforts of the Mandate authorities, particularly their improvements in the infrastructure of the country.[69]

In our examination of the Zionist preparations for the trip, we saw how important it was for the Zionist Organization to present Einstein with a well-coordinated position on plans for the Hebrew University. However, apart from this specific issue, I have not found any evidence in the available sources for correspondence between the Zionist Organization in London and the Zionist Executive in Palestine pertaining to detailed preparations of Einstein's itinerary. In-

Figure 18. Albert and Elsa Einstein with British high commissioner Sir Herbert Samuel, Lady Samuel, Norman Bentwich, Mrs. Bentwich, and others, Government House, Jerusalem, February 1923

deed, I have found only one document that provides us with some insight into the local behind-the-scenes arrangements concerning Einstein's visit. Prior to his arrival in Tel Aviv, Ginzberg informed Mossinson that the distinguished visitor "does *not* want to give any lectures, except in Jerusalem." The foremost purpose of the tour was defined by Ginzberg as follows: "For him, and I think for all of us, the main thing is not that we will enjoy his presence but that he will enjoy his visit to Palestine and will leave with an impression of the beauty and the work which will connect him to the land."[70] As we have seen, the Zionist Executive was successful in providing Einstein with this impression during his visit. Yet, as we will see presently, "enjoy[ing] his presence" *was* a major factor in the Yishuv's reception of Einstein.

Despite the virtual lack of documentation pertaining to detailed preparations, it is patently clear that Einstein was presented with the mainstream Zionist narrative about Jewish colonization efforts, the needs of the Yishuv, and relations with the local Arab inhabitants. This is most evident in the itinerary of the visit, which presented the achievements of the Jewish community in the best possible light. We have already seen how internal conflicts were downplayed. There was also a clear avoidance of visiting centers of Arab nationalism such as Jaffa, where violence had occurred less than two years previously. Thus, apart from some discussions with representatives of the British Mandate authorities on the situation in Palestine, Einstein had minimal exposure to alternative narratives besides the Zionist one.

The Zionist Organization evidently viewed Einstein's visit as an event of utmost national importance. When Einstein arrived in Palestine, Weizmann sent him the following cable: "Hearty welcome to Erez Israel whose regeneration you will largely contribute feel sure that happy impressions you will receive will link you still closer to our hopes [your visit will be great inspiration to Yischuw]."[71] Thus, Weizmann bestowed on Einstein a national role he had not previously ascribed to him, that of national regenerator, both in Palestine and in the Diaspora.

The Zionist establishment in Palestine was also well aware of Einstein's national significance, and its senior representatives made a concerted effort to highlight this during his tour. This is clearly evident in the official welcoming speeches of the leaders of the Yishuv. The glorification of Einstein in these addresses bordered on an apotheosis of the distinguished visitor. In his speech, David Yellin bestowed the title of "Gaon" (genius) on Einstein, claiming that he deserved it "more than anyone else."[72] Furthermore, as we have seen, Ussishkin ascribed a quasi-Messiah-like role to Einstein when he called on him to mount the stage "which has been waiting for you for 2,000 years." Perhaps the most hyperbolic statements can be found in a speech made by a member of Haifa's Jewish community council, who ascribed prophetic and almost superhuman traits to the distinguished guest: Einstein "is equal to Moses and sixty Rabbis. . . . We stand before an incomprehensible wonder, as we bow and prostrate ourselves before the secret *torah* which this genius disseminates." In the same speech, Einstein's national role was consciously highlighted at the expense of his international one: "This international man whose theory is universalistic and whose light is world-encompassing, he is national in the Jewish sense." And the speaker expressed the hope that "we'll soon see the day and he'll settle here and spread his theory from Zion and that will glorify us in the eyes of other nations." The perception of Einstein as a national Jewish icon was also underlined by Boris Schatz, who asked the visitor for a portrait for the planned national museum at the Bezalel Art Academy.[73]

As was planned in the preparations for the visit, the tour was utilized for propaganda purposes. However, this use was far more extensive than the original limited plans of propaganda for the Hebrew University. The propaganda was particularly blatant in Mossinson's speeches, in which he made bold claims about Einstein learning Hebrew and intending to settle in Jerusalem, which were not truthful. This can be seen as a cynical exploitation of Einstein's ignorance of the Hebrew language.

From the point of view of the Zionist movement in general, it is important to differentiate between the perspective of the Zionist Organization in London and that of the Zionist Executive in Palestine. The tour was initially organized by the Zionist Organization, and its original limited goal was to make Einstein acquainted with the situation on the ground in regard to the Hebrew University. However, once the visit got under way, it took on its own dynamic, and the two agents that had the greatest impact on the actual itinerary, course, and agenda of the tour were the British Mandate authority and, most crucially, the Zionist Executive.

From the perspective of the Zionist establishment in Palestine, Einstein's visit had a twofold propagandistic role: vis-à-vis Einstein and the Yishuv. Their goal was to give Einstein a most favorable impression of their efforts, and in this they were highly successful. As for the Yishuv, the functionaries could use Einstein's lack of knowledge of Hebrew to further their propagandistic aims: claims could be made about Einstein that would enthuse the masses but were not necessarily true.

Einstein's British hosts saw his visit as a state occasion. His lodging at Government House, the official receptions, and the gun salutes were all clear signs of this. There may have been various reasons for this: the high regard in which Einstein was held in British government circles, the influence of his close associate Lord Haldane (with whom he had lodged in London in 1921), or merely a personal decision on the part of Herbert Samuel.

It is not clear whether there was any planned coordination between Einstein's hosts concerning his visit. There is no documentation of such in the Zionist archival material at our disposal. Yet there was a clear division of labor between the two camps that indirectly reveals some coordination. Even though Einstein was initially welcomed to Palestine by a Zionist delegation in Lod, the British authorities took over in Jerusalem. After a mixed (and informal) British and Zionist agenda during the first weekend of the visit, the official Zionist agenda took over on 5 February and remained dominant for the remainder of the visit, with the exception of a few social events organized by the British and Einstein's second lecture in Jerusalem, in which the German Foreign Ministry was also involved.

Even though Einstein's lecture on Mt. Scopus was organized by the Zionist Executive, the presence of the high commissioner and other senior British officials made the event not only national in character but also statelike. In his speech, Herbert Samuel expressed his view of the historical significance of the festive occasion: "He pointed out the amazing fact that a member of the same nation that created the most ancient *torah* in the world now created the most modern theory in the world."[74] Thus, the occasion of Einstein's visit elicited from both his Zionist and his British hosts analogies between ancient and modern wisdom emanating from Jerusalem.

It is interesting to note that the British Mandate authorities were actively involved in some of the Einstein-related events in both Jerusalem and Haifa, but apparently abstained from participation in the festivities in Tel Aviv, thus reflecting the authorities' varying relationship toward the cities with mixed Jewish and Arab populations as opposed to those towns obviously perceived as being exclusively Jewish. It is also clear that the festivities in Tel Aviv studiously ignored and avoided the neighboring Arab town of Jaffa, which is particularly striking in light of the riots there less than two years earlier.

Intriguingly, some of the individuals with whom Einstein most closely associated during his visit had double affinities to both the Jewish Yishuv and the Mandate authorities. The most notable examples of this were Einstein's official Zionist "shadow," Solomon Ginzberg, and the high commissioner's daughter-in-law, Hadassah Samuel.

The Yishuv's reception and perception of Einstein can best be gleaned from the coverage of the visit in the local press. The superlatives lavished on Einstein were full of hyperbole: on the eve of his arrival, he was greeted as "the greatest in the Nation and of this generation." The adulation of Einstein was boundless: the moment of his "feet touch[ing] the ground of the nation's homeland" was anticipated as an occasion for great "rejoicing." He was also seen as "the golden link in the chain of the greatest lights of the world of our generations."[75]

Einstein's visit was perceived as being particularly unique: even though "there have been many visits from many great men . . . this week is a greater occasion than any, for to-day Jerusalem contains within its walls the greatest man of science of the day."

Einstein fever in the Yishuv reached a frenzy at the central event of his visit, the inaugural scientific lecture of the Hebrew University on Mt. Scopus. The daily *Ha'aretz* termed the festive occasion "a national festival and a scientific festival." Einstein's mounting the stage to deliver his lecture was interpreted by the press as no less than a sacred moment. Torah was once more being dispensed from Zion. Moreover, the distinguished visitor was perceived as the high priest in the new "temple of science that will be built in Zion." Einstein's opening his lecture in Hebrew was seen as the realization of "Ben-Yehuda's vision," a reference to Eliezer Ben-Yehuda, the "reviver" of the Hebrew language.[76] If the leaders of the Zionist establishment hinted at an apotheosis of Einstein in their speeches, the journalists in the Hebrew press presented their readers with a full-blown version. The daily *Doar Hayom* saw his lecture as a sign of divine intervention: "Shouldn't we see this as a sign from Heaven? . . . His voice will be like a shofar and every Jew and every Gentile in the world will hear his great voice . . . his words are the words of the living God."[77]

Reports in the more sober English-language press echoed the connection made by Herbert Samuel between Judaism's ancient torah and Einstein's modern theories:

> It was from Mount Moriah, which faces Mount Scopus, that the great theory of Monotheism and Humanity was propounded, and now thirty centuries later, another Jew who has also concentrated his great genius on the study of the heavens, but in its relation with the necessities of earth, is giving forth to Israel his triumphant discovery.

The writer hoped that the lecture would inspire the Diaspora Jews

> to realize at last the great scheme of raising a building on the site acquired for a Hebrew University, with Professor Einstein at its head—not as a German, not as an Internationalist, but as a Jew, whose only purpose is the enlightenment of Humanity.[78]

In its reception of Einstein, the Yishuv thoroughly adopted Einstein as one of their own—as their "great brother" and national icon. Clearly, both Einstein and Palestine saw each other in a highly idealized manner.

However, Einstein's visit did not occasion universal celebration. There was some criticism, which was directed at the organizers of his lectures rather than at Einstein himself. There was considerable discontent among some of the journalists concerning overcrowding at Einstein's lectures: "Though this acclamation testifies to the enthusiasm awakened throughout Palestine by the visit of Professor Einstein, it was not a little disturbing to the ticket holders, who had to fight their way through the surging mob. A little self-control on such occasions would not be amiss." The *Ha'aretz* reporter complained about the lack of order in the hall at the Mt. Scopus lecture: "Visitors such as [Zionist ideologue] Ahad Ha'am did not have a place to sit. The ushers caused terrible overcrowding in the small hall by giving out more tickets than there were seats." There was also resentment that access to the receptions in Jerusalem and Tel Aviv was severely restricted: "the petty officials and caretakers . . . ignore the presence of the people of the writing and of the book." *Doar Hayom* reported that it had received complaints from merchants in Jerusalem that, in contrast to Tel Aviv, where a popular reception was organized, in Jerusalem "the same 'scholars' and their wives, etc. are invited and the other strata do not get a foot-in, not even to a slight extent. . . . [A]ll of us are equal vis-à-vis Einstein's theory (namely in our lack of understanding)."[79]

This quote also hints at the initial reception of relativity in the Yishuv. The myth of the alleged incomprehensibility of the theory of relativity, so widespread in other countries,[80] was readily accepted in Palestine: "There's no possibility to render in the paper the lecture or chapter headings of it. . . . [O]nly few of those in the audience could grasp its meaning."[81] Furthermore, an indication that the crucial significance of Einstein's lecture lay more in its merely being held than in its actual content was confirmed by mayor Meir Dizengoff. At a reception in Tel Aviv he claimed he "had been all the way to Jerusalem but I admit without shame that I have not fully understood Einstein's system and therefore hesitate to tell you what his greatness is. From his entire lecture I understood one thing—that the huge audience did not understand anything." [82] At the same time, the English and Hebrew press seized the occasion of Einstein's visit to publish various in-depth articles on Einstein's scientific theories, thus contributing to the popularization of relativity in the Yishuv.[83]

The tour was used by some in the Hebrew press to mock the local Arab population. At the reception welcoming Einstein to Jerusalem, a journalist was allegedly asked by two Arab passersby who Einstein was and whether he was "a Russian or one of the Rothschilds." When they heard that he was "an expert on stars and astrology, they lost interest and walked off." The absence of the Muslim dignitaries at the lecture was interpreted in two (contradictory) ways: that "these people are probably still far removed from the world of science" and, alternatively (and mockingly), that "they are experts in the theory of relativity" and therefore did not need to attend the lecture.[84]

However, the visit was also seen by some outside of Palestine as an opportunity to subtly attack Zionist aspirations. *Ha'aretz* brought its readers the following quote from the *Egyptian Gazette*: "We hope . . .

that when he is there [in Jerusalem], he will apply his theory of relativity to the relative importance of the Zionist Executive."[85]

Historians of travel to Palestine have distinguished in their works between three main types of visitors: pilgrims, travelers, and tourists. The goal of pilgrims was to establish "a link between earth and heaven," travelers focused on their own individual experience of the land, and tourists visited en masse in organized tours. Historians have also distinguished between travelers, who are "inclined to compare civilizations, often to the detriment of [their] own," and tourists, who are "committed to [their] own civilization[s]." Another focus of analysis has been the issue of whether visitors arrived in Palestine "already conditioned by the values of the society from which they [have] come" or whether they "accept[ed] local habits or customs without criticism."[86] This crucial issue has also been focused on in specific studies on German Jewish and German Zionist travel accounts to Palestine. In his study, German historian Wolf Kaiser claims that the visitors' "deep-rooted religious and political orientation" influenced the "selection of what is seen" and the "emotional attitude towards it." Thus, the travel accounts need to be seen against a "double background": "the social and cultural situation of German-speaking Jewry [and the] state of Palestine's Jewish population." Furthermore, the travelers visited the country with preconceived ideals which they saw confirmed or contradicted by the reality they experienced, or, alternatively, they disregarded the reality and "project[ed] their ideal values onto the people of Palestine."[87] Zionist travel to Palestine has also been viewed in the more general context of the cultural debate on Orientalism and the multiple (and often conflicting) perceptions of the East. Because Palestine was located in the Orient and was being settled predominantly by Eastern European immigrants, the Zionists had to grapple with the European perception of the Orient and Eastern Europe as "the twin counterpoints to the idea of a western, or European, civilization." Some Zionists subscribed to the view of a backward Orient "into which western civilization must be imported." Others were convinced that Zionism's encounter with the East was fundamentally different from that of Europe's colonial empires and could even lead to a blend of Western and Eastern cultures.[88]

In the context of these insights and theories of historians of travel to the Middle East, the following questions can be posed: Was Einstein's trip to Palestine in 1923 that of a pilgrim, traveler, or tourist? How did his ideological values and opinions affect how he perceived the reality he encountered in Palestine and were these values and opinions influenced by his visit in any way? Did he perceive Palestine as a backward country into which European culture had to be imported, or did he subscribe to the views of those Zionists who believed that their efforts were fundamentally different from those of the colonial powers?

It can definitely be concluded that the narrative in Einstein's travel diary was secular in nature—the few allusions to biblical figures do not render his trip to the Holy Land a religious pilgrimage. But was it a secular pilgrimage, or was he a traveler or a tourist? Apart from the fact that he insisted on traveling second class on the train to Lod (even though a sleeping car had been reserved for him),[89] there are no indications that he intended to endure hardship during his visit. He lodged at Government House in Jerusalem and in comfortable quarters in the other cities. There is also no indication that he wanted to establish "a link between heaven and earth" through his trip. The only manner in which his visit can be termed a *secular* pilgrimage is that he was looking for confirmation of his beliefs about the Jewish colonization efforts he had formed prior to his visit.

As for whether Einstein can be described as a "traveler" to Palestine, his tour of the country was definitely not independent enough to qualify as such. There are no indications that he made any autonomous arrangements during his tour of the country. Einstein's independent-minded views of the people and places he encountered are the only aspect of his visit that would fit the definition of a traveler. Even though he was highly enthusiastic about the achievements presented to him by his Zionist hosts and was strongly committed to assist them in furthering their goals, he did not become entirely enthralled by their efforts, and he did not completely identify with their goals. In this, as with his relationship to Zionism in general, he kept a critical distance. We have seen how this was mainly due to his perception of Palestine as a mere instrument for advancing goals of greater significance to him: the curing of the "sick Jewish soul" and the improvement of the status of the Jews in the eyes of the Gentiles.

Therefore, Einstein's visit can be seen as that of a Zionist tourist, or, more precisely, as that of a tourist *in* Zionism. He was presented with a Zionist narrative, given a highly organized and carefully coordinated tour of the country, shown only the places his hosts wanted him to see, not informed of internal Jewish conflicts, and kept away from the more controversial individuals and locations. It would seem that a large part of the responsibility for this would lie with Einstein himself, as I have found no evidence that he asked to meet with more radical representatives of the Arab population or to tour areas that may have been less welcoming.

There is also no indication that Einstein's preconceptions about Palestine and Zionism underwent a significant transformation as a result of his visit. Palestine was of secondary importance to him prior to his trip and it remained so after his visit. As with other tourists before him, the reality on the ground as he perceived it confirmed Einstein's views and opinions about the colonization efforts. At the same time, he was deeply impressed by the achievements of the Yishuv, and his visit confirmed his belief that their efforts were a highly worthy cause to which he would continue to devote considerable time and energy. Particularly his positive perception of the (mostly) Russian settlers was confirmed and increased his preexisting sympathy for young Jews from Eastern Europe.

Yet how can we categorize Einstein's perception of Palestine in the context of the historiography of travel to the region? Israeli historian Yeshoshua Ben-Arieh has developed a typology of various perceptions by visitors to the Holy Land.[90] Einstein's perception fits three of these categories. To a limited extent, he perceived Palestine as "the land of the Bible and the holy places." We have seen that this was not of great importance to him. To a considerable extent, he saw the country as an "exotic, oriental land"; this was especially the case with the Old City of Jerusalem and his reaction to the desert landscapes and their inhabitants. Yet above all, Einstein perceived Palestine as "a land of new beginnings." We have seen how impressed he was by the modern urban facilities in Tel Aviv and by what he perceived as the positive effect the agrarian colonization efforts would have on the pioneers (*halutzim*) as "complete human beings." Perhaps it was his rural Jewish origins in Swabia which heightened his enthusiasm for the members of the rural settlements and his sympathetic attitude towards these "new Jews."

An additional important issue is how Einstein perceived Palestine in relation to its location as part of the Levant. He acknowledged its quality as "strangely Oriental" during his tour of Jerusalem's Old City, but he seems to have perceived the Yishuv (without explicitly saying so) as quintessentially European. Practically all the aspects he admired in the Yishuv were of European origin: Richard Kaufmann's garden suburbs in Jerusalem, Boris Schatz's faux-Oriental artwork at Bezalel, the "congenial Russians" on the moshavim and kibbutzim, the display of gymnastic exercises performed by the pupils of the Herzliya high school, the modern facilities and factories in Tel Aviv and Haifa, and many more such instances.

As to the question of whether Einstein believed that Palestine needed to import European culture or blend in with its Levantine environment, he definitely advocated the Europeanization of the country. In this, Einstein was following a Zionist tradition that began with Theodor Herzl himself. Historian Robert Wistrich has pointed out that "Herzl's vision of Israel was thoroughly European, rational, liberal-utopian and modernist. It was to be a transplant of the best Western European culture . . . to the Eastern Mediterranean."[91] And indeed, Einstein believed that Zionist efforts would be of great benefit to the peoples of the region. One can even see the focal point of his visit, his lecture on Mt. Scopus, as the most blatant symbolic representation of this. His second lecture in Jerusalem, delivered in German, was also of special significance in this regard. Yet there is a strong irony here: we have seen elsewhere that Einstein was highly critical of and even derogatory toward the "horrid Europeans."[92] This is a contradiction that can only be resolved by understanding that Einstein rejected European politics, but did not by any means reject European culture. However, this is a contradiction of which he himself does not seem to have been aware.

The question remains to what extent Einstein saw Palestine through Zionist eyes during his visit. As Einstein had an exceptionally independent mind, we can surmise that he saw the country through his own eyes. However, in light of the propaganda tour he was given, one could argue that while he was there he wore heavily Zionist-colored glasses. Intriguingly, once Einstein returned to Berlin, there was no more discussion of his original plan to revisit the country for an extended stay in Palestine.

By the spring of 1923, Einstein had undertaken two major trips as a consequence of his association with the Zionist movement. Were there any similarities or differences between Einstein's trip to the United States and his visit to Palestine? One similarity was that in both instances he was completely in the hands of his Zionist hosts. Nearly all of the time he was dependent on these escorts to receive all pertinent information and determine his itinerary. The only exception to this in the United States was when he broke out of Weizmann's camp and obtained information from Justice Brandeis independently. In Palestine, there were no instances of his traveling alone; he was either with his Zionist escorts or with his British hosts. In the United States he was part of a visiting delegation, in Palestine he was not. Another major difference is that in Palestine (as opposed to the United States), there was no major conflict within the Zionist movement in which Einstein was involved.

His brief tour of Palestine brought Einstein face-to-face with the reality on the ground, as it was presented to him by his hosts. This reality does not seem to have clashed considerably with Einstein's vision of a Jewish homeland. Perhaps this was because he was not overly invested in the reality of the Jewish colonization efforts as a consequence of his ideological priorities. However, what would happen once his vision for the specific Zionist project he cherished the most and in which he was most invested, the Hebrew University, collided with the mundane reality of the everyday activities of running an academic institution?

CHAPTER 5

THE "BOTCHED UNIVERSITY"

Einstein's Involvement in the Hebrew University, 1924-1929

"If at this year's annual meeting of the Board of Governors it proves impossible to wrest control of the institution from the Philistines in Jerusalem ... I will resign with an uproar from the administrative bodies of the University with no regard for dealing a heavy blow to its cause. Better dead than so alive."[1] With these extremely harsh words written to a colleague in 1928, Einstein threatened to divest himself of all involvement in the Hebrew University, the same institution of higher learning that had brought him to associate himself with the Zionist movement nine years earlier. Merely three years after the young university's festive opening in 1925, Einstein had become completely disillusioned with his pet project and was prepared to inflict major political damage on an enterprise he had repeatedly declared was "close to my heart." How did this dramatic chain of events come about, and what impact did this disenchantment with his favorite cause have on Einstein's support for Zionism?

Most historians of the Hebrew University have paid only cursory attention to Einstein's role in the contentious battles over the young school's development during the first decade of its existence. He has been viewed as primarily playing an auxiliary role to Zionist leader Chaim Weizmann in his acrimonious conflict with the university's first chancellor, Judah L. Magnes, over the administration of the newly founded institution. In general, Einstein has been seen as the chief advocate among the main figures involved in the establishment of

181

the university for the separation of its academic and administrative branches. In addition, he has been perceived as "the personification of uncompromising scientific standards."[2]

Biographers of Einstein have also noted in passing his role in the intense disputes over the direction of the Hebrew University, particularly his very personal conflict with Judah Magnes. Einstein's difficult relationship with the executive bodies of the university has also been seen as a typical instance of the physicist's difficult relationship with organizations in general.[3]

By the end of 1923, a new stage in the development of the Hebrew University had begun. After years of intense planning and fundraising, initial concrete steps were being taken to establish the administrative bodies which would govern the future university. The first indication that Einstein would be directly involved in these concrete steps can be found in a letter from October 1923. Weizmann expressed his "joy" at learning from Leonard Ornstein that Einstein had accepted the presidency of the planned board of governors (*Kuratorium*), the university's principal governing body.[4]

Yet various proposals regarding the exact responsibilities of the board and the other major administrative bodies were being disseminated at the time. One of these, which was drawn up by the German-Zionist Palestine University Committee, was presented by German Zionist Otto Warburg to the Zionist Organization "on behalf of Prof. Einstein." It is not clear how much of this document was actually authored by Einstein himself, but the essential tenets laid out in this document corresponded largely to Einstein's own ideas regarding the scope of authority for the university's central institutions. According to the proposal, the university should have "a central administrative and academic body," consisting of "a few independent and influential individuals." As Jerusalem was not yet deemed to be a suitable location for this proposed board of trustees, its headquarters should be located in London. On academic matters, this board would consult with a number of committees of Jewish scholars in various countries. These committees would elect representatives to the "Academic Council" of

Figure 19. Einstein addressing the Conference of Jewish Students in
Germany at the Logenhaus, Berlin, 27 February 1924(?).

the board. Representatives from relevant disciplines would advise the
board on academic appointments. The faculty members in Jerusalem
would also elect a committee to be heard by the board.[5] Thus, this
early proposal clearly gave control over administrative matters at the
university to a select group of academics based in Europe.

The German plan was apparently well-received by the Zionist Executive. In his reply to Warburg, Leo Kohn, then secretary of the
Zionist Organization's Central Office in London, announced the
Zionist Executive's intention to establish the planned board of governors. He also clarified more explicitly who would belong to the
planned body: not only "some of the few leading personalities of contemporary Jewry, such as Professor Einstein and Ahad Haam," but also
"some of the main philanthropists who have made notable contributions to the University, such as Baron Edmond de Rothschild."[6] Thus,
there was a clearer emphasis on the part of the Zionist Executive on the
need to involve major donors in the administration of the university.

In early 1924, British Zionist David Eder officially invited Einstein on behalf of the Zionist Executive to join the Board of Governors. Eder also sent Einstein a revised version of the original German plan for the central bodies of the university: the board would be responsible for the university's general administration, its external affairs, and its general fund. Its initial members would be Einstein, Ahad Ha'am, James de Rothschild, Alfred Mond, Lord Rothschild, and "a few" leading American Jewish scholars and philanthropists. The Academic Council would serve as an advisory committee to the board in all academic matters such as the creation of new institutes and the appointment of professors. There would also be administrative bodies for the individual faculties, which would consist of experts and the directors of the institutes. The model for these faculty committees would be the American Jewish Physicians Committee. They would determine the institutes' curricula and propose appointments to the Academic Council. All these bodies would have representatives from Palestine, yet ultimate responsibility would lie with the board and the Academic Council, which would predominantly consist of members from outside Palestine. Einstein's positive response to the invitation is not extant, though referenced in Eder's reply. Eder also welcomed Einstein's proposal to invite a few more scholars to join the board.[7] Thus, Einstein's strong preference for scholars to have a decisive say in the administration of the university was once more underscored.

In July 1924, following a meeting with Magnes, Einstein found that he "essentially" agreed with him on the university. Einstein asserted that the board should not number more than ten members and expressed his opposition to "political entities" exerting too much influence on its decisions.[8] This confirms Einstein's tendency toward elitism in academic matters, as well as his antipathy toward the overpoliticization of academia.

Following years of planning by the Zionist Organization, the Hebrew University was officially opened on 1 April 1925.[9] Einstein did not attend the festive opening, as he had made previous plans to visit South America. Instead he attended a rally at the Theatre Coliseo in Buenos Aires organized by the local Jewish community at which some four thousand people celebrated the opening of the nascent institu-

Figure 20. Sir Herbert Samuel addressing the festive opening
of the Hebrew University, 1 April 1925

tion. He also published an article on the occasion of the opening.
Einstein expressed his view that it was one of the "noblest tasks" of
"our educational institutions . . . to keep our people free from nationalistic obscurantism and aggressive intolerance." In previous chapters
we have seen how Einstein warned against what he perceived as the
perils of such tendencies. The article also clarified the varying roles
Einstein saw the university playing in the Diaspora and in Palestine:
it would inspire international Jewry "with assurance and pride *vis-à-
vis* the entire world," and it would provide "our laboring class" in
Palestine access to "the treasures of culture."[10] Yet this advocacy of a
more populist approach to higher education would prove to be
fleeting.

In the meantime, the first meeting of the university's board was
held in Tel Aviv in April 1925. Still in South America, Einstein was not
present. The constitutive members of the board of governors were
Ahad Ha'am, the Hebrew poet Chaim Bialik, Einstein, Magnes, Sir
Alfred Mond, James de Rothschild, Nahum Sokolow, Felix Warburg,
and Weizmann. Magnes was appointed chairman of the board. It was

at this meeting that Weizmann transferred responsibility for the university from the Zionist Organization to the board of governors. As a portent of the crucial differences between them in future discussions on the university, Weizmann and Magnes expressed their divergent opinions on funding for the university. Weizmann favored a "public exchequer" to be funded by the Jewish Agency, analogous to a state grant to British universities. Magnes opposed "state universities, where [the] freedom of science was imperiled." Weizmann retorted that he supported a "state-aided, not a state-controlled University."[11] These differences made it clear which models of university administration these major protagonists favored: Weizmann supported a British model, whereas Magnes preferred an American one.

The first signs of Einstein's displeasure with developments at the Hebrew University can be found in early July 1925, merely three months after the institution's opening. In his reply to a letter by Weizmann, who had informed him that he was resigning both as president of the Zionist Organization, out of strong opposition to the Jewish Agency, and as a member of the university's board of governors, Einstein pleaded with Weizmann not to resign from the board. Yet he did suspect "that you too are dissatisfied with the attempt by the wealthy American donors to exclude Europe."[12] This was another instance of Einstein's antipathy toward what he viewed as undue influence by American Jewish philanthropists on academic matters, which took root during his trip to the United States in the spring of 1921.

The United States was emerging as only one of the fronts on which Einstein (and the Zionist Executive) began to wage battle over the direction of the university. Palestine soon emerged as another. A plan proposed at the first meeting of the board to transfer the administration of the university to Palestine was opposed by the Zionist Executive. The Executive's resistance to this move was "strongly influenced by a communication from Prof. Einstein expressing his disapproval of the proposal."[13] Thus, at this point in time, the Zionist Executive clearly supported Einstein's position on where the locus of control in university matters should be. It is also evident from Eder's letter how important Einstein's opinions were for the London head-quarters of the Zionist movement.

In reaction to the attempts by the board members from Palestine to move the focus of control over the university to Jerusalem, London informed Magnes that an executive committee would be established for the university consisting of the board's president and two members from outside Palestine. There would also be a Jerusalem Executive, which would consist of the board members from Palestine. However, the "real authority" would rest with the board as a whole. The rationale given for this decision was the following: as the university was "the intellectual property of the entire Jewish people," the Zionist Executive felt obliged to empower the "most significant intellectual agents of all of Jewry" with a decisive influence on the university's administration. In the Zionist Executive's opinion, this was the only way "to guarantee a high intellectual standard" for the institution "and to achieve for it the confidence of the entire Jewish world." The Zionist Executive also saw the raison d'être of the university as the central intellectual institution of the Jewish people. These perceptions—of the university as the central intellectual hub of the Jewish people and of the university needing to acquire the respect and trust of world Jewry—were very closely related to Einstein's own perceptions of the roles of the Hebrew University. To clearly stress that academics would have the decisive say in the university's administration, four additional members were proposed for the board, all of them renowned European scholars and three of them close colleagues of Einstein: Leonard Ornstein, Sigmund Freud, Rudolf Ehrmann, and Edmund Landau. Not surprisingly, Einstein welcomed both the Zionist Executive's approach and the additional members proposed for the board.[14]

However, the tug-of-war between the university's representatives in Jerusalem and Europe had only just begun. In response to the proposals presented by the Zionist Executive, Magnes and the other Palestinian members of the board stated that they agreed to London's "formulation of the functions of the Board of Governors," yet objected "most emphatically" to the proposal for a non-Palestinian Executive. In their opinion, "the center of gravity of the Hebrew University of Jerusalem must be in Jerusalem." They welcomed the cooperation of "all savants, Jewish and non-Jewish," yet they believed that two executive offices would cause "waste and duplication and [would lead] to friction."[15]

Two weeks prior to the second meeting of the board of governors, Einstein explained to Magnes why he opposed moving the focus of control to Jerusalem. In his opinion, it would be "extremely perilous ... were the appointment of scholars and determination of the budget and the work plan of the University to be effected from Jerusalem." He justified this in a manner which must have seemed offensive to Magnes and the faculty in Jerusalem: "The intellectual competence of the scholars currently residing in Palestine cannot be sufficient to cope with these serious tasks due to the sparse settlement of the country."[16] The seeds of this perception of the poor academic level in Palestine were laid during Einstein's visit in February 1923. As we saw, he was far more impressed with the country's colonialization efforts than with its efforts in the field of academia, mainly because at the time, there were initial academic preparations only at the Technion. Einstein's dismissive view of his colleagues in Jerusalem reveals that his perception had not changed in the interim since his visit, even though more serious academic efforts were already under way.

In late September 1925 the second meeting of the board of governors took place in Munich. This time Einstein participated in the meeting, yet Weizmann did not, as he was in the United States. According to the official minutes of the meeting, two new bodies within the board of governors were to be established: a Presidium in London consisting of two members of the board, whose main tasks would be to represent the university externally and to review material for the further development of the institution, and an Executive in Jerusalem, whose task would be to execute decisions of the board in Palestine. Einstein and Weizmann were elected the two presidents of the board of governors. Magnes and British Zionist Norman Bentwich were elected members of the Palestinian Executive and would be known as the chancellor and vice-chancellor, respectively. Einstein was also elected chairman of the Academic Council. In addition, nineteen new members (most of them European scholars) were elected to the board of governors.[17]

However, there was considerable displeasure about the outcome of this meeting among some of the prominent board members who had not been in attendance. Weizmann was concerned that so many new members, "some of whom are incompetent," had been elected to the board of governors and that English university titles had been bestowed on Magnes and Bentwich, which were not "titles devoid of significance." Indeed, as he noted, the vice-chancellor was the "highest-ranking administrative official" at British universities. David Kaliski, chairman of the American University Committee, informed Weizmann of the American perspective on the Munich meeting. Felix Warburg, the most influential member of the committee, was opposed to the "overwhelming Zionist complexion" of the expanded board. Warburg and Julian Mack viewed "the Hebrew University as a matter to be removed as far as possible from Zionist concern and influence, not only here in America but also in Europe." Perhaps fearful for the success of his plans for an alliance between Zionists and non-Zionists in the establishment of the proposed Jewish Agency, Weizmann agreed that the university should "not be regarded as a Zionist institution and that it must remain free from all political influences."[18]

It was at the meeting in Munich that there seems to have been a sea change in Einstein's opinion of Magnes. He told Zionist Executive official Leo Kohn that it was Magnes himself who had proposed the title of chancellor to the board and that he would never forget "such a mortifying experience." However, it was Magnes' actions *following* the session regarding the minutes of the Munich meeting that incensed Einstein the most. Indeed, a major controversy broke out between them over the board's protocol. According to Einstein, Magnes had sent out his own version of the minutes and had changed crucial decisions taken at the meeting. Einstein expressed his ire over this in a highly combative (and rather arrogant) tone: "Mr. Magnes has to be told explicitly . . . that for all intents and purposes he is the executive organ of the Board and that he may only coordinate matters which have been decided upon by the Board (aside from minor ad hoc matters)." And Einstein issued a warning: "If he continues to behave badly, then money or no money, there will be nothing but to throw him out . . . I do not fear the Americans to whom I will explain the matter, should it be necessary."[19] At this stage, Einstein evidently still believed he could wield sufficient influence over developments at the university so as to both make a decisive impact on events on the ground in Jerusalem and to change the opinions of the American board members.

In a programmatic letter to Magnes, Weizmann expressed his support for Einstein's view of the university's administration and for his version of the events at the Munich meeting. Furthermore, he clearly delineated what he perceived as his "European" approach to the administration of the university and contrasted it to his definition of the American approach: he was opposed to establishing an "office of such scope and power as that of an American University President or Chancellor." In his opinion, the university should be governed "by committee," not by "powerful executive officers." Its structure should be "dispersed . . . as in other Zionist enterprises." Similar to his conflict with the Brandeis faction over the Foundation Fund in 1921, he advocated control of the university by the Zionist Organization as a "democratic institution" and not by donors. These benefactors should have "some influence in the affairs of the University, and not merely because of their money." Yet he did not believe that "the control of individuals [was] necessarily less political than that of an organization."[20]

The latent conflict between Einstein and Magnes turned into an open one in December 1925, when Einstein sent Magnes an angrily worded letter in which he protested against a second protocol of the Munich meeting being sent out by Jerusalem. Einstein insisted that only the first protocol was the official one. Privately, Einstein expressed himself even more extremely about Magnes: according to the Jewish philosopher and publicist Jacob Klatzkin, Einstein termed him a "swindler" who had "forged" the minutes of the board sessions.[21]

Einstein continued in his efforts to convince the American members of the board of the correctness of his approach to the university administration. In a letter to Felix Warburg, Einstein reiterated his view that the academics in Palestine lacked the "necessary experience and the intellectual caliber ... to be able to establish and lead a university worthy of the entire Jewish people." Until such time the institution's intellectual leadership needed "to remain centered where the intellectual focal point of Jewry lies, namely outside of Palestine." In his opinion, the board and the Academic Council had to provide what was missing in such a new country: "the intellectual atmosphere, the collaboration of scientific minds, that intellectual aura upon which the intensity of all scholarship depends and which can normally only develop gradually." [22] This strikes at the core of why Einstein opposed ceding control of the academic affairs at the Hebrew University to the Jerusalem Executive. If what he perceived as the gradual process of establishing suitable conditions for scholarship could not be guaranteed by the faculty in Jerusalem, then control of the university's academics had to continue to rest with those Jewish intellectuals in the Diaspora who were willing to contribute to its development. Only thus would it be possible to attract "the best minds from all of Jewry as teachers and students" and thereby become "for all of Jewry an intellectual center on a par with the universities of other cultures." However, there is an interesting contradiction here. Elsewhere, such as in his statements on anti-Semitism, Einstein stated that the Jews need not be concerned about the views of non-Jews. Yet Einstein clearly did care about the opinions of Gentiles, especially when it came to academia. Indeed, this is another indication that non-Jewish institutes of higher learning constituted Einstein's most important frame of reference for issues relating to the Hebrew University's administration. Furthermore, it is also clear that despite his rhetoric on behalf of the education of laborers in the Jewish community of Palestine, he was hardly concerned with the needs of the local population. Indeed, in his opinion, the university must never become "merely a provincial Palestinian university," and there would actually be no sense in its existence as such. [23]

In an obvious attempt to reach a compromise over the university's administration, Magnes proposed a "balancing of forces" between the main foci of power. The board of governors would be constituted by three main factors, "the Jewish Agency (or Zionist Organization), Jewish scholarship, and the material and moral support that we may be able to get from committees such as the one constituted in America." The Academic Committee would have three sections, European, American, and Palestinian.[24]

There is no direct statement by Einstein on this proposed compromise. Indeed, when Einstein learned that Magnes had appointed American chemist Israel Kligler head of the university's microbiology department without assent from the board, he grew even more incensed by his actions: "I view the University as lost . . . if it continues in this way. . . . If the Board does not decide on a radical purge [*Säuberung*], I will no doubt abandon the cause completely, as it is just a pity about the expended time and effort."[25] This was the first dramatic statement on Einstein's part in which he threatened to no longer have anything to do with the university. And it marked the continuation of his employment of language that pertained to sanitation—we have seen previous instances of such language during his trip to the United States.

In the meantime, Einstein was not satisfied with Magnes's explanation for sending out a corrected version of the Munich minutes. Magnes claimed he did not believe the Munich resolutions were binding, owing to previous decisions that had been made at the first meeting of the board of governors in Tel Aviv. Einstein therefore intended to issue an ultimatum to Magnes at the next meeting of the board, as he had concluded that "things are not going to work out with him." However, after consulting with his close colleague Rudolf Ehrmann, he decided to be more conciliatory: "No combative stance against Magnes, but rather constant pressure and non-approval of all steps to which he is not entitled."[26]

Although Einstein did not refer directly to Magnes's compromise attempts, it does seem that he was prepared for some kind of accommodation with other proposals for the administration of the university: in regard to the planned medical institute, he and Ehrmann favored a discussion "in how far the various European methods could be combined with those pursued in America."[27]

However, possibly encouraged by Weizmann's intention to resign from the Zionist Executive and dedicate himself to the university, Einstein sent a harshly worded letter to Magnes, protesting his refusal to retract the revised minutes and declaring that it was "futile to continue negotiating with you."[28]

One month later, Einstein received a firsthand account on the state of the university from Weizmann, who had just visited Palestine, exactly a year after its official opening. The Zionist leader was not at all happy with his impressions of the young institution: he complained that Magnes was interfering in the Jewish studies department, there was "no science to speak of," and the chemist Andor Fodor was a poor administrator who had chased away young talented co-workers. Weizmann declared his intention to break Magnes's "autarchy," but was at a loss as to how this could be achieved. He informed Einstein that he would like to take up a position in Jerusalem but did not know whether he would be "allowed" to.[29]

Yet, as in previous instances, such intentions did not lead to any concrete action on Weizmann's part. Einstein concluded therefore that Zionist politics were too important for Weizmann to forgo for the sake of the university. Consequently, he felt he could no longer believe the Zionist leader's announcements about his plans for direct involvement in the university. He declared the university could be salvaged only "if a man of scholarly ability and commercial and psychological disposition resides as a permanent representative of the Board [in Jerusalem]." At this point in time, he still saw Magnes as "decent and well-meaning, but narrow-minded and stiff-necked." He did not presume there were any significant differences between himself and Weizmann in their approach to the university. The most important goal was that by the next meeting of the board, an individual will be identified "whom we can place in Palestine to [carry out] the rehabilitation. Maybe this can be achieved without inflicting a fatal blow on Magnes. I would prefer this, in spite of the fact that he has committed grave formal and factual errors." Thus, at this stage, Einstein still believed the board could wield sufficient influence on the appointment of a replacement for Magnes. This statement also marked a significant shift in the solution proposed by Einstein: he apparently no longer believed it was necessary to control matters in Jerusalem remotely via the board. And he was adamantly opposed to the solution proposed by Magnes and his supporters: that the Jerusalem Executive wield the decisive control over the university's administration. Instead, he advocated a solution that would involve the import of a sufficiently talented individual to Palestine. Thus, the introduction of European academia into the Hebrew University, which began with Einstein's own inaugural lecture on Mt. Scopus, found its logical conclusion. In line with his support for the university as a solution for young Jewish academics, Einstein now advocated the hiring of "competent young individuals" who would be sufficiently independent "of foreign countries," as "no aspiring person will move to Palestine if such a dependent role is ascribed to him."[30]

However, merely a few weeks later, matters deteriorated again. As veteran German Zionist Otto Warburg expressed his support for Magnes, Einstein declared that he would refrain from recommending his dismissal to the board. Yet he reiterated firmly that he could no longer collaborate with Magnes and stated that "as long as Mr. Magnes is directing the affairs in Palestine, I will not be involved actively with the concerns of the University."[31]

In response, Weizmann stated that the Zionist Executive would "do [their] utmost to prevent your resignation." He reassured Einstein that he was convinced that his diagnosis that "sooner or later one must definitely get rid of [Magnes]" was "absolutely correct." He informed Einstein that he had written to Magnes, drawing "his attention to the general indignation over his autocratic administration and constant reliance on the American moneybags, and I told him quite plainly that it would be more dignified for the University not to accept such donations rather than be perpetually dependent on the whims and threats of the donors." Rather optimistically, he surmised that when Magnes received letters from himself and Einstein, "he may resign." This was apparently Weizmann's only hope, as "[he] cannot personally lead the attack against Dr. Magnes in view of [his] political responsibilities."[32] Indeed, this reveals the major weakness of Weizmann and Einstein's opposition to Magnes's rule in Jerusalem: Weizmann's inability (and unwillingness) to dedicate all his energy to the university's affairs and his constant prioritizing of Zionist politics over the university. Seen historically, this was merely a continuation of Weizmann's approach to the university during their joint trip to the United States in 1921, when his efforts on behalf of the Keren Hayseod had a negative impact on the Zionist delegations' mission to the United States in regard to the university. In this specific instance, the "moneybags" with which Magnes had excellent relations were the same Jewish philanthropists the Zionist Organization needed to fund its settlement projects in Palestine. At the time, Weizmann hoped to achieve this goal by establishing a grand alliance between Zionists and non-Zionists in an expanded Jewish Agency.[33]

Einstein seems to have been only too aware of this weakness. Merely two weeks after Weizmann's letter, he recommended to the board to place the "management of the affairs in Jerusalem into other hands." He justified this proposal by claiming that Magnes had "systematically ignored the resolutions taken by the Board in Munich." He also issued an ultimatum to the board that until such a change took place, he could no longer continue to be involved in the university's affairs. He assured them that he would not make this decision public so as not to inflict damage on the cause of the university.[34]

Einstein remained intransigent in this stance, even after Magnes visited him personally. He refused to discuss these matters with him so as not to introduce "confusion into these disputes." He refrained from taking a position on the planned constitution for the university, as he did not want to influence participants in the upcoming London meeting of the board. He particularly wanted to avoid "that any decisions be taken in a biased manner out of consideration for my person." He reassured Weizmann that he was "completely unaffected by personal [attacks]" and that "no incident whatsoever" could detract from his "devotedness to the cause of the University."[35] Yet it seems that Einstein thereby put the board in quite a difficult situation. A compromise could not be reached if Einstein was not willing to discuss specific points of contention. Moreover, he claimed that he wanted matters to be decided in a "businesslike" manner. However, his attacks on Magnes were to a great extent ad hominen in nature. One is also struck by the contradiction between his claim that he was "completely unaffected by personal [attacks]" and the emotional intensity in his reactions to the events as they unfolded.

Following the board's London meeting in August 1926, Einstein remained unresponsive, and Leo Kohn was sent to Berlin to prevent his resignation. Kohn learned that Einstein had decided to "sever all connections with the University." He succeeded in convincing Einstein to remain a member of the board, yet could not prevent him resigning as chairman of the Academic Council.[36] This step once more reveals the primacy for Einstein of academic matters over administrative ones—it was evidently what he perceived as the failings of the academics at the university that distressed him the most.

His resignation from the Academic Council led to Einstein's being less involved in the university's affairs. Six months later Kohn sent him an update on the major developments. Einstein was informed there was practically no influence of the bodies outside Palestine on the running of the university: "The entire management of the University is virtually concentrated in Jerusalem." The appointment of Kligler as head of the Hygiene Institute had proven to be a "serious mistake." Appointments by Kligler had not been brought to the board for approval. At the Jewish Studies Institute, Magnes refused to employ a professor whose permanent appointment had been confirmed by the board and for whom there was sufficient funding. Weizmann was very pessimistic about the situation and intended to resign as chairman of the board. Kohn surmised the only solution was to entrust the university's administration to the hands of a diligent younger person "who really understands something about the administration of a university and is knowledgeable about the specific problems of this University." To his mind, the Russian British mathematician Selig Brodetsky would be such a suitable candidate.[37]

As Einstein had realized that it was mainly through funding for the university that Magnes was wielding his power at the university, he now attempted to exert influence on how monies would be spent. When asked to assist in soliciting funds for scholarships from a Portuguese Jewish philanthropist, he demanded that he (and another person of his choosing) be given direct control of the funds.[38] This was a blatant departure from the German model of university funding, whereby a board of trustees kept a watchful eye on how monies were spent by the scholars. It is difficult to escape the conclusion that, by this stage, exercising control over the university had become a more important issue for Einstein than following ethical standards of administration.

After receiving an update on the Hygiene Institute from the Polish Jewish bacteriologist Arthur Felix, Einstein was even more incensed about developments at the university. The transfer of Felix, the university's "best scientist," to the Lister Institute in England was seen by Einstein as a huge loss and a disgrace. He attacked the director of the Hygiene Institute, Kligler, as a "schemer" and threatened (once again) to resign from all of the university's administrative bodies if the agreement Magnes had recently concluded with the medical organization Hadassah for the affiliation of its staff with the university was implemented. In Einstein's words, according to the stipulations of this agreement, "all the ineffectual dabblers who are employed at Hadassah, are simultaneously to be 'professors' at the University."[39] Thus, the blatant disregard for both the authority of the board and proper procedure was what angered Einstein the most.

Two weeks later, there were additional reasons for Einstein to be distressed. Leo Kohn sent him details of Magnes's proposals for significant changes to the university's constitution. All aspects of the university's administration would now be in the hands of the University Council, which consisted of Magnes and the local professors in Jerusalem. The Academic Council was to be abolished, to be replaced by foreign academic advisers to the University Council. The chancellor would be both the administrative and academic head of the university. Kohn did not believe that Weizmann would intervene to prevent these developments. Therefore, he claimed there was only one way out: that Einstein write a completely open letter to power-broker Felix Warburg in which he would ask Warburg whether keeping Magnes on as "dictator of the University" was a conditio sine qua non for the continued funding of the university by the Americans.[40]

Once again exhibiting an unwillingness to be overly cooperative, Einstein refused to write to Warburg, stating that he had already clarified his position to him and had not received any reply. He informed Kohn that if the agreement with Hadassah was accepted or if Magnes's proposals for changes in the constitution were adopted in part or in their entirety, he would resign from the board and justify this step "without reserve in the Jewish press." He was doing so convinced that "it well-nigh no longer constituted any actual harm if the University were to be compromised in the eyes of the Jewish public."[41]

This represented a further radicalization of his stance. It also revealed the degree of Einstein's frustration about the direction in which his "problem child" (*Schmerzenskind*)[42] was headed. He was now apparently demoralized to such an extent that he was no longer concerned with the public image of the university (and of the Zionist Organization, for that matter) and even hoped that a frank and public discussion about its future would bring about the changes he wanted to achieve.

The following month, Einstein received another firsthand account of developments in Jerusalem, this time from the prominent German Jewish physician Felix Danziger (who was not a member of the Jerusalem faculty). He reported that his support for Arthur Felix had been perceived as "German meddling" and that the "antithesis German-American was increasingly coming to the fore" at the university. Danziger viewed this development as "psychologically understandable, as all disciplines ... have had to resort to individuals with a German cultural background time and again and as a matter of course." In reaction, those individuals had been "systematically removed" from the Jerusalem board and an "Anglicization" of the entire executive had taken place. Einstein did not relate explicitly to this dichotomy between German and Anglo American academics, yet it is significant that yet another of his correspondents stressed these differences. Danziger also attacked Kligler's "unprecedented arrogance," his "protectionism," and his suppression of the opinions of others. He criticized the alleged insignificance of his scientific achievements and highlighted the helplessness of the rest of the faculty in light of his funding from the American Jewish relief organization, the Joint Distribution Committee.[43]

The case of Arthur Felix apparently had a profound and far-reaching impact on Einstein. He told Andor Fodor that "men of scientific competence are being kept at bay or being driven away in disgust" from the university. As he felt it was "practically impossible" for him "to fight against the existing conditions with any measure of success," he had decided to withdraw officially from the board and the Academic Council. He declared he could not share the responsibility "for that which occurs there any longer." In another letter to Fodor a month later, he cited the "arrogance of American money" as the reason for his inability to effect any changes at the university. He only hoped that by his own resignation (and that of others), the mistakes of the current leadership would become clear to all and their situation would become "untenable."[44]

Einstein expressed his extreme demoralization to Weizmann as well. He was especially exasperated that the board was merely leading "an illusory existence." If genuine attempts were made to appoint an academic head for the university and Magnes's role were restricted to administrative matters, he would refrain from taking any public steps for the duration of one year. If not, he would openly sever all his ties with the university. And he expressed his displeasure in a particularly dramatic fashion: "It is by far preferable to postpone the establishment of a Hebrew University for an entire generation than to build up a botched university now under the pressure of external circumstances."[45] It seems that Einstein was thereby referring to the local conditions in Palestine, which, as we have seen, were of little consequence to him. He does not seem to have been referring to the desperate situation of young Eastern European academics, as it is highly unlikely he thought their plight could have been ignored for an entire generation. It is in this context that we should note that, parallel to his efforts on behalf of the Hebrew University, Einstein had pursued (and continued to pursue) other options for Eastern European Jewish academics, such as the introduction of special university courses for Eastern European Jewish students in Berlin and a planned Jewish University in Kaunas in Lithuania.[46] This multilateral approach to solutions for the predicament of Eastern European academics was analogous to Einstein's approach to Jewish nationalism in general. If Palestine could not provide all of the necessary solution, other options could and should be found.

Weizmann expressed his support for Einstein's position. He also informed Felix Warburg that he would follow Einstein should he resign from the board. However, it seems that Weizmann was seeking an excuse to withdraw from the board of governors in any case. In late January 1928, he asked for Einstein's opinion on his intention to resign from the presidency of the Zionist Organization and take up a position in Jerusalem as the university's academic head. Einstein welcomed the idea and stated that he thought the university would "blossom" under Weizmann's leadership, yet he told him that he did not actually lend credence to his plan, as he had already proposed this before.[47]

Shortly after receiving an apparently nonextant letter from Einstein, Magnes showed himself more conciliatory. He was ready to yield on the matter of an academic head if the university's administrative structure was simplified and the authority of the university itself was strengthened. In these first formative years, a "lay head should be the actual executive head." Until the board (or he himself) believed there was a better lay leader than himself, he would continue in his current position. Even though Einstein was still very suspicious of Magnes's intentions, he had renewed hopes for the university, especially if Selig Brodetsky were to be appointed academic head.[48]

However, Magnes's reaction to the proposal to appoint Brodetsky was unenthusiastic. He doubted whether Brodetsky had sufficient prestige, and held that other candidates should be considered. He also believed it was "premature to fix dogmatically the final form the Academic Headship is to take. . . . The University should be given the chance to accumulate its own experience, and its constitution should be flexible enough to enable it to adapt itself to its own vital needs."[49]

The same month, Einstein received the most explicit details yet on the acrimonious rivalries at the university from the head of the chemistry department, Andor Fodor. In his bitter report, Fodor claimed it was his adherence to the "German model" he had applied to his work in the department as "director of the Institute" which led to attempts by the university's administration to terminate his employment. He also accused those assistants who did not support him of being "openly abusive of German characteristics." And Fodor summarily condemned the character of the university as comparable to that of an "upper secondary school."[50]

Even though Einstein was aware that Fodor was "an agitated and not very tactful person" and made no explicit reference to the accusations of anti-German sentiments at the university,[51] Fodor's report clearly had a strong effect on him. It was in reaction to the information he received from the Hungarian chemist that evoked from Einstein the unbridled threat to dissociate himself from the entire enterprise of the university that was quoted at the beginning of this chapter. This was certainly, once more, an extreme reaction, which can only be explained by the bitter disappointment he felt over developments at the university. Yet his willingness to witness the demise of the whole project merely because it was not developing in the manner he wanted was quite irrational. This can only be explained by the deep significance this project had for Einstein and his intense emotional involvement.

Even though Einstein sympathized with Fodor's plight, he did not approve of the next step he took: the chemist presented his allegations of the "persecution" of German scholars at the university to the German consul in Jerusalem. Einstein objected strongly and remarked, "It is always unwise and destructive to involve strangers in a family row. One's self-esteem alone should prohibit one from doing so." This was once more indicative of Einstein's ethnocentric perception of the Germans as being strangers and not to be involved in internal Jewish affairs. Yet it was perhaps slightly absurd for him to criticize Fodor over his appeal to the German consul at the same time that he himself was deliberating whether to resign from the university with an "uproar." Fodor subsequently clarified that he had only conferred with the consul to obtain a recommendation for German universities.[52]

In late May, on the eve of the annual meeting of the board of governors to be held in London, Einstein made a decisive and dramatic move. He sent Weizmann two letters to be presented at the meeting. The first letter, which consisted mainly of Einstein's program for the restoration of the university and his advocacy of the need for an academic head, was to be presented at the beginning of the meeting. The second, in which he asked Weizmann to propose Selig Brodetsky for the position of academic head in both their names, was to be submitted when the board was to elect the head. Einstein added that if an academic head was not elected, it would be "far better to close down the whole shebang. Because if matters continue in this manner, not only will the University itself have become a laughing stock in a couple of years' time but indirectly it will also tremendously harm the prestige of Palestine and of all those who are working on its behalf in the eyes of the whole world."[53] This is possibly Einstein's most sincere statement on the university and his involvement in its affairs. It seems that here he finally revealed that it was his own prestige which he was most concerned about. We must remember that in 1921 he had gone out on a limb with his non-Jewish and non-Zionist scientific colleagues (such as Lorentz and Haber) on behalf of the Hebrew University. He had forgone participation in such a supremely prestigious conference as the Solvay Congress to raise funds for the university. If the whole project failed, it would reflect detrimentally on Einstein himself. It is difficult to escape the impression that this is what troubled him the most in regard to the direction the university was headed.

Einstein informed Weizmann that the election of an academic head at the London meeting was a conditio sine qua non for his continued involvement. This had to be decided "immediately," and there had to be an exact determination of the head's functions.[54] This was a tall order, especially in light of the fact that Einstein would not himself be present at the meeting. Yet he still expected the board to dance to his tune, thereby exhibiting little understanding for the workings of organizations in general and committees in particular.

From his letter to the board, it is obvious that Einstein's main concern was the academic standards of the university. As the issue of the introduction of teaching for beginning students was the central topic to be discussed by the board, Einstein clarified his position on this matter: teaching should only be instituted once research had reached a certain level; students should only be accepted if they had a willingness and capability for scholarly activity—"mass production of academics in title only" was "absolutely out of the question"; courses should only be taught by professors and lecturers and not merely by unqualified instructors, e.g., "schoolmasters." For this reason and also for the development of the university in general, it was fundamentally important to appoint an academic head. As the university was self-governing, he deemed the English model of an academic vice-chancellor to be most appropriate, especially as the university did not yet have a representative council of lecturers. The academic head should not merely be a one-year honorary president like the German *Rektor* but rather a long-term "functionary" who would direct the academic development of the university. Furthermore, he had to be fully independent of the university's administration.[55]

When Felix Warburg, without whose support there was no chance of swaying the board in Einstein's favor, received these letters on the eve of the meeting, he was not at all pleased. He stated that he was "very unhappy that these bombs have been thrown into the sessions" and asked Einstein to amend those passages in his letters "which are so unjust and of an aggressive nature."[56] Yet no reply of Einstein's is extant.

Following the sessions in London, Weizmann sent Einstein a detailed report. Prior to the meeting, Weizmann had conferred with Warburg and Magnes, who expressed their agreement to the appointment of an academic head, but only for a limited period and with curtailed functions. However, at the meeting itself, there was such opposition to Einstein's proposal, especially from some of the European scholars that it was decided to transfer the issue to a committee, which would present its conclusions at the board's next meeting. Einstein was proposed as chairman of the committee. Knowing that this could hardly be described as the result Einstein had hoped for, Weizmann asked him not to take any further steps until they had met.[57]

Yet, as was to be expected, Einstein was extremely dissatisfied with the outcome. In his view, "all attempts to find a healthy basis for the development of the University [have] failed for the time being." He was very blunt in his judgment of the sessions: "I think nothing of the resultant compromises. Under these circumstances I have decided to de facto withdraw entirely from this whole enterprise." Yet he had changed his mind about making this decision public: "I will refrain from an official resignation from the Board and the Academic Council, in contrast to my original intention. I do not want to contribute seemingly or actually to the failure of the development of the University by discrediting it in the eyes of the public."[58] As in previous instances, Einstein was opposed to the compromises reached. Even though some progress was achieved at the London meeting, he was not prepared to continue to try pushing the university in the direction he wanted. Yet it is interesting that Einstein went back on his original (and repeated) threat to publicize his resignation from the board. This would make it seem that he had not been entirely serious in his previous threats and they were part of a complex game of brinkmanship.

Einstein's unwillingness to continue negotiating with those involved in the university's administration extended during this period even to the leader of the Zionist movement. He asked that Weizmann and Brodestky refrain from visiting him, ostensibly for medical reasons, though this may have merely been a convenient excuse. He also requested that his name be struck from the list of members of the board and the Academic Council.[59]

He gave a more explicit explanation to Fodor of the reasons for his refraining from making his resignation public: "I am now convinced that every minute I continue to dedicate to this foul enterprise is a waste of time. On the other hand, I do not want to resign from the Board and the Academic Council with an uproar so as not to provide ammunition to the *goyim* and the anti-Zionists." It seems that a month after his initial disappointment in the outcome of the London meeting, he had gained some perspective on the proceedings: "You must bear in mind that there is much good will behind the enterprise, even if the administration has fallen into the wrong hands. . . . Needless to say, I will dissuade everyone from going to the University in Jerusalem."[60] It can therefore be concluded that it was more important to Einstein to protect the image of the Zionists and their most prestigious project vis-à-vis the Gentiles and anti-Zionists than it was for him to publicize his profound disappointment over the university's direction. Furthermore, he was prepared to acknowledge that at least some of those involved were well-intentioned. And perhaps most significant, he had reached a stage in his involvement with the university at which he was no longer willing to view it as an attractive option for young Jewish academics, the original motivation for his getting involved in the university project in the first place.

Reports from visitors to Einstein in the aftermath of his dramatic step reveal that his resignation may have been tactical in nature—at least in part. American physician Emanuel Libman reported to Weizmann that Einstein "believed that his resignation would lead to certain changes, —and that then he could again become openly interested." He also stated that "if anybody attempted to make capital of his position, he was to be informed, and that he would make a statement for the press which would show his interest and confidence in the development of the University." Thus, it seems that following a brief period, in which he was furious over the nonacceptance of his demands, he returned to some degree of loyalty to the university. Meanwhile from Jerusalem, Magnes condemned Einstein's course of action, especially his attacks against him "from his Olympian heights" and his refusal to confront him and the board in person.[61]

Even though he had officially resigned from the university's administrative bodies, Einstein still remained interested and committed to this institution. In a letter to Brodetsky, he was perhaps most sincere and self-reflective about his dramatic decision: "it was especially kind of you that my obstinate behavior in the matter of the University has not deterred you from being in contact with me. Among our eloquent

Jewish brothers I possess the status of a savage who can only make himself understood convincingly by means of gestures. This is how you should view my step and not out of lack of internal solidarity." At this stage he apparently saw himself as a Moses-like figure who may never enter the Promised Land of a reformed university: "Even if I do not live to see the day on which I can rescind my decision I will not cease to view the Jerusalem University as a matter of the heart. I believe I can best serve the cause by following my instincts without too much reflection. So far this has been the best policy." However, he apparently did see some merit in a greater preparedness to compromise: "Yet at the same time I do think that a more conciliatory stance is preferable for more politically skilled figures such as you and Weizmann."[62] It is intriguing that even Einstein himself viewed his own behavior as "obstinate." There was also some extent of helplessness vis-à-vis the board on Einstein's part, particularly in regard to his communication with its members. Nevertheless, Einstein's emotional commitment to the university remained high, despite the great disappointments he had experienced. It is also revealing that Einstein viewed his own thinking on a matter as cerebral as the university as primarily instinctual. This seems to be more of a myth he told himself (and others) about himself, part of his philosophy of trying to adhere to the "natural," which may have had its origins in his rural Jewish heritage. There is also a slight indication that Einstein saw himself as morally superior to "political" figures such as Brodetsky and Weizmann.

His intention not to refer young academics to the university did not last for long, as can be deduced from his informing Leo Kohn of a young colleague who had studied under Marie Curie: "One should get hold of him as soon as possible for Palestine. . . . In general, it would be preferable to get a few such eager young people there who could inspire each other."[63]

Around this time, Einstein also became disappointed with the other major educational project of the Zionist Organization in Palestine he was involved in, the Technion. His method of dealing with the developments he was not pleased with was identical. He threatened that if Technion board members Shmuel Yosef Pevzner and Aharon Czerniawski were not removed, he would resign and publicize his reasons. To his mind, the situation there was "worse" than at the university.[64]

Although Einstein's resignation was not publicized in the press, "broad circles" knew about it. Weizmann tried to avoid the impression that his step "derived in any way from animosity toward the Zionist leadership in general." He therefore appealed to him to speak at the opening of the upcoming Zionist Congress in Switzerland. Even though he had originally planned not to attend the congress, this appeal may have swayed him to change his mind, as he did travel to Zurich.[65]

Einstein's speech was greeted with wild enthusiasm by the attendants of the congress.[66] He restated the importance of the Zionist enterprise in Palestine for all of world Jewry, and gave homage to the movement's founder and current leader:

> We must . . .not forget that the development of Palestine is not only a great task for all Jews but that this challenge is also a great gift to us. For that we are indebted not only to the two great leaders [i.e., Herzl and Weizmann], but also to that courageous and enthusiastic minority that calls itself Zionist. We, the others, owe these men our national solidarity, and I believe we should be mindful of the fact that they have the moral right to continue to exert the strongest influence on the work that we wish to carry out."[67]

Einstein's wording in this speech is highly ambiguous, perhaps intentionally so. "We, the others" could be a reference to "we, the other members of the Zionist movement," or it might be a reference to "we, the non-Zionists." It depends on what he meant by the phrase "these men." Was he referring to Herzl and Weizmann or to the Zionist leadership in general? Unfortunately, we cannot be certain as Einstein often referred to the Zionist leadership by using the term "these men"—in the male-dominated world of Zionist politics, the women in the movement did not figure prominently in Einstein's mind. In my opinion, in his speech, Einstein was giving the congress a subtle hint that he continued to consider himself a non-Zionist, yet one who was willing to utilize the Zionist cause as long as his own goals were being advanced.

Einstein was clearly impressed by the major accomplishment of the Zurich congress, the enlargement of the Jewish Agency, the quasi-governmental body that served the administrative needs of the Jewish community in Palestine. In its expanded form it would encompass both Zionists and non-Zionists, a goal Chaim Weizmann had been striving toward for years. Recognizing this achievement, Einstein sent a postcard from the luxurious hotel he was staying at in Zurich to Weizmann's brother-in-law: "On this day the seed of Herzl and Weitzmann [*sic*] has ripened in a wonderful manner. None of those present remained unmoved."[68]

The congress in Zurich would turn out to be one of two major developments that affected Einstein's position on Zionism in the year 1929. Within a fortnight, even more momentous events would unfold in Palestine that would have a decisive impact on Einstein's views on the Zionist movement.

CHAPTER 6

"A GENUINE SYMBIOSIS"

Einstein on the 1929 Clashes in Palestine

"If things continue in this vein, no decent Jew can be a Zionist any longer."[1] This was Einstein's reaction on learning of a call by Menachem Ussishkin, head of the Jewish National Fund in Jerusalem, to transfer the Arab population from Palestine in the spring of 1930. Einstein was appalled by the Zionist leader's remarks on "the Arab question" and characterized them as a "scandalous speech." In reaction to the violent clashes in August 1929 that had led to hundreds of fatalities, Ussishkin had stated, "If the land is free of inhabitants—good! And if there are inhabitants there, then they will just have to be transferred to some other place wherever that may be, but *we* have to take over the land. We have an ideal which is greater and nobler than taking care of several hundred [thousand] *fellahin*."[2]

The violence that erupted in Palestine in late August 1929 in which 133 Jews and 116 Arabs lost their lives[3] proved particularly testing to Einstein's continued commitment to the Zionist cause during this period. While he was becoming increasingly entangled in the feuding over the direction in which the Hebrew University was heading, Einstein continued to concern himself with pressing general Zionist issues as well.

The internal inconsistencies between his support for Zionism and his antipathy toward a particularistic brand of nationalism remained one of the complex issues Einstein had to deal with. In early 1925 he wrote, "A Jew who strives to impregnate his spirit with humanitarian ideals can call himself a Zionist without contradiction."⁴ This was another attempt by Einstein to resolve his basic universalistic worldview with his ethnocentric approach in the context of Jewish nationalism. These incongruities are even expressed in the language he uses: the statement is impersonal; it may refer to himself, yet it also may not.

Time and again he had to contend with the core questions of whether he was a Zionist or not and whether he was a nationalist or not. In 1926, he allegedly told Jacob Klatzkin,

> You know, I am a Zionist but not a nationalist. Admittedly, as long as humanity is divided into peoples, the Jewish people must also have its place in the community of peoples. But the ideal situation would be if all peoples as such, including the Jewish people, would vanish as nations.⁵

We cannot be certain of the authenticity of this statement as it is based on Klatzkin's own notes of his series of conversations with Einstein (on another occasion, his subject supposedly told Klatzkin that he "had learnt quite a bit from Klatzkin, but he has learnt very little from me: for I understand him, but he does not understand me."⁶ This is not exactly an ideal endorsement for the authenticity of Klatzkin's notes.) In any case, similar to his position on Jewish nationalism in the early 1920s, it illustrates how Einstein continued to divest Zionism of its normative identity as a form of nationalism and present his utopian vision of a world without nations. Moreover, this passage once again demonstrates Einstein's instrumental approach to Zionist ideology.

During the mid-1920s, Einstein was involved in propaganda efforts on behalf of the Zionist cause both within and beyond the Jewish community. In 1926 the German Pro Palestine Committee (Deutsches Komitee Pro Palästina) was established. Its goal was to garner support on the German political scene for Jewish settlement in Palestine. Einstein was regarded as one of the most prominent non-Zionists to serve on the committee. Other distinguished members included the senior German politicians Graf Bernstorff and Katharina von Oheimb and the high-ranking diplomat Carl von Schubert.⁷

Within the Jewish community, Einstein continued to be instrumental in Zionist fundraising among world Jewry. At the beginning of

Figure 21. Einstein and Weizmann and members of the German Pro Palestine Committee: Graf Bernstorff, Carl von Schubert, Katharina von Oheimb, Oskar Wassermann, Kurt Blumenfeld, 1928

a year that would prove to be a fateful one for Palestine and for Einstein's enduring commitment to the Zionist movement, 1929, he issued a call to raise funds for the United Jewish Appeal. In a rather hyperbolic statement, he expressed his lofty vision for the role Palestine would play in elevating Jewish morale: "Palestine already represents a wonderful living monument to the Jewish collective spirit. See to it that it becomes our pride and joy and our rallying center, as was the Temple in Jerusalem for our ancestors."[8] We have already seen how Einstein (and others) established a connection between the ancient Temple and plans for the Hebrew University during his inaugural scientific lecture there in 1923. Now this analogy was being extended to encompass the entire Zionist settlement enterprise. This was quite a tall order for an impoverished country in a region with a slowly developing infrastructure. However, we have already noted that for Einstein, the ethical and cultural roles of Palestine far outweighed the economic and political ones. We also need to take into consideration that this statement was intended for fundraising purposes, and this would have had an impact on Einstein's choice of analogies.

The immediate trigger for the August 1929 violent clashes in Palestine was the issue of access by Jews to the Western Wall in Jerusalem. Following demonstrations by Revisionist Jews, which were perceived by the more radical Arabs as provocative and inflammatory, a week of violence ensued, initially in Jerusalem. Yet it soon spread to other locations in Palestine, most notably to Hebron and Safed. Most of the Jewish deaths were caused by Arab rioters; most of the Arab casualties were the result of British military and police actions, although there were also some acts of retaliation by Jews.[9]

Einstein's first public reaction to the events was a message sent to a rally held by Jewish organizations in Berlin in solidarity with the Yishuv. In his letter, Einstein was at pains to stress that "such setbacks" would not adversely affect "our resolve to continue the peaceful development" of Palestine. In his opinion, the repetition of such events in the future could only be avoided by two measures, first and foremost by creating "such a form of coexistence with the Arab people" that would resolve frictions between the sides by means of "organized cooperation." Einstein identified the lack of everyday contact between Jews and Arabs as a cause for fostering "an atmosphere of mutual fear and mistrust which promoted such regrettable explosions of passion." In his analysis of the causes for the clashes, Einstein chose to focus on the psychological factors of the interethnic strife and advocated that lessons be learned from Jewish history: "It is imperative for us Jews to demonstrate that in the difficult periods of our past we have gained sufficient understanding and psychological experience that we can cope with this psychological and organizational problem." He was convinced that the differences between Arabs and Jews were not irreconcilable. And in line with his usual predilection to focus on his own ethnic group's prejudices rather than on the other side's biases, he defined "blind chauvinism among our own ranks" as the greatest danger in the current situation. He warned against the Jews succumbing to the belief that "reason and understanding could be replaced by the bayonets of the English police." The other necessary measure was for the Jews to demand from the mandatory authorities that "institutions for the safety of people working peacefully" should be found that would be fair toward the dispersed Jewish settlements and could work toward bridging the national gap between the ethnic groups. He was quite vocal in his criticism of the British being remiss on that count. In accordance with his strong opinion that particularistic goals had to be subject to universalistic ones, he restated that "we must not forget for a moment that our national task is at its core a supranational one, and that the power of the entire movement is based on its ethical value and stands and falls with it."[10]

This was not the first time that Einstein had expressed his belief that everyday interaction between the ethnic groups could play a beneficial role in lessening friction between the two sides. We have seen that six years earlier, during his visit to Palestine, he had asserted that "[t]he common folk knows no nationalism," thereby implying that day-to-day contacts could alleviate the interethnic tensions between Jews and Arabs. Following his 1923 tour, he also stated that "most of the difficulty" in regard to nationalism "comes from the intellectuals—and at that, not from the Arab intellectuals alone."[11] Einstein clearly believed that, similar to the situation in Germany during World War I, the more educated classes had a responsibility to guide the general populace toward more tolerance and not to encourage nationalistic fervor. Yet he had been disappointed with the role German academics had played in that conflict. Was he about to be disappointed again?

A week after the riots, Einstein made an important private statement on Zionist calls for the removal of the Arab population from Palestine. In a letter to his close friend Michele Besso, he wrote, "Displacing the Arabs from their soil is completely out of the question. The country is sparsely populated in relation to its potential."[12] This assertion is particularly revealing in light of remarks on the future of the region he had allegedly made six years earlier during his visit to Palestine. We saw then that he had reportedly stated that "the country will be ours."[13] However, it was not clear whether this meant he believed the land would be exclusively Jewish or merely predominantly Jewish. But by now it is apparent that even in the wake of the violent clashes, he was strongly opposed to any part of the Arab population being removed from Palestine.

Einstein was also concerned with the effect of the violence on British public opinion and its potential for an unwanted change in British foreign policy. An open letter bearing his name as signatory was therefore published in the *Manchester Guardian* on 12 October. It is a rallying cry intended to express concern at the criticism of Zionist policies in Palestine in reaction to the clashes and to garner support for the Jewish colonialization efforts. However, it is a peculiar document: its style is not at all typical of Einstein, and some of its content would seem to contradict his positions on Zionism and on the clashes in Palestine. Unlike the statement for the rally in Berlin, the letter completely lacks references to what is required of the Jewish population to prevent further deterioration of the conflict. There has therefore been some speculation that the document was not actually written by Einstein.[14] I would definitely agree that he did not author the letter as it was published. Passages referring to "a martyrdom of centuries" and "innocent youths ... butchered in cold blood" do not appear elsewhere in Einstein's published or unpublished reaction to the clashes and do not sound as if they were authored by Einstein.[15] However, there are some similarities to the message he sent to the Berlin rally, such as a reference to "British bayonets," and a few months later Einstein did refer to the missive as "my letter."[16] It seems very likely that a draft version prepared by Einstein was reworked by the Zionist Organization in London to render it a more effective tool for Zionist propaganda. In any case, no actual documentation proving either Einstein's direct authorship or any reworking of the letter by the Zionist Organization has come to light in the available sources.

Two months after the events, Einstein attempted once more to establish an ethical vision for Zionism that would distinguish it from other forms of nationalism. In a rare instance in which Einstein actually initiated an exchange on Zionism, he wrote to the former state president of the German province of Baden, Willy Hellpach, who had written an article on the Zionist Congress in Zurich that Einstein had attended, about his association with the Zionist movement. Einstein informed Hellpach that his initial motivation for supporting Zionism had been his perception of the erosion of self-esteem among Jews. In his opinion, this could only be remedied by establishing a "common enterprise" around which all Jews could rally. He did concede that this was somewhat linked to nationalism. However, he made this important distinction: "it is a nationalism that does not strive for power but for dignity and recovery. If we did not have to live among intolerant, narrow-minded, and violent people, I would be the first to reject every form of nationalism in favor of a universal humanity."[17] Even though he did not explicitly mention the August riots in his letter, it is safe to assume that he had them in mind, as they had occurred only two months earlier. The escalation of violence in the region had apparently confirmed Einstein in his belief that Zionism should be qualitatively different in character from other forms of nationalism. He once again defined Palestine as a "center" (*Zentralstelle*) for the realization of Zionist aspirations rather than a "homestead" (*Heimstätte*), thereby possibly advocating a political entity that would be even more removed from a nation-state than a homeland. *not clear what that practically means*

In the aftermath of the clashes, Einstein entered into important exchanges on the outbreak of hostilities with two prominent individuals. The first exchange was with Hugo Bergmann, whom Einstein had initially encountered in Prague as a representative of that city's Zionist and philosophical associations and with whom he had subsequently correspondended in his new role as a Zionist functionary in London. He had then visited Bergmann during his 1923 Palestine tour in Jerusalem, where the latter had settled as director of the Jewish National and University Library. In the meantime, Bergmann had also become one of the central figures in the pacifist Brith Shalom movement, which strove for the peaceful coexistence of Jews and Arabs in Palestine.[18]

In the first extant letter from their correspondence, Einstein introduced his vision for a peaceful coexistence in the region: "The events in Palestine seem to me to have proven again how necessary it is to create a genuine symbiosis between Jews and Arabs in Palestine. This means the existence of permanently functioning joint administrative, economic and social organizations." He again emphasized that the separate existence, "side by side" of the two groups, "must lead from time to time to dangerous tensions."[19] Intriguigly, the one concrete measure he suggested to Bergmann was an educational one: all Jewish children should learn Arabic. The riots obviously led Einstein to conclude that a nonviolent future could not be secured by the two communities living segregated from one another, each with its own institutions. In order to guarantee that the violence would not escalate, it was imperative to establish collaborative organizations.

In his reply, Bergmann stated that Einstein's use of the term "symbiosis" was a precise formulation of the goal Jews and Arabs should strive toward. However, in his opinion, psychological barriers, particularly among the Jews, would make the realization of this goal quite difficult. He admitted that in recent time the "willingness for peace" among the Arab leadership was not "tangible," owing to the strengthening of the Arab national movement. However, he opined that ever since the Balfour Declaration, the Jews had "immersed themselves in the dream of a Jewish state or at least of a state in which the Jews formed the overwhelming majority." But this dream had been "shattered by reality." This had led to serious psychological obstacles among Jews in Palestine and the Diaspora and hindered their willingness to strive for peaceful coexistence with the Arabs. He therefore saw little hope for Einstein's plan to establish joint Arab-Jewish institutions.[20]

Nevertheless, Einstein reacted with enthusiasm to Bergmann's reply and described it as an "exquisite characterization of the required policy on Palestine." He therefore requested permission to publish the letter, to which Bergmann agreed.[21] However, one week later Einstein apparently changed his mind about the style of the letter and proposed that Bergmann adopt a more positive tone. Bergmann agreed to the suggested changes and sent a revised version.[22] Einstein then tried to get Bergmann's letter published in the German Zionist organ *Jüdische Rundschau*. However, its editor in chief, Robert Weltsch, who privately supported Bergmann's stance on Arab-Jewish affairs, refused to publish the letter, as he did not agree with Bergmann's public propaganda on behalf of Jewish-Arab understanding. Surprisingly, Einstein agreed with Weltsch that Bergmann's letter should not be published.[23] Perhaps he was not satisfied with Bergmann's changes, or perhaps he now believed that a less public stance on these issues should be advocated. However, merely four days later, Einstein forwarded Bergmann's letter to Weizmann in response to a letter from the Zionist leader in which he had attacked the Arab leadership. Einstein described Bergmann's letter as "hitting home at the essentials."[24] The same day he informed Bergmann that he agreed with the changes he had made to his letter and authorized him to make any use of it he deemed fit. However, he also stated that he did not think it was the suitable time to publish the letter in the press. Einstein was obviously quite conflicted as to the correct path to follow: "As soon as I have gained some clarity, I will write to you again."[25] Einstein's various reactions to Bergmann's letter are intriguing in that they clearly demonstrate how often he could change his mind about one and the same issue. They also show how he could adopt various tactical stances on a given issue, depending on the target audience.

Bergmann did not object to Einstein's decision not to publish his letter. He informed Einstein that in the wake of his recent political comments, his name was "on the list now proscribed by the Zionists," and almost all the Zionist press was branding him as a "traitor" and calling for his resignation as director of the National Library. However, he did believe that Einstein should make public his own views on peaceful Arab-Jewish coexistence as soon as possible.[26]

It was at this point that Einstein's exchange with Bergmann became a matter of increasing concern to his "handler" in Berlin, the German Zionist Kurt Blumenfeld. He was very concerned that Bergmann would undermine Einstein's trust in the Jewish nationalist movement and have an adverse effect on his commitment to Zionism in general. He expressed this to Bergmann in a typically denigrating manner: "Your correspondence . . . has merely had the result that Einstein, who is quite mercurial and easily susceptible and who places his name at the disposal of both good and bad causes, is gradually becoming an opponent of Zionism." In a clear attempt to demoralize Bergmann, Blumenfeld warned him against becoming "unfaithful" to the Zionists and separating himself from "the fate of his own people." He also disputed Bergmann's right to "appear as a legitimate representative of the Zionist movement."[27]

There followed a four-month interval in the correspondence between Einstein and Bergmann. In late April 1930, Bergmann appealed to Einstein to intercede on behalf of the thirty indviduals (twenty-nine Arabs and one Jew) sentenced to death by the British Mandate authorities as a consequence of the violence of August 1929. He stressed the importance of demonstrating that public opinion around the world opposed the planned executions and asked Einstein to convey this to the British high commissioner in Palestine, Sir John Chancellor.[28] Robert Weltsch also conveyed such an appeal against the executions that had been initiated by another German member of Brith Shalom, the prominent philosopher Martin Buber.[29] It is intriguing to note that Weltsch had by now apparently assumed a more important role in Einstein's contacts with the German Zionists, perhaps even ousting Kurt Blumenfeld from his central position. It is highly likely that this shift was caused by Einstein's greater affinity for Weltsch's than for Blumenfeld's position in regard to the issue of Arab-Jewish coexistence in Palestine. In any case, Einstein subsequently sent the telegram to Chancellor, urging that the executions be commuted to life sentences, using Buber's proposed text.[30] In early June, after all but three of the death sentences had been commuted, Bergmann informed Einstein that his telegram had featured prominently in the Arab press. He regretted the fact that high-ranking Zionist officials, most notably Chaim Weizmann, had chosen not to appeal for commutation of the sentences. In his opinion, this would have been a "golden opportunity" for the Zionist leadership "to change the relationship between the two peoples by means of an energetic measure." In any case, he now thought it imperative "not to squander new opportunities for approchement." He therefore advocated a "radical change in the relationship between the two peoples" and demanded that the Jews "side with the Arabs with total awareness." He advocated a "program of Semitic cooperation" and presented a concrete proposal for the future of the region which, if adopted, would have far-reaching consequences for Zionist policy. Bergmann's plan included the following points: the "irrevocable renouncement of the Jewish state"; regulation of Jewish colonization so that no fellahs are dispossessed of their land; Jewish aid for the fellah; alignment of Jewish immigration to the economic capacity of the country; Jewish support for an Arab federation consisting of the three countries under the British Mandate: Palestine, Transjordan, and Iraq, whereby the special status of Palestine would be preserved and Jews would acquire the right to settle in Transjordan and Iraq (both of which would remain outside the Jewish homeland); and the establishment of a democratic parliament in Palestine.[31] This plan clearly placed Bergmann far beyond the national Zionist consensus: up to then, all Zionist factions had agreed on the establishment of a Jewish nation-state in Palestine. The proposal for a drastic reduction in Jewish immigration was also beyond the Zionist consensus.

In reply to Bergmann's plan, Einstein stated that he "fully agreed with everything [he] had written" except for one point: "A unified parliament can have catastrophic results for the minority in the case of a conflict. A way has to be found to secure our minority existence and cultural autonomy." He warned that if the Jews did not realize that "only direct cooperation with the Arabs can create a dignified and secure existence," then "the entire Jewish position within the cluster of Arab countries will gradually become more and more untenable. *I am saddened less by the fact that the Jews are not clever enough to grasp this but that they are not fair-minded enough to want it.*"[32] Here again, Einstein's basic ethical stance determined his perception of the Arab-Jewish conflict. His agreement with all of Bergmann's points (except for one) was a far-reaching step. Einstein had thereby clearly placed himself outside the national consensus. On the other hand, his rejection of a fully democratic parliament owing to the demographic balance in Palestine was obviously based on his deeply entrenched Jewish ethnocentrism.

In his reply, the final letter in their exchange on the 1929 clashes, Bergmann reassured Einstein that he did not think their respective positions on the issue of an elected parliament were very far apart. He was also convinced that an accommodation could be achieved with the Arabs in regard to the establishment of a joint parliament that would protect the cultural autonomy of the Jewish community. He was strongly opposed to the establishment of two separate parliamentary bodies, as most major issues needed to be dealt with between the two ethnic groups as joint inhabitants of the country. Bergmann also expressed his regret that the executions "have once more reduced the hopes for [achieving an] understanding for an extended period of time." He proposed publishing Einstein's letter of 19 June in Hebrew in the organ of Brith Shalom—in his opinion, the letter would have a positive effect, particularly its last sentence.[33]

Einstein's second significant exchange in the wake of the riots was with Azmi El-Nashashibi, the twenty-six-year-old editor of the Arab newspaper *Falastin*, published in Jaffa. In early 1930 Einstein contaced El-Nashashibi in reaction to an article that had appeared in the newspaper a few months previously entitled "Relativity and Propaganda."[34] This article had been written in response to Einstein's letter in the *Manchester Guardian*, which had been reprinted in the *Palestine Bulletin*.[35] Although in his opinion the article in *Falastin* included some "insulting remarks about the character of the Jewish people," Einstein was willing to ignore these, as "Unfortunately, it has become a habit of quarreling peoples to belittle the other nation." Here again, Einstein took pains to mention the psychological factors involved in the Arab-Jewish conflict. However, he was most interested in focusing on his vision for the future of Palestine. In reaction to the doubt expressed in the piece that the Jews "only want peaceful relations with the Arab population of Palestine," Einstein reassured the editor that as someone who "for decades has been of the conviction that future humanity must be based on an intimate community of nations and that aggressive nationalism has to be overcome," he can "only imagine the future of Palestine in the form of peaceful cooperation between the two peoples residing there." This was another reiteration of his previously expressed position that his views on Zionism and Jewish settlement of Palestine were based on his wider universalistic perception of the peaceful interaction of all nations. He clearly wanted this position to be held by the other side in the conflict as well: he expected "the great Arab people to better appreciate the Jews' need to reconstruct their national home in the ancient Jewish homeland." He even anticipated that the Arabs would assist the Jews in "jointly finding means and ways to enable the expansion of Jewish settlement of the country." He expressed his conviction that "the loving devotion of the Jewish people to Palestine will benefit all the inhabitants of the country, not only materially, but also culturally and nationally." Einstein even believed that Jewish support could benefit the entire Arab population in the Middle East, not merely in Palestine. These are interesting statements, as they seem to imply a somewhat colonialist attitude on Einstein's part, if an enlightened one. One could even say that Einstein was thus describing the Jewish man's burden in the Levant.[36] In any case, he was of the firm opinion that "the two great Semitic peoples" had made "everlasting contributions" to European culture and could have "a great future together" by mutually supporting each other's national and cultural aspirations. Einstein ended his letter by referring to the "sad events of last August," which he regretted for two reasons: they had revealed human nature "in its most debased manifestation" and they had alienated the two peoples from each other.[37] Einstein's letter was subsequently published in the Arabic and English editions of *Falastin* on 1 February 1930.[38]

El-Nashashibi sent Einstein two copies of the English-language issue in which his letter was published. He claimed that his paper was the only one in Palestine "which advocated peace, and published several articles on peace and understanding by prominent personalities." He informed Einstein that Magnes and Bergmann had also published essays in it but that "their articles have been condemned and severely criticized by almost all the official Zionist organizations the world over." He invited Einstein to write more on the Palestine crisis and promised to publish an article by him.[39]

Einstein accepted the invitation and replied in late February. He took pains to emphasize that this was a personal letter and that he bore "*sole* responsibility for it."[40] He also informed Robert Weltsch that he had sent his reply to El-Nashashibi "of my own accord." He added, "As this reply includes concrete proposals I did not want to burden your conscience with it, as otherwise you will be placed in an awkward position vis-à-vis the Zionist Organization."[41] This would seem to imply that Weltsch and the Zionist Organization were involved in the publication of Einstein's first letter in *Falastin*. It also demonstrates that Einstein was very much aware of the tensions within the Zionist Organization over his statements and those of other intellectuals who were advocating peace and conciliation.

In his second letter, Einstein stated that El-Nashashibi's reply proved to him that "there is goodwill on your side for a solution of the existing difficulties worthy of both our peoples." He reiterated his opinion that the complexities were "more of a psychological than substantial nature and that they would be solved if both sides exhibit sincere goodwill." He decried the fact that "Jews and Arabs face each other as quarreling parties vis-à-vis the mandatory authorities." He viewed this state of affairs as "unworthy of both nations" and believed it could only be rectified by finding a way "among ourselves" toward proposals to which both parties could agree. He then proceeded to present a concrete proposal for interethnic cooperation in the region. He stressed that this was his personal opinion, which he had not discussed with anyone: "A Privy Council is to be formed to which the Jews and Arabs shall each send four representatives, who must be independent of all political parties." The four representatives from each side would be a doctor, a lawyer, a working man's representative, and an ecclesiastic. They would each be selected by their representative bodies. These eight council members would meet once a week. They would undertake not merely to represent the interests of their profession or their nation "but rather to consider the well-being of the entire population of the country to the best of their knowledge." Whenever decisions gained the support of at least three members from each side, they were to be made public as the decisions of the entire council. If one of the members dissented, he could resign, but he would still be sworn to secrecy. The ultimate goal of the council was to form a joint representation of the populace vis-à-vis the Mandate that would "rise above the politics of the day."[42]

The council was clearly intended to be a committee of technocrats. It is worth noting that Einstein sought a nonpolitical (and nonpartisan) solution to the highly sectarian conflict in Palestine. We have seen elsewhere that Einstein preferred nonpolitical resolutions to political conflicts. As a technocratic solution, this would also not be a democratic means to solve the interethnic dispute. As we saw in his correspondence with Bergmann, Einstein was clearly fearful that a strictly representational parliament would be disadvantageous to the Jews in Palestine because of their smaller share of the total population. Intriguingly, the concreteness of this plan and its sophistication illustrate that Einstein was far from being naive when it came to practical political proposals.

This second letter was published in the 15 March edition of *Falastin*, accompanied by an editor's note by El-Nashashibi. The note reveals that the letter constituted a real breakthrough in the vicious circle of mutual mistrust between the two sides:

> Prof. Einstein has made a practical contribution towards the solving of the present difficulties. Taking into account the sincerity of purpose which must be credited to whatever he may have to say, it becomes a duty of Arabs and Jews in Palestine to give their opinion on the proposal put forward. It may be that it may require some modification. It may also be that the sum total of goodwill may not amount to a quantity sufficient to make this proposal practicable at present. On the other hand, it would be *criminal* not to explore all channels which contain a chance of making the two communities co-operate. That the proposal by Prof. Einstein contains such a chance is evident. At the present unfortunate state of tension it is a piece of good fortune to get Prof. Einstein's name attached to a proposal, which otherwise may be open to suspicion.

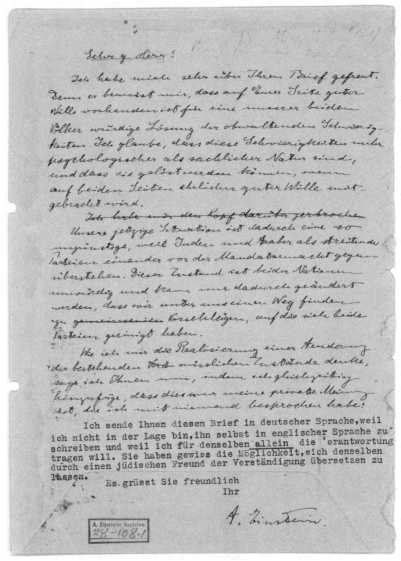

Figure 22. Letter to Azmi El-Nashashibi, published in *Falastin* in 1930

This ended the exchange between Einstein and El-Nashashibi. The German Jewish physicist had obviously gained the Arab editor's trust. It is certainly a testimony to Einstein's credibility among such moderate Arab intellectuals as El-Nashashibi that he could convince him of the sincerity of his proposals. The question subsequently became

Es wird ein „geheimer Rat" gebildet, zu dem die Juden
und die Araber je vier Vertreter senden, welche nicht
von einer politischen Instanz abhängen dürfen.
Zusammensetzung auf jeder Seite

1 Arzt, gewählt von der Ärztegesellschaft
1 Jurist gewählt von den Rechtsanwälten
1 Arbeitervertreter gewählt von den Gewerkschaften
1 Geistlicher gewählt von den Geistlichen.

Diese 8 Leute kommen einmal per Woche zusammen.
Sie verpflichten sich, nicht Interessenvertreter ihres
Berufes und ihrer Nation sein zu wollen, sondern
nach bestem Wissen und Gewissen das Gedeihen
der ganzen Landesbevölkerung ins Auge zu fassen.
Die Besprechungen sind geheim und es darf über
sie absolut nicht berichtet werden, auch nicht
privatim.   über irgend etwas)
    Wenn eine Zustande gekommen ist,
dass mindestens drei Männer von jeder Seite zustimmen,
so kann in der Beschluss der Bekanntgabe
gefasst werden, die aber im Namen des ganzen Rates
geht. Ist einer nicht einverstanden, so kann er
aus dem Rate austreten, aber er wird von der Pflicht
der Geheimhaltung dadurch nicht entbunden.
Wenn eine der gewählten wählenden Genossenschaften mit einer
Resolution des Rates unzufrieden ist, kann sie
ihren Vertreter durch einen anderen ersetzen.
    Wenn dieser „geheime Rat" auch keine bestimmten
Kompetenzen hat, so wird er doch dazu führen
können, dass allmählich die Differenzen ausgeglichen
werden, und dass eine gemeinsame Vertretung der
Landesinteressen gegenüber der Mandatarmacht
entsteht, welche über die Tagespolitik erhaben ist.

whether this breakthrough on a personal level between two prominent individuals could be extended to the leaders of their respective
communities or even to larger groups among Jews and Arabs.

A possible impediment to reaching out to wider circles was the
lack of trust on both sides of the conflict. Such mistrust may have led
Robert Weltsch to warn Einstein off from establishing contact with
Mohammed Nafi Tschelebi, a Muslim student from Syria living in
Berlin who had founded the Islam Institute there. Weltsch informed
Einstein that Tschelebi "seemed to be a political agent" and cautioned
him against being in contact with him.[43]

Einstein's support for a more conciliatory approach toward the Arab
population led to some serious antagonism between him and the most
distinguished Zionist leader he was associated with, Chaim Weizmann. Three months after the August riots, Einstein issued a dire
jeremiad-like warning to the Zionist leader about the future of Jewish
settlement in Palestine that constituted somewhat of an indictment of
his policies: "If we do not succeed in finding the path of sincere cooperation and negotiating with the Arabs, then we will have learned
nothing from our two-thousand-year-old ordeal and will deserve the
fate that will await us."[44] Weizmann seems to have been quite offended
by Einstein's criticism of the current Zionist policy. He was particularly aggrieved that opponents of Zionism had begun to quote Einstein (alongside Bergmann and Magnes) "against us."[45] Einstein was
taken aback by Weizmann's reaction and quicky sought to reassure
him that he had no intention of waging a battle against the Zionist
leadership. He agreed with Weizmann that negotiations should not be
conducted with the "[Arab] agitators." However, he was clearly not
satisfied with an analysis of the events that merely focused on nationalistic incitement and ignored the need for "sincere cooperation,"
without which there would be "no peace and no security."[46]

The reaction of Weizmann and the Zionist mainstream to the riots
led to a decisive deterioration in Einstein's support for Zionism. By
late 1930 Einstein had actually clearly reached a nadir in his support
for the Zionist movement. In an interview with a correspondent of the

Austrian newspaper *Wiener Tag* he was quoted as saying, "The continuing conflict in Palestine repels those more superior members among the Jewish people for whom the work there is a cause based on idealism; they would be far more likely to collaborate if the political issues could be solved by means of benevolent agreement."[47] This passage, if accurate, once again demonstrates Einstein's primary preoccupation with the effect of Zionism on Western Jewry (as opposed to developments on the ground in Palestine). And, needless to say, by "those more superior members among the Jewish people" he primarily meant himself.

In any case, Einstein's comments in the Viennese newspaper evoked Weizmann's ire, and he responded in a defensive manner by blaming the Arabs for the conflict: "It is not *we* who do not desire peace with the Arabs but rather the Arabs (at least their leaders) and I am fearful of the English (at least of their people in Palestine) and the last White Book of the government makes this very clear." He was apparently even personally offended by Einstein's remarks: "I am all the more hurt by your interview and your statement that decent people will soon turn their backs on Zionism etc. etc."[48]

The fallout from the interethnic violence had an adverse effect not only on Einstein's position with respect to the Zionist leadership but also on his stance toward the Hebrew University. When Hugo Bergmann and Judah Magnes—two leading figures in Brith Shalom and at the university—were subjected to acrimonious public attacks by their more belligerent colleagues and students,[49] Einstein was quite dismayed, and this had a negative impact on his opinion of the Jerusalem faculty. Thus, once again he was deeply disappointed with the reaction of a group of academics with whom he felt an affinitiy in regard to a political issue. Fifteen years earlier he had been dismayed by the response of German academics to the militarization of German society. Now he was disillusioned with the reaction of the faculty at his favorite Jewish academic institution, the Hebrew University, over their attacks on their colleagues who advocated Arab-Jewish coexistence. Needless to say, it is highly ironic that Einstein backed Magnes's position on the riots, yet in regard to all other university matters remained firmly convinced that Magnes had to be removed as that institution's chancellor.

The serious erosion of Einstein's support for Zionism had an additional consequence. He began to express his backing of a non-Zionist settlement project outside Palestine. In July 1930 he demonstrated his interest in a plan to settle Eastern European Jews in Peru. He found the project "quite remarkable" and saw it as a real possibility to obtain a healthy existence for a large section of the Jewish people.[50] It seems fair to speculate that his support for alternative colonization efforts at this stage was at least in part due to his disillusionment with the Zionist movement. It is even possible that his strong disenchantment with the Hebrew University also played a role in this: if Palestine could not provide a setting for his pet project, then maybe other venues would suffice just as well.

Yet another indication of Einstein's weakened support for Zionism during this period can be found in Arthur Ruppin's diary. When visiting Einstein at his summer house in Caputh outside Berlin, the prominent German Zionist asked him why he had aligned himself with Zionism. Einstein responded that he saw Zionism as an idea "which has enabled Jews to remain a community with dignity." Ruppin then asked him "whether he would recommend Jewish nationalism for German Jewry if it were possible for them to become assimilated into German society with their dignity intact and join the German nation." In response, Einstein replied evasively that "it was good that this question was merely hypothetical as he would find it difficult to provide an answer."[51] If this account is accurate, it seems to suggest that (at this stage at least), the perceived rejection of German Jewry by the majority society was Einstein's ultimate motivation for his support of Zionism.

The 1929 clashes represented a major watershed moment in Einstein's affiliation with the Zionist movement. It tested his sympathy for Zionism in the extreme. The outbreak of violence actually led Einstein to a substantial reassessment of his stance toward the movement and its leaders. In the aftermath of the violent events, he offered his vision of future peaceful existence in the region and a concrete solution to the Arab-Jewish conflict in Palestine. Yet by doing so he placed himself outside the mainstream Zionist consensus and created a significant ideological distance between himself and the Zionist leadership. As the situation in the region became increasingly more complex, clear fault lines began to appear between Einstein and the highest echelons of the Zionist movement.

Yet in spite of his severe disillusionment with events in the Zionist arena, on the ground in Palestine and at the Hebrew University, Einstein continued to concern himself intensely with the ongoing struggle over administrative and academic issues at that institution in Jerusalem. I now return to consider his positions in the acrimonious battles in that (bloodless) conflict.

CHAPTER 7

THE "BUG-INFESTED HOUSE"

Einstein's Involvement in the Hebrew University, 1930–1933

"It is completely clear to me that tampering with the constitution will be of no use as long as downright vermin continue to play the leading roles in the Executive and faculty in Palestine."[1] In such dramatic language did Einstein express his despair to Chaim Weizmann over developments at the Hebrew University in May 1933.

During his final years in Berlin, events at Einstein's favorite Zionist project continued to unfold in a dramatic manner. And even after his departure from Europe, the struggle over the administration of the university headed toward a remarkable showdown.

In July 1930, an intense correspondence began between Einstein and the young mathematician Adolf Abraham Fraenkel, from whom Einstein had requested reports on the university a year previously, as "there is no more objective and pure judge than you." Fraenkel, who had just been elected head of the Jerusalem faculty,[2] was to become Einstein's most important informant on the university over the next three years.

From his correspondence with Fraenkel, it is clear that despite his disassociation from its official bodies, Einstein viewed his role as contributing indirectly to the "thriving" of the university "that is dear to both of us." In this context, he was willing to give Fraenkel advice on appointments for the university's physics department.[3]

During this period, Fraenkel functioned as a mediator between Einstein and Magnes. The chancellor proposed that Einstein take the

230

Figure 23. Adolf Abraham Fraenkel, professor of mathematics
at the Hebrew University, early 1930s

initiative and talk unofficially with suitable scholars from various countries to search for an academic head. His only condition was that Weizmann not be included on the search committee for the moment. However, Einstein did not think he could be effective, as "the people with the clout, i.e., those who wield the money, don't listen to me." Moreover, after an apparent deterioration in Fraenkel's relationship with Israel Kligler and with the British professor of philosophy Leon Roth (who both exerted considerable influence on Magnes), Fraenkel told Einstein he would probably leave the university in the summer. By March 1931 he had confirmed his imminent departure and had also informed Einstein that at its upcoming meeting, the board of governors would discuss plans for the appointment of a rector (i.e., academic head), yet in a form that would diverge substantially from that envisioned by Einstein and himself.[4]

In a further move toward once more adopting an official role in the university's administration, Einstein held discussions with Julian Mack and Felix Warburg. He subsequently told Fraenkel that the

Americans would welcome "a solid plan" and a concrete proposal for the position of academic head. He asked Fraenkel whether he could propose any candidates.[5]

In the meantime, Fraenkel continued to wage his own battles in Jerusalem. He now saw Kligler's influence on Magnes as "the greatest danger" for the development of the university. He had garnered support from ten members of the University Council for his proposal for an academic head, yet there were three weighty opponents: Magnes, Kligler, and Roth. He stated that because of his precarious situation, he could not propose a candidate for the position from Jerusalem. He believed that the European members of the board could do so more easily.[6]

At this stage, Einstein believed that Fraenkel himself would be a suitable candidate for academic head if he were given the necessary powers. In May 1931, Weizmann informed Einstein that Magnes was now convinced that an academic head was necessary. He also asked Einstein for his frank opinion on his plan to move to Jerusalem and take up the "academic administration" of the university. Einstein supported the idea, yet thought it could only serve as a temporary solution until a more permanent appointment could be made. At that time, he expected Weizmann to resume his involvement in political affairs. He informed Weizmann that he was prepared to rejoin the board of governors "if an end is put to the current ghastly situation as a result of your appointment." The following month, Otto Warburg suggested that Einstein raise the issue with Felix Warburg. He did so, asking him to propose Weizmann to the board once he had consulted with Magnes. Felix Warburg was hesitant about the proposal, however, and seemed to prefer that Weizmann take up a professorship in chemistry in London rather than run the risk of becoming involved in political entanglements at the Hebrew University. Einstein was disappointed with this outcome and told Fraenkel that he was glad "that he no longer directly has anything to do with the affair."[7]

Following this disappointment, Einstein's involvement in university affairs appears to have been on hold for almost a year. He did not renew his active participation until September 1932, when Fraenkel informed him of the recent board of governors meeting that had taken place in London. It had been decided that Weizmann would take up a position at the university beginning in the autumn of 1933. Furthermore, Fraenkel continued, "the Kligler regime seems to have decidedly passed its apex and lacks substantial support even in America, with the possible exception of [prominent American educator] Abraham Flexner." Einstein welcomed this news and confirmed that he had not heard anything about it previously. In a second letter, Fraenkel gave further details of the meeting. Weizmann had attended the board meeting for the first time in a long while. And from the United States, "gentlemen interested in the matters at hand" were present, not merely the "money people" or their representatives. The initial funding for the new physics institute was approved. Therefore, Einstein's involvement was all the more necessary, as his expertise was required to address the difficult issues to be decided regarding appointments and the direction of the institute. Perhaps most important, a commission to deal with the central organizational issues of the university was to be established, to be led by Weizmann. Einstein was encouraged by this report yet preferred to continue not to be involved officially: "it is certainly tactically preferable if I continue to absent myself from the administration. I can be consulted at any time if necessary. I fear that the situation will not improve substantially until a genuinely competent Head is in place."[8]

Weizmann also provided Einstein with a detailed report on the recent board meeting, but additionally informed him of some of the background to the meeting. He had threatened Felix Warburg with his resignation from the board if the course of the university was not changed. This led to a meeting between Weizmann and Magnes at which the Zionist leader accused the chancellor of "completely strangling the exact sciences." He demanded that the disciplines of physics and chemistry be developed properly before work started in biology and medicine. Weizmann concluded that as he had succeeded in raising funds for the physics and chemistry departments from the American Jewish Physicians Committee, the matter was secured. The other main issues discussed were the "nebulous constitution" of the univer-

Figure 24. Judah L. Magnes, chancellor of the
Hebrew University, early 1930s

sity, the role of the board of governors, and pressure from the American donors. Magnes and the board agreed to the establishment of a Structure Committee to examine these issues, to change the constitution, and to restructure the board. This committee would consist of Norman Bentwich, Philip Hartog, Herbert Samuel, and Weizmann himself. In Weizmann's opinion, the university had entered a new and improved stage, and he wanted to remind Einstein of his promise, made eighteen months earlier, that he would rejoin the board if Weizmann were to join the university.[9]

Einstein replied that he was extraordinarily interested and pleased by Weizmann's letter regarding the "'rehabilitation'" of the university: "With this state of affairs, I am pleased to rejoin the Board." However, within a brief period, once it had become apparent that Weizmann would not be joining the Jerusalem faculty after all, Einstein became demoralized again and reiterated his intention not to have anything to do with the university until Magnes was removed from office. Eventually, a Survey Committee to conduct an independent investigation into all administrative matters at the university was established in late 1933 (around the time that Einstein left Europe for the United States). The committee was led by Sir Philip Hartog and submitted its findings in April 1934. Its report criticized "almost every aspect of the university's activity" and recommended that Magnes's authority be radically curtailed and that an academic head be appointed. Consequently, the position of chancellor was abolished in 1935, and Magnes became the first president of the Hebrew University, with strongly limited executive powers. In addition, Hugo Bergmann was appointed the university's first rector. As all his demands for radical change at the university had been met, Einstein was prepared to return to the university's board of governors in late 1935.[10]

This study has now followed Einstein's involvement in the administrative and academic affairs of the Hebrew University throughout its first decade. What conclusions can be drawn from his intense and at times highly dramatic and even acrimonious engagement in the university's early development? In order to analyze Einstein's association with the university, I will place his activities in the context of recent historiography of both German academia and the Hebrew University itself.

Recent studies on the early history of the Hebrew University have mainly focused on the following debates pertaining to its fundamental character: Which of the widely divergent schools of thought were proposed for adoption in the early development of the university? Was the university to be primarily a national popular institution or a universalistic, elitist research center? Was it to be primarily a university of the local Yishuv or of the entire Jewish people? And which activity should be the primary focus of the university, research or teaching?[11]

The various academic traditions to which the major proponents of the university adhered have been widely discussed in recent studies. Historians have noted that advocates of the British, German, and American academic models tried to influence the direction of the university. This jockeying for control was exacerbated by the absence of a "universal framework of accepted procedures and regulations." David Myers has pointed out that the projection of "an institutional model conceived in Europe onto the *terra incognita* of Palestine" was a "difficult enterprise."[12] This transfer was compounded by the fact that the state authority in Palestine, the British Mandate, was not directly involved in the establishment of the university. Instead, the Zionist Organization in London played this role. However, this also changed after 1925, once ownership of the university had been transferred to the university's board of governors.

Yet historians of the Hebrew University seem to agree that the battle over the various academic models was mainly limited to the deliberations of the administrative bodies of the young institution. When it came to the *academic* structure and content at the university and the central authority of its professors, these have been seen as deriving primarily from the German research university tradition. Myers has noted that "virtually all of the first-generation Jerusalem scholars, those from both Eastern and Central Europe, had studied in Germany where they were exposed to the methods and standards of the validation of [the German concept of] *Wissenschaft*."[13]

It was certainly this adherence to a German research ethic that guaranteed Einstein's respect for those scholars at the university who excelled in their fields on an international level. Yet it is interesting to note that no instances in which Einstein referred to the German research model explicitly as "German" have been found. This may have been because he would have felt that any accentuation of the German character of such a research ethic constituted excessive Germanness or because he viewed the German concept of research (*Forschung*) he had inculcated in his student days as not particularly (or exclusively) German. A third possibility is that he took the German character of this research ethic for granted to such an extent that he saw no need to stress its Germanness.

In this context, it is also beneficial to take a look at recent studies on the German university. Some of these have noted the coexistence at these universities of heterogeneous organizational principles: they were simultaneously partly charismatic, traditionally corporative, and bureaucratic. This was especially true from the late nineteenth century on, when new research institutes came to the fore.[14] Therefore, the perception of the research tradition described in studies of the Hebrew University simply as "German" in character does not do justice to the complex nature of this tradition's background and its impact on the Jerusalem university.

This is even more relevant in light of the fact that new research has cast doubt on the universal existence of the archetypal German professor. As the German historian Rüdiger vom Bruch has stated, "There were professors in the *Kaiserreich*, but not *the* German professor." Recent studies have emphasized significant differences among the various disciplines, regions, and localities.[15]

In this context, rather than the classical Humboldtian university (which has been seen by some scholars as having a decisive influence on the early development of the Hebrew University), I believe that Einstein's preferences for the administrative and academic structures of the university were primarily influenced by his experiences as director of the Kaiser Wilhelm Institute for Physics. Toward the end of the *Kaiserreich*, corporate ties within German universities were weakening and were being replaced by both growing state influence and new affiliations with financiers and entrepreneurs. This was especially apparent at the newly founded Kaiser Wilhelm Institutes. The directors of the institutes held very high status in German academia and society and had direct contact with both senior government officials and wealthy financial backers.[16] Einstein fostered crucial ties with both social democratic ministers of culture such as Konrad Haenisch and Carl Becker and the main funder of the Kaiser Wilhelm Institute for Physics, the prominent Jewish banker and entrepreneur Leopold Koppel. It seems fair to say that Einstein would have preferred a similar setup at the Hebrew University: strong support from state functionaries (in this case from the Zionist Organization) and from financiers who respected the expertise of scholars in all matters academic.

It is also important to note that the classical university, dominated by full professors (the *Ordinarienuniversität*), was under siege at the time Einstein was receiving his university education and struggling as a young lecturer. The "decline of German academia" was proclaimed regularly in contemporary literature. In polemical articles, the associate professors and private lecturers (*Privatdozenten*) were attacked as a "swarm . . . of dabblers"[17] We have seen that Einstein employed the same term, "dabbler" (*Pfuscher*), to describe what he perceived as the unqualified Hadassah doctors who were incorporated as faculty members on Magnes's initiative. This indicates the complexity of Einstein's relationship to German academia. Although he had considerable disdain for most of his colleagues at German universities (especially in the humanities) because of their political beliefs,[18] he clearly adopted at least some of their views on what they perceived as less qualified colleagues and the threat they constituted to the status of the full professors. Yet, as with most things with Einstein, this issue was a complex and multidimensional one. At the time of his strong criticism of the direction the Hebrew University was headed in, he was a well-established full professor (*Ordinarius*) himself and a prestigious member of the Prussian Academy. Yet he was also haunted by memories of his days as a young alumnus from the Zurich Polytechnic and his failure to secure a position as an assistant, the first rung on the academic ladder. As we have seen, this was the driving force that motivated him to seek out viable alternatives for young, up-and-coming colleagues. However, we need to ask whether his advocating the transfer of such young colleagues to the Orient was subconsciously a means for him to dispose of talented rivals. Even though there were several instances after World War I in which Einstein admitted he had nothing new to contribute to science, he does not seem to have ever considered vacating his own positions for the sake of younger colleagues. Indeed, it has been pointed out by some Einstein scholars that despite his enormous scientific achievements, he did not create an Einsteinian school in the discipline of physics. The only young academics he fostered were the numerous assistants who assisted him in his mathematical calculations.[19]

Einstein's experiences with the Kaiser Wilhelm Institute also influenced his preferences for the organizational structure of the Hebrew University. Once responsibility for the university was transferred from the Zionist Organization to the board of governors, it became a private university. Some historians have determined that the university's model of governance at this stage was that of an American autocratic presidency. This may have been Magnes's preference, yet the European members of the board (and most prominently among them Einstein) were accustomed to a system of academic self-governance.[20] This explains Einstein's outrage and exasperation at Magnes's open and repeated defiance of the board's authority.

Yet it is particularly intriguing that when it came to what he perceived as the university's most crucial necessity, the appointment of an academic head, Einstein preferred the British university model and its position of a vice chancellor. Here again, this choice may have been motivated by his highly ambivalent relationship to German academia. On the other hand, he may just have been influenced by Weizmann, who also advocated the British model. Another possible explanation is that he was so favorably impressed by Selig Brodetsky that he came to support the system espoused by the mathematician from Leeds.

As for the American university model, we have seen how Einstein was confronted with this model of higher education during his trip to the United States in 1921. The exposure of European (and especially German) scholars to the American model was a general trend from 1900 on. They began to visit the United States on exploratory trips, and American scholars traveled to Europe as visiting professors.[21] Yet in the framework of the Hebrew University, Einstein was confronted with the American model in a distorted manner. These distortions were multifold. The university was not in the United States but in Palestine. The American model had to compete with rival academic models such as the German and the British. In addition, Magnes's import of the American system was idiosyncratic, especially in light of the fact that, as his critics quickly pointed out, the former rabbi from New York lacked prior administrative and academic expertise at an American university—or anywhere else, for that matter.[22] As for the issue of American financial backing for the university, it is at least somewhat ironic that Einstein opposed what he perceived as the undue influence of American Jewish financiers, in light of the close relations he himself fostered with German Jewish bankers and entrepreneurs such as Leopold Koppel and the Warburg brothers.

In the context of the influence of various university models on the early development of the Hebrew University, we should also note that recent studies on the history of German universities have pointed out that the three models usually distinguished in the historiography of European universities, the German research model, the French training model, and the English personality model, seem insufficient when compared with the actual reality of European higher education. New studies on universities in Russia and in northern and Eastern Europe have demonstrated that these three models were not valid in all of Europe. Recent research has also shown that the classical Humboldtian university did not serve as a "modernization foil" for other universities in Germany. Rather, there were numerous paths to reform.[23] Here again, the newer trend in the historiography of higher education toward dismantling the monolithic view of the German (and other) university models needs to be adapted more systematically to the historical study of the Hebrew University.

Issues discussed in recent studies on the history of the Hebrew University can also help to contextualize Einstein's positions. In the debate over whether the university should be a popular institution or an elitist research center, Einstein was one of the most ardent advocates of the university fulfilling the role of a superior center for research. Einstein expected the standards of research at the university to be top-notch almost from its inception, and this raises the question of how realistic this expectation was. Einstein's support for an elite research center was tied to his adamant claim that the university was not capable of running its own affairs. As a consequence of his 1923 trip to Palestine, he was convinced that the Yishuv and the university lacked the necessary infrastructure to support such first-class efforts. Therefore, all major administrative and academic matters had to be guided by the board of governors in London. However, as more institutes were established in Jerusalem over time, Einstein apparently found this position less and less tenable. This is most likely the reason why, in June 1926, he changed his position and began to advocate hiring an academic head from Europe.

The issue of the character of the university as a popular or elite institution is closely related to the debate on whether the university would be primarily an institution of the local Yishuv or of the entire Jewish people. Historiographers of the university have seen the conflict between the Zionist Executive in London and the university's executive body in Jerusalem as a struggle between the center and the province. They have noted that this conflict was a "microcosm of the larger issue of a newly assertive Yishuv challenging the hegemony of the diaspora-located Zionist executive." In the specific case of the university, recent studies have pointed out that such external agendas on the part of Magnes and Weizmann exacerbated the already severe clashes over the university.[24] In Palestine, Magnes's perception of the university within the general Zionist context concurred with that of most of his Jerusalem colleagues: once they moved to the university, they advocated the local University Council having the decisive say in matters of administration and resented attempts by London to control the running of the young institution. Himself an import from the United States, Magnes's support of local needs and interests has been seen as ironic by some historians.[25]

For Weizmann, on the other hand, the university was "one of a myriad of Zionist projects he directed"[26] and could not be realized without direct and strong intervention by the Zionist Organization in London. Myers has described this attitude of Weizmann and others as a "paternalistic scepticism in the ability of the locals to guide their own affairs."[27] Einstein seems to have fully subscribed to this paternalistic attitude.

The debate over the role of the university as primarily a global Jewish or a local Yishuv institution was closely tied to the issue of the role of politics at the university. For Weizmann, it was to be the crowning achievement of the Zionist movement. For Magnes, the university was to be "above politics," and he did not perceive it as a project of any specific faction within world Jewry. In this, he concurred with the most influential figures involved in the university in the United States, such as Felix Warburg and Louis Marshall.[28] Ironically, in light of his antipathy toward Magnes, Einstein concurred with him on the issue of the university being above politics. In principle, he was against the "overpoliticization" of the university.[29] Yet he also concurred with Weizmann's view that the Zionist Organization was the most useful instrument to establish the Hebrew University. It seems that this apparent contradiction can be resolved once one realizes that Einstein did not define the Zionist Organization as a political institution, at least not in any narrow sense of the term.

In addition, as we have seen elsewhere, Einstein viewed the political with much the same disdain as his fellow full professors did. The maxim of being "above politics" advocated by the majority of German university professors in the *Kaiserreich*[30] was a trait Einstein had absorbed during his own early years as a young lecturer at German-speaking institutions of higher learning. Much has been written about just how much of a political dissident Einstein was during this period. In spite of his initial courage in signing the antiwar declaration in 1914, Einstein's primary mode of action during World War I seems to have been one of inner exile. This modus operandi continued even after the war had ended. Days after the dissolution of the *Kaiserreich*, he summed up his attitude towards politics by recommending to the prominent social democrat and former private lecturer Leo Arons, "Keep your trap shut!"[31]

This was another instance of Einstein seeing himself primarily as a servant of science (*Wissenschaft*), a common attribute among many other German scholars. Historians of the German university have differentiated among three main groups of professors in regard to their relationship to politics: the "apolitical academics," who constituted the great majority; those who saw themselves as "political" only in regard to their responsibility for the "common good"; and a small minority who actively identified with a specific political party.[32] Even though Einstein was viewed as a dangerous dissident in some quarters of academia, he avidly avoided any direct association with specific political parties.

Yet what was Einstein's position on the prospective national role of the Hebrew University? We have seen that during his early involvement with plans for the university, it was very important for Einstein to stress its national role. Prior to its opening, he saw its main function as being a potential haven for Eastern European Jewish academics. In this, he concurred with Weizmann, for whom the university was a means for rescuing Jews from anti-Semitism. This position was advocated in an even more radical manner by the leader of Zionist Revisionism, Vladimir Jabotinsky, who supported the establishment of a mass Jewish university to rescue Jews from Eastern Europe.[33] However, in contrast to Jabotinsky, both Weizmann and Einstein were primarily concerned with establishing a small research center of qualitative excellence that would absorb the elite of Jewish scholars.

Moreover, once Einstein became embroiled in the power struggles over the university, he seems to have lost sight of his initial primary goal of rescuing the talents of young Eastern European Jewish scholars. They ceased to figure in his correspondence on the university in any prominent way. Indeed, the academic standards of the young institution became his overriding concern in relation to the university.

With regard to the local role of the university, Einstein does not seem to have had much understanding for or patience with the difficulties inherent in transferring European plans for the university to the reality of the Orient. It is clear that he expected the Yishuv to adopt a European institutional model regardless of the local conditions and challenges. In this, one could claim that Einstein adhered to the mandarin intellectuals' insular ideology of being primarily preoccupied with colleagues, then with state authorities (in this case, the Zionist Organization), and only then with ordinary people.[34] Similar to his reaction to Palestine during his visit just a few years earlier, it was the European efforts that impressed him. Now, in the case of the university, he demanded that the same research standards he was accustomed to be implemented at the Hebrew University. Magnes's autocratic rule at the university certainly made it easier for Einstein to attack the direction the young institution was headed, yet it seems fair to speculate that he would not have been satisfied with its development even if Magnes's managerial style had been less tyrannical.

There were other local Palestinian issues of which Einstein apparently had little or no awareness. One of these was the chasm between the effete scholars in Jerusalem and the hegemonic culture of Labor Zionism, which has been described as the "sense of alienation between the 'Mountain'—Mount Scopus—and the 'Valley'—the Jezreel Valley." This has also been seen as primarily a dichotomy between professors of Central European (and to a lesser extent of Anglo-Saxon) origin and agricultural settlers of Eastern European origin.[35]

Einstein also seems to have had little interest in the scholars who moved to Palestine. As Myers has pointed out, most were marginal figures who were "never fully at home either in Europe or in Palestine."[36] A marginal figure himself, Einstein might have been expected to have had more empathy for the professors in Jerusalem. Yet such empathy on Einstein's part seems to have been limited to a few individual cases, such as Arthur Felix, Abraham Fraenkel, and, to a more restrained extent, Andor Fodor. Needless to say, all of them had been trained in the German research tradition and excelled in their fields academically.

Even though Einstein expected the university to have a beneficial effect on "all the peoples of the East," there is no indication that he expected local Arab students to enroll at the university. In this, he was no different from most other supporters of the young institution. As one historian has pointed out, although officially the university was open to "all races and creeds," because instruction was to be in Hebrew, there was no expectation that Arabs would attend.[37]

One local event of which Einstein was very much aware was the violent clashes of August 1929. These had a decisive impact on the university. As Israeli historian Michael Heyd has stated, if there had been any hopes that the university would also serve students from the non-Jewish population, they were dashed after the flare-up of ethnic violence.[38] This book has shown how the events of August 1929 (and their aftermath) had a profound impact on Einstein. He was sincerely shocked by the way in which proponents of peaceful coexistence with the Arabs were treated by their colleagues and students at the university. This contributed greatly to his disenchantment with the university at the time.

The issue of the status of the Hebrew University within the wider context of British colonial efforts has been raised by researchers. There is no evidence that Einstein was aware of these endeavors to establish institutions of higher learning in the region. It has been noted that the British plan to establish a college in Palestine initiated by the governor of Jerusalem, Ronald Storrs, pushed the university into opening sooner and probably also propelled its administrators toward introducing teaching earlier than planned.[39]

Einstein's position in the debate over the emphasis on research versus teaching at the university was a direct function of his advocacy for its development as an elitist research center. He was adamantly opposed to the introduction of teaching in any given field at the university before research in that field had reached a high standard of excellence. In his support for research over teaching, he concurred with Baron Rothschild and Weizmann, who favored the establishment of small research centers modeled on the Pasteur Institute in Paris.[40] Yet, as with Weizmann, there were internal contradictions in Einstein's logic: they both envisioned the university as an elitist research institute, yet also as a refuge for young Jewish academics and as a means to educate the "Jewish workman and farm laborer."[41]

Like Weizmann,[42] Einstein supported the primacy of research over teaching for practical reasons as well: to foster research into some of the practical problems of the country such as local diseases and experimental forms of agriculture. This was the only concession Einstein was prepared to make to the needs of the local population in Palestine.

Einstein's demand for an academic head should also be seen in the context of sociologist of science Joseph Ben David's notion of *Forschung* (research) as a charismatic act.[43] In his search for an academic chief for the university and also in his considerations on the competency of candidates for faculty positions at the university, this was the key element he was looking for in prospective nominees.

Einstein's advocacy of a research university of first-class excellence also needs to be seen in the context of another important characteristic of German universities noted by scholars: the creation of a system of "entrance," "advancement," and "end-of-the-line" universities.[44]

Seen in this context, Einstein's demand for an elite research center can be interpreted as an impatient demand on his part for an "end-of-the-line university." He wanted the university to be top-notch from its inception and was not prepared to give it time to work through early, less demanding stages of its development.

We have seen that the local needs of the Yishuv in general and the local faculty in Jerusalem were not a high priority for Einstein. He was not concerned with the increasing demands for a full-fledged teaching institution that would bestow undergraduate degrees. Studies on the Hebrew University have noted that the Yishuv and the University Council demanded this type of institution as they feared the lack of undergraduate teaching would force Palestinian youth to study abroad.[45]

In the context of Einstein's adamant advocacy for the primacy of the research function of the university (as opposed to its teaching role), it is interesting to note that David Myers has pointed out that the idea of a small research center as the core of the university was "largely the province of non-Zionist supporters," such as Baron Rothschild, [Paul] Ehrlich, Cyrus Adler, and Felix Warburg. However, Myers fails to mention that the notable exception in this context was Weizmann. Myers claims that the involvement of the non-Zionists was in part "a gambit to control potential Zionist excesses." Similar to what we have seen with Einstein, these non-Zionists "sought to ensure that European standards of scholarship [would] prevail in Jerusalem."[46] So, what does this tell us about Einstein's relationship to Zionism? This study has stressed that the primary concern for Einstein once the university opened its doors was its academic standards. In this, he concurred with the non-Zionists cited by Myers. Like them, the needs and interests of the *Yishuv* and its students do not seem to have concerned him to a great extent. This was another clear expression of Einstein's affinity for a Jewish nationalism that was not Palestinocentric. Yet he does not seem to have opposed the idea of teaching at the university per se. His central condition for its introduction in a particular discipline was that research in that field first reach a certain standard of excellence. So, once again, as we have seen elsewhere, Einstein occupied a middle ground between Zionists and non-Zionists, adapting those characteristics from each ideology that suited his personal needs.

But where did Einstein's advocacy of the primacy of research at the university come from? Recent studies on the history of German academia may provide us with at least a partial answer. Traditionally, there had been a fusion of research (*Forschung*) and teaching (*Lehre*) at German universities. Yet toward the end of the nineteenth century, cracks began to appear in this unity as a result of the growing trend of scientific innovation being carried out in semi-autonomous institutes, supported by foundations like the Kaiser Wilhelm Society.[47] Einstein was influenced by these developments, especially as he was personally involved in just such an enterprise as the first director of the Kaiser Wilhelm Institute for Physics.

Einstein's advocacy of research over teaching also derived from his position on what he perceived as the inherent perils of the creation of a "mass production of academics in title only." Einstein was certainly influenced in his views on the universities' social role in granting degrees to its graduates by the dramatic changes in German-speaking academia in the late nineteenth and early twentieth centuries when he was himself a student at the Zurich Polytechnic and then eventually a young lecturer at various institutions. By 1914, the most serious issue in German universities was overcrowding following a massive increase in the number of students. The internal structure of the universities had not kept up with the changes, and the status of the full professors was increasingly under attack. Nontenured faculty (associate professors, private lecturers, and assistants) were being exploited. The specter of an "academic proletariat" had actually haunted the German universities since the mid-nineteenth century. It had a significant impact on Prussian educational policy from the 1880s on.[48] This would partly explain Einstein's wariness of the potential creation of an "academic proletariat" in Palestine. As we saw in the analysis of Einstein's visit to Palestine, another reason might have been his support for the reproductivization of the Jewish people. Ironically, the prominent physicist who came to epitomize eruditeness was opposed to any step that would increase the overintellectualization of the Jews.

Einstein's involvement with the Hebrew University reveals a great deal about his organizational behavior. He often did not participate in crucial meetings of the board of governors (or of any other university bodies) yet expected others to be convinced of the verity of his opinions and points of view, which he had expressed in writing. He also assumed that he could guide events from afar, thus exhibiting a fundamental lack of understanding of the workings of organizations and committees.

We have seen that unofficial information obtained from onsite testimonies was far more important to Einstein than official information from the university authorities. Actually, apart from occasional letters from Magnes, Einstein received very little direct information from official university sources in Jerusalem. However, he was in regular contact with the leading officials at the Zionist Organization in London who dealt with the university's affairs. As has been noted, he also came to perceive the Jerusalem Executive as untrustworthy; an example is the second version of minutes from the second board meeting in 1926. In the first few years, the Zionist Organization's University Committee was the most important source of information for Einstein. However, with time, direct testimonies from disenchanted or embattled faculty members in Jerusalem (or those who had recently left Jerusalem) constituted Einstein's main source of information. It is clear that these reports had a definitive impact on Einstein's perception of developments at the university. It can hardly be argued that he received a balanced picture of these developments. Yet he probably would not have been very open to hearing reports that contradicted his point of view.

Academic prestige on an international level was clearly the most important criterion for Einstein when he received firsthand testimonies from scholars at the university. If a scholar had substantial achievements in his field, Einstein viewed his reports as authentic.

What he perceived as personal integrity also played a role, yet one cannot escape the impression that a large part of what made up a scholar's personal integrity for Einstein was his academic track record. The adherence to German research standards on the part of those academics he deemed trustworthy also meant, conversely, that it was those scholars who did not adhere to such standards whom Einstein suspected of being incompetent. This explains his disdain for Kligler and the doctors at Hadassah whom he described as "dabblers."

Einstein's correspondence concerning the university also provides us with fascinating insight into the evaluation of scholars by their peers. It reveals which criteria Einstein and his colleagues employed to make judgments regarding the qualifications of candidates for positions.

As we saw in the analysis of both Einstein's induction into the Zionist movement and his trip to the United States, the relationship between Einstein and Weizmann was highly complex. The intense struggle over the Hebrew University tested their difficult relationship to the limit. The Zionist leader's right-hand man on university affairs during this period, Leo Kohn, played an important role in this context. He served as a vital conduit of communication between the two, informing each of the positions and plans of the other. It is intriguing that they needed such an intermediary in their relationship. Yet he seems to have been a necessary adjunct in their delicate dance around each other.

As biographer Ronald Clark has noted, Einstein's organizational behavior vis-à-vis the Hebrew University was not limited to that institution. He displayed similar behavior in regard to the League of Nations' Intellectual Committee on International Cooperation. This was another case in which he frequently resigned and rejoined institutional bodies.[49]

Another aspect of his organizational behavior was the language Einstein employed in his attacks against his opponents at the university. As we saw at the outset of this chapter, he termed Magnes and his supporters "downright vermin" (*ausgesprochene Schädlinge*). On another occasion, he called the university a "bug-infested house" (*verwanztes Haus*).[50] This use of biological terminology by Einstein reveals the intense emotion he felt in the matter of the Hebrew University. He often described the institution as a "matter close to my heart," and his large emotional investment made his disappointment and frustration at the project not developing in the way he wanted all the greater. Yet this terminology also shows us the limits of Einstein's humanism. Faced with opponents who did not see matters the way he did, he did not refrain from dehumanizing them and comparing them to vermin.

Einstein's intense involvement over the years in the early development of the university raises the question of how much influence he actually had on the university during this period. The establishment of the Hartog Committee, its acceptance of his major criticisms of the university, and the implementation of the committee's major recommendations would seem to point to Einstein wielding considerable influence on the direction of the university. However, it could also be argued that the rigidity of his positions, his unwillingness to compromise on major issues, and his lack of interest in local needs and interests on the ground in Jerusalem actually diminished the concrete effect he had on the course of the young institution. Other reasons have been mentioned by historians for the reining in of Magnes in the wake of the Hartog Report (apart from a desire to placate Einstein): contributions from his American supporters dropped drastically after the 1929 crash, and increasingly more faculty members grew tired of his autocratic regime.[51]

In a wider context, it seems that three main factors had a detrimental effect on the university's early development: Magnes's autocratic regime, Weizmann's consistently setting Zionist politics before the university's affairs, and the negative impact that Einstein's emotional entanglement had on his ability to negotiate effectively within such a complex organization as the Hebrew University.

What effect did Einstein's experiences with the Hebrew University—the project that had brought him into the Zionist fold and about which he felt most passionate—have on his relationship to Zionism in general? Disenchanted by both the intense power struggles over the university's administration and the harsh treatment of faculty members who supported coexistence with the Arabs (such as Hugo Bergmann) by their colleagues and students, Einstein felt compelled to seek out academic alternatives to his pet project. Yet, as we have seen in the case of the Peruvian Jewish settlement project and the interview he gave to a Viennese paper in 1929, there are indications that Einstein's disillusionment with the university also led to a reassessment of his relationship to Zionism in general. Einstein clearly always kept his emotional energy invested in more than one cause or project. I assume he did so to protect himself from the anguish that came with disappointment. Yet even though Einstein kept his interest alive in alternative projects of higher learning for young Jewish academics, it was the Hebrew University that provided him with his most distressing and bitter disappointments in his involvement in Jewish affairs during this period. It had this effect on him because it was obviously of such great significance to him. In no other matter did he use such strong language of despair. The prosaic daily reality of the Hebrew University could not live up to the lofty vision one of its most ardent supporters had of it—at least not in the first decade of its existence.

As we have seen, Einstein often employed religious terminology to refer to the secular activity of conducting scholarly research. In January 1933, he stood on the brink of a new chapter in his life, forced to emigrate from his country of birth. Einstein now had to decide whether he could continue to view the Hebrew University—the "problem child" (*Schmerzenskind*) that had caused him such painful disappointments—as a suitable secular Jewish framework for the "sacred" activity of research or whether he would prefer to seek out an alternate venue for what he perceived as a "holy" endeavor.[52]

CONCLUSION

We have come to the end of our examination of Einstein's association with the Zionist movement during his European years. It is now time to return to the questions we posed at the outset of this study and draw some general conclusions in light of the evidence I have gathered.

It has been a fascinating journey. We have seen how Einstein's family background and early socialization in Germany shaped the formation of both his Jewish identity and his political outlook. His family's specific German Jewish subculture instilled in the young child a fledgling sense of pan-Jewish solidarity, an antipathy toward Jewish religiosity, and a wariness of the German Gentile majority. Rigid discipline at school fostered in the adolescent a budding oppositional defiance. Yet the informal education he received from a young Eastern European Jewish student evoked both a profound enthusiasm for intellectual pursuits and an admiration for his ethnic brethren from the East. Einstein's Swiss years had a deep impact on his role as a dual outsider—as a Jew and as a foreigner. The liberal views he was exposed to during his final school year in Aarau further reinforced his religious tolerance and political progressiveness. His initial difficulties in finding permanent employment and the long delays in establishing his academic career affected his ideological preferences. The difficulties in his personal life, initially with his first wife and then with his own offspring, further affected his value system.

Einstein's first encounter with a group of Zionists in Prague during his tenure as a professor there prior to World War I provoked a strong antipathy to their representatives and their views. Shortly after the war, he was contacted by Zionist functionaries in Berlin who were striving to expand the movement's appeal to prominent Jewish non-Zionists. Einstein responded with initial skepticism toward the validity of their cause and their movement's ideological tenets. By the spring of 1919, however, he had become convinced of the veracity of some of their views, and began to display genuine enthusiasm for some of the goals of the movement, especially in regard to the planned establishment of the Hebrew University in Jerusalem. This keen interest eventually led to his agreeing to be utilized by the Zionist Organization for propaganda purposes, most notably as part of a delegation to the United States in the spring of 1921. While in the United States he became embroiled in the bitter dispute between Weizmann and Brandeis over funding for settlement in Palestine and received a sobering lesson in Zionist politics. He then toured the land of his forefathers in the winter of 1923 to see for himself the colonization efforts he had supported over the previous four years. This generated a substantial amount of enthusiasm in Einstein for the achievements of the Jewish community in Palestine but did not fundamentally change his views on the limited practical impact the country would have on his "ethnic comrades"—he saw its potential worth primarily as a moral and spiritual center for Diaspora Jewry.

This book has also traced Einstein's involvement in the project that brought him into the Zionist fold in the first place, the Hebrew University. His disillusionment with developments on the ground at the university in Jerusalem began soon after its official opening in 1925 and eventually led to his resignation from its official bodies. The mounting interethnic tension in Palestine between Jews and Arabs, which erupted into violence in 1929, had a profound impact on Einstein's view on Zionism, and he proposed some far-reaching solutions for the future of the region. The manner in which members of the pacifist Brith Shalom movement, with whom Einstein felt a great deal of affinity, were treated by their peers at the Hebrew University also affected his views on the young institution and the Zionist movement itself. Finally, we saw how he continued to be involved with the development of the university even after he had removed himself from its central institutions. He eventually resumed his official ties once the reforms he had demanded had been implemented.

At the outset, this study asked how Einstein saw himself as a Jew and as a member of the Jewish community. This is a deceptively simple question. While it is clear that by the early 1920s Einstein had formulated well-defined opinions on what constituted Jewish identity, it is particularly tricky to pinpoint when Einstein originally adopted those views. There is a fundamental methodological problem here: we do not have sufficient contemporary evidence from the very period that Einstein later claimed had been critical to formulating his views. If we take Einstein's narrative on the genesis of his own Jewish identity at face value, he became aware of a "sense of being different" while at elementary school in Munich. Presumed physiognomic idiosyncrasies, verbal and physical attacks by classmates, and instances of anti-Jewish sentiment expressed by teachers instilled in the young Jewish child a "vivid sense of strangeness." The encounter with other Jewish students at secondary school fostered a feeling of ethnic solidarity that further distinguished the minority members from their German counterparts. From the time he dropped out of gymnasium, he had distanced himself from an official association with the Jewish community by repeatedly declaring himself to be "without religious affiliation."

In contrast to his explicit testimony in regard to his schooling in Munich, we have very few direct statements by Einstein from his Swiss and Prague years on his Jewish identity. I have therefore had to rely on accounts by relatives and biographers to gauge the impact of Swiss anti-Semitism on his growing sense of alienation from his Gentile environment and to ascertain the influence of the unique situation of Prague Jewry on his feelings of solidarity with his fellow Jews. Einstein later attributed his interest in public affairs related to Jews to his return to Germany in 1914, but I believe his brief sojourn in the Czech capital was more instrumental in this regard.

In any case, by the time Einstein gave public expression to his views on Jewish identity in the early 1920s, his definitions were primarily ethnic and cultural, not religious, in nature. It is patently clear that he rejected a religious concept of Jewish identity. Moreover, we have seen time and again that his preferred term for referring to his coreligionists was "ethnic comrades" (*Stammesgenossen*), thereby illustrating the primacy for him of the ethnic bond with his fellow Jews. For him, the central characteristics of the Jews as a nation were ethnic lineage, "a sense of being different," and "predominantly" nonreligious traditions.

Einstein's views on the Jews of Germany and of Eastern Europe were closely linked to how he defined Jewish identity. Himself the product of Southern German rural Jewry, he saw the assimilationist strivings of the urban, bourgeois majority of German Jewry as "undignified mimicry." He particularly disapproved of what he perceived as their pathological lack of self-respect, their allegedly futile efforts to combat anti-Semitism, and their unwillingness to express ethnic solidarity with their fellow Jews in Eastern Europe. In contrast, he was profoundly impressed by what he perceived as the ethnic authenticity and cultural achievements of the Ostjuden and was deeply concerned about the discriminatory measures that adversely affected their young academics.

Einstein's own status in Germany underwent two contradictory processes during the period under study. As a prestigious member of the academic establishment, his counsel was sought out by colleagues and government officials alike. Yet as a target of anti-Semitic attacks, he grew increasingly uncomfortable with residing in Germany, did not feel "rooted" there (in contrast to his Gentile colleagues—at least in his perception), and often sought refuge in extended travels abroad. Occasionally he considered emigrating from Germany altogether.

His relationship to his own German identity was also fraught with ambivalence. From the repudiation of his Württemberg citizenship at the age of fifteen on, Einstein repeatedly denied his official affiliation with the German state. Even after his return to Germany and his appointment as a German civil servant, he continually claimed he was a Swiss national. Only the awarding of the Nobel Prize in 1922 eventually led to governmental clarification that he had indeed reacquired

German citizenship upon his appointment to the Prussian Academy in 1914. Yet despite these repeated denials, he became a somewhat reluctant ambassador at large for German foreign relations. And though never explicitly acknowledged by Einstein, he felt a great deal of allegiance to German culture, and even more to the German scholarly ethic.

The same ideological preferences that led Einstein to forswear any official membership in the Jewish community or the German state profoundly affected his concept of nationalism. The strict, rigid military-like discipline of the Bavarian educational system was perceived by Einstein as mind-stultifying and spirit-crushing. The unbridled militarism that swept through German society during World War I, engulfing the overwhelming majority of intellectuals and academics, was a grave disappointment for Einstein and further reinforced his negative perception of nationalistic fervor. He also reacted with strong disdain to the jingoism and anti-German zeal of the Entente countries. In the immediate aftermath of the war, he was thoroughly disgusted with the "horrid Europeans" in general. This anti-European sentiment also resulted in Einstein defining himself (and the Jews) as non-European. This was a key turn that had a critical impact on his concept of nationalism.

We have seen that Einstein's perception of nationalism as an ideology was a highly negative one. I have found numerous statements in which he refused to define himself as a nationalist. For him, the European example of nationalistic fervor was such an abhorrent one that he seems to have been unable to conceive of a more humanistic variety of nationalism.

However, we are confronted with a gaping discrepancy between Einstein's revulsion toward European forms of nationalism and his willingness to collaborate with the proponents of the Zionist movement and embrace some of their ideological tenets and endeavors. He had to perform some deft intellectual acrobatics to overcome this profound contradiction. He did so by concluding that Zionism could be divested of its character as a nationalistic ideology altogether. Whether consciously or not, he employed his Jewish ethnocentrism, ingrained in him from a very early age, to conclude that Zionism was qualitatively different from other forms of nationalism and therefore worthy of his support. Hand in hand with this assumption, he also surmised that it was more ethical than other nationalistic enterprises. This led to a momentous consequence: Einstein imposed extremely high ethical standards on the Zionist movement.

But how did Einstein become affiliated with a political organization that strove to establish a Jewish state in the British Mandate for Palestine? And what transformation did he undergo in his transition from fervent disdain vis-à-vis Zionism merely a few years earlier to willingness to give limited support to some of its tenets and causes? This process took place in two distinct stages: during World War I, a number of key developments made Einstein susceptible to the appeal of Zionism, and immediately after the war, the Zionist movement reached out to Einstein in an effort to mobilize him for their cause.

Numerous factors that played a crucial role in leading Einstein to become *engagé* on behalf of Zionism have been overlooked in previous studies. As a consequence of his increasing antipathy toward Germany, he was willing to acknowledge science (*Wissenschaft*) as the only patria to which he felt any allegiance. The wanton destruction of the war also brought about an increasing disenchantment with the future of Europe. In addition, he became severely disillusioned with the younger generation of German academics because of their extreme militarism. He was also greatly dismayed at the deaths of young, talented scholars as a consequence of the war. Major setbacks in his relationships with his sons rendered him progressively more uncomfortable with the emotional demands he had to contend with as their father. In fact, he ceased to see them as his "temporal successors." In addition, he was convinced that his scientific output was diminishing, and his poor health made him increasingly aware of his own mortality. All these factors served to heighten Einstein's acute desire for the establishment of a new (emotion-free) organizational framework in which the intellectual talents of young academics could be fostered.

This susceptibility on Einstein's part for such a new institutional framework was synchronous with a decisive attempt by the leadership of the German Zionist Federation to broaden their organization's appeal to prominent Jewish non-Zionists, and their most celebrated recruit was Albert Einstein. It was primarily the project to establish a Jewish university in Jerusalem that drew Einstein into the Zionist fold. The Zionist Organization's initial plans for the Hebrew University seemed to fit exactly Einstein's vision of a new educational institution in which the intellectual talents of young Jewish academics would not go to waste. This was extremely important for Einstein, as back in his days in Bern he had sworn a "solemn oath" to aid gifted young scholars. The university thus fulfilled a crucial role for Einstein, both intellectually and emotionally.

The Hebrew University was not the only Zionist project Einstein was involved in. Other notable enterprises were the establishment of the Institute for Technology (the Technion), the Bezalel Arts Academy, and the National Library. He lent his name to numerous appeals on behalf of Jewish colonization efforts in Palestine. He also participated in endeavors to propagate the Zionist cause in Germany beyond the narrow confines of the Jewish community by serving on the Pro Palestine Committee.

The benefits Einstein saw in Zionism and Palestine were closely related to how he defined what ills were afflicting the various communities among the Jewish people. For Western Jews, he hoped the movement would enhance their self-respect and that Palestine would provide a spiritual center. For Eastern Jews, he anticipated that Zionism would offer a partial solution as a physical refuge for the young and dynamic sectors among the Ostjuden. For Einstein himself, Zionism provided a renewed sense of purpose. It also afforded him a sense of ethnic belonging precisely at a time when he felt rootless and alienated from the German nation.

The amount of time and effort Einstein was prepared to dedicate to Zionism waxed and waned with the fluctuations in his enthusiasm for this ideological cause. In the heady phase of his initial involvement, he was even prepared to forgo the most prestigious international physics conference, the Solvay Congress, to accompany Weizmann on his tour of the United States (although admittedly, there were other factors that led to his decision to be absent from the conference and that motivated him to cross the Atlantic Ocean). Soon after his return from that trip, he started to complain about the demands the Zionists made of him. This was a clear sign that his initial enthusiasm was waning and that he was beginning to ask himself whether he should heed their call in every instance. However, he was prepared to visit Palestine on his return trip from the Far East. But his planned extended sojourn in the country never materialized. One cause he almost always seemed to have time for was the Hebrew University. Yet at the height of his disillusionment with this institution's executive body, he maintained his distance even from this, his pet Zionist project, at least temporarily.

However, in accordance with his significant ambivalence toward Zionism, Einstein did not place all his ideological eggs in one basket. Parallel to his involvement with Zionist enterprises, he continued to pursue non-Zionist Jewish projects. This eclectic approach led him to support Jewish colonization initiatives in such far-flung places as Peru and educational undertakings in Lithuania.

A crucial question in determining the degree of Einstein's engagement on behalf of Zionism is the extent to which Einstein accepted the central tenets of this ideology. The first issue that had to be clarified in this regard (and that previous studies have largely ignored) was Einstein's exposure to Zionism in a very specific historical context. As he attended the Zurich Polytechnic he did not share what seems to have been the defining experience for the future leadership of German Zionism—academic anti-Semitism during the university years and membership in Zionist student associations. However, it was precisely the most notable figures in the "second generation" of German Zionists who played a critical role in initiating Einstein into the Zionist movement. These prominent leaders were slightly younger than Einstein. It is striking that he was more attuned to some of the central tenets of the previous, older generation, most notably their ethnic definition of Jewish identity and their pan-Jewish solidarity with the Jews from Eastern Europe. He shared the younger generation's alienation from German society, its outright rejection of the assimilationist stance, and its perception of the futility of combating anti-Semitism. However, he was clearly indifferent to the younger generation's interest in the renaissance of a distinct Jewish and Hebrew culture. He also rejected the young Zionist leaders' negation of life in the Diaspora and their espousal of immigration to Palestine as a personal solution for every member of the Zionist movement.

So, how far did Einstein's commitment to Jewish nationalism extend? Did he convert to Zionism? Did he become a full-fledged Zionist? Looking at the entire period under discussion, I have found no evidence that Einstein underwent a genuine conversion to Zionism. He maintained his distance from many of the central tenets of Zionist ideology. He also did not adopt a Zionist identity for himself, and he did not identify fully with the movement: "the Zionists" always remained "them" for Einstein, they did not make the all-important transition to "us."

Einstein's ambivalence toward Zionism can be deduced from the language he employed to describe his relationship to the movement. He repeatedly used phrases and statements that introduced a certain distance between himself and Zionism. An example of this tendency can be found as early as 1920, when he declared that "My Zionism is not such a terribly serious matter," which does not clarify at all the degree of his commitment to the Zionist cause. Five years later, he remarked rather impersonally, "A Jew who strives to impregnate his spirit with humanitarian ideals can call himself a Zionist without contradiction."[1] And another five years on, he warned that "[i]f things continue in this vein, no decent Jew can be a Zionist any longer," thereby leaving us in doubt whether this was a reference to himself or not.[2]

How, then, can one define Einstein's stance toward Zionist ideology? I have found that his relationship to Zionism was primarily emotional and instrumental in nature. Somewhat surprisingly for an individual who regarded himself (and is regarded by others) as acutely cerebral, Einstein's Zionism was, to a large extent, a "Zionism of the heart." It was his deeply rooted ethnocentrism that made him susceptible to the appeal of Zionism in the face of his intense disdain for all other forms of nationalism. Time and again, Zionist-related events and experiences tugged at his heart-strings, whether it was his encounter with the "Jewish masses" in the United States, his enthusiasm for the young pioneers in Palestine, or his initial excitement about the establishment of the Hebrew University.

Yet his approach to the Zionist cause was also highly instrumental in nature. Indeed, he was very selective about the Zionist causes he would support. We have seen repeatedly that he would only lend his name to those specific Zionist tenets, goals, and actions that would further his agendas in other general, not specifically Zionist, spheres, such as academia and pan-Jewish solidarity. These goals were of far greater importance to Einstein than the Zionist cause itself. Therefore, his was a "pick-and-choose" Zionism.

My conclusion, therefore, is that, despite his enthusiasm for some of the goals advocated by the Zionists, Einstein did not "convert" to Zionism and did not become a fully committed member of the movement.

In light of this conclusion, we must ask ourselves whether Einstein can even be termed a Jewish nationalist. I actually do not believe that he can be defined as such. He did perceive the Jews in favorable ethnic terms, which means that he meets the central criterion of a Jewish ethnicist, that is, someone with "a state of mind of positively valued self-awareness" in regard to the ethnic attributes of the Jews. Yet to qualify as a Jewish nationalist he would have had to subscribe to "the principle that the ethnic unit, generally now called the 'nation', and the polity should be congruent, a principle the full satisfaction of which requires political independence within the ethnic unit's own territory."[3] Even though there may be ideological differences among nationalists as to how to realize the ultimate goal of creating a nation-state, the national consensus encompasses all those who advocate such a solution.[4] However, during the period covered by this investigation, Einstein remained opposed to the creation of a Jewish nation-state. His ideological position was thus located outside the national consensus. He was also quite indifferent to whether a sufficient number of Jews would actually amass in one geographic location (whether this be in Palestine or Peru) to be able to establish a socially and economically viable national entity. Furthermore, "nationalist" was a term at which Einstein bristled repeatedly.

By the same token, I do not believe that Einstein can be defined as a Jewish cultural nationalist. As we have seen, he was indifferent to the development of a specific Jewish or Hebrew national culture. He was not opposed to his ethnic brethren developing their own national culture, yet he believed this could be achieved anywhere in the world. He did not expect that doing so would require them to give up any aspects of their existing cultural identity in their host countries.

Einstein's affiliation with the Zionist cause brought him into contact with the top echelons of the German and international leadership of the movement. This study has taught us much about the nature of the relationships he established with these prominent individuals.

Einstein began his contacts with the Zionists on a modest scale. He first met with local leaders of the German Zionist Federation, then progressed to higher-level functionaries in the hierarchy of the Zionist Organization in London. He subsequently encountered Chaim Weizmann and other major figures of the Zionist Executive on the voyage to the United States; next, he became acquainted with the leadership of the Zionist Organization of America; finally, he met with all the major figures of the Zionist Executive in Palestine during his visit in 1923.

His contacts with German Zionist leader Kurt Blumenfeld formed perhaps the most significant Zionist relationship for Einstein. To some extent, Blumenfeld was the face of Zionism for Einstein. After substantial efforts at convincing him of the justness of the cause, Einstein became susceptible to the manipulative charisma of the former head of Zionist propaganda.[5] However, he remained seemingly unaware of the degree to which Blumenfeld did not trust the movement's "prize-winning ox" and to what extent he filtered the information Einstein received and tried to have his public utterances muzzled. Yet this study has also made clear that Einstein did not fully trust Blumenfeld either and would occasionally subject him to his own manipulations. Eventually, in the aftermath of the 1929 riots in Palestine, the editor of the *Jüdische Rundschau*, Robert Weltsch, seems to have supplanted Blumenfeld as Einstein's favorite German Zionist. This was no doubt the consequence of a greater affinity in their positions vis-à-vis the Arab-Jewish conflict. Thus, Einstein shifted his allegiance from Blumenfeld, a member of the erstwhile radical faction among German Zionists, to Weltsch, a member of the radical circle within Brith Shalom.[6]

Einstein's relationship with Chaim Weizmann was even more intricate. They seem to have found dealing with each other directly quite difficult. Therefore, they often needed a mediator to act as a conduit of information between them. In the early stages of their relationship, this role was fulfilled by Blumenfeld. Later on, when most of their dealings pertained to the Hebrew University, the Zionist functionary Leo Kohn assumed this role. Intriguingly, both Blumenfeld and Kohn were German Jews. Thus, it seems that despite their mutual admiration (and wariness of each other), Einstein and Weizmann needed a German Jew to act as an intermediary. Perhaps Weizmann, the brilliant chemist turned wily politician, was not the ideal Ostjude for Einstein, and the distinguished physicist may not have been the ideal Westjude for Weizmann.

This study has also demonstrated how Einstein's role was defined within the Zionist movement. Throughout the period under discussion, the Zionist Organization repeatedly referred to Einstein as a non-Zionist. The movement was particularly proud that such a celebrated and illustrious individual who was not a Zionist agreed to lend his support to (select) Zionist enterprises. Einstein gave the Zionists hope that other prominent Jewish non-Zionists would come to advocate their cause. However, Einstein's case does seem to be unique. Other prominent non-Zionists (such as Sigmund Freud) may have lent their occasional support to an individual project (such as the Hebrew University), but no other Jews of Einstein's stature consistently allowed their names to be utilized for Zionist purposes.

In light of all this, how should Einstein's role in the Zionist movement be defined? I do not believe he was sufficiently naive to fit the description of a fellow-traveler or "hanger-on." He was far too independent-minded and critical to be described as such. He was definitely sympathetic to the cause, yet maintained his own idiosyncratic views on the central issues. However, he was also something more: he was a Zionist icon, lionized in the United States, Palestine, and eventually even in Berlin. Historians of nationalism have long recognized the crucial role played by national icons: "It is through the lionized hero that the perceived genius and redemption of the nation is realized, embodied and worshipped."[7] Yet there is a supreme irony here: Einstein was a Zionist icon without actually being a Zionist. And this is another factor that renders Einstein unique in the Zionist movement.

Two high points of Einstein's affiliation with Zionism were his encounters with the Jewish masses in the United States and with the modern Jewish community in Palestine. He was struck by the enthusiasm of the (mostly Eastern European) pro-Zionist crowds that greeted him and Weizmann at every major Jewish port of call during their American tour. And he was deeply impressed by the energetic vitality of the "Russian" pioneers on the kibbutzim and of the laborers in the urban centers of Palestine. His positive perception of Eastern European Jewry was further reinforced by these encounters. And he was confirmed in his belief that Palestine could form a healthy environment to cure some of the social ills that afflicted modern Jewry.[8]

Einstein's tour of Palestine also brought him face-to-face with its Arab population. Yet as we have seen, his Zionist hosts played it safe and introduced their guest only to moderate leaders among that sector of the populace. There is also no evidence that Einstein went out of his way to meet with more radical Arab representatives. In light of the violent clashes less than two years before his visit, he seems to have been quite complacent at this stage about the dangers of the Arab-Jewish dispute in Palestine. Some of his statements in the wake of his tour can also be interpreted as rather dismissive of the Arab population. In this, his position was similar to that of the leaders of German Zionism.

However, the interethnic violence that erupted in August 1929 brought about a sea change in Einstein's approach to the burgeoning conflict. He became disillusioned with mainstream Zionist policy on

Arab-Jewish relations in Palestine and placed himself even farther away from the national consensus. Serious fault lines opened up between Einstein and the highest echelons of the Zionist leadership. In correspondence with prominent figures in the Brith Shalom movement (most notably Hugo Bergmann) and with the editor of the Arab newspaper *Falastin*, Azmi El-Nashashibi, he outlined his proposals for peaceful coexistence between Jews and Arabs in the region. He also made it patently clear that he opposed the establishment of a Jewish nation-state and preferred an interethnic solution.

Einstein reserved his greatest hopes—and despair—for his pet Zionist project, the Hebrew University. This institution represented the quintessence of his expectations from the Zionist movement: academic brilliance, personal integrity, ethnic solidarity, and the pure pursuit of scholarly truth. Yet as great as his hopes were for this enterprise, as deep was his disillusionment. His relationship with the university was also affected by the fallout from the 1929 clashes. His perception of the intolerance demonstrated toward supporters of Brith Shalom at the university further alienated him from developments on the ground in Jerusalem. This seems to have strengthened him in his resolve to seek out and seriously consider alternative solutions both in academia and in settlement efforts.

Einstein's involvement in the affairs of the Hebrew University demonstrated an intriguing discrepancy in relation to his identity as a European. By the end of World War I, Einstein was defining himself as a non-European and denouncing those "horrid Europeans." Yet when it came to the university and other educational and cultural enterprises in Palestine, he strongly advocated the import of European norms and standards and had little awareness or concern for the concrete needs of the local population, whether Jewish or Arab.

To a large extent, this study has concerned itself with Einstein's profound "sense of strangeness." An intense awareness of being different accompanied him from a very early age, and he came to view that alterity as intricately linked to his Jewishness. That basic perception was further reinforced through subsequent experiences of anti-Semitism and xenophobia. We can therefore appreciate the loftiness of the hopes with which he began his association with the Zionist movement. Einstein sincerely longed for Zionism to alleviate his deep feelings of personal and national alienation. However, the "little patch of earth on which our ethnic comrades will not be considered aliens"[9] was clearly not intended to solve the issue of Einstein's own personal alienation. At best, the emigration of others to this small yet qualitative entity could somewhat alleviate his own "sense of strangeness." To a certain degree, this was Einstein's own doing: he plunged into enterprises for which he had not prepared himself adequately and which were rife with internecine struggles. He also preferred to maintain the familiar emotion of distancing himself from organizations and causes and utilize Jewish nationalism for his own instrumental purposes rather than enter the precarious waters of full immersion in a movement and an ideology to which he could not fully adhere. Thus ultimately, Zionism could not relieve Einstein's feelings of alienation. The "sense of strangeness" prevailed, even (ironically) toward the Zionist movement itself.

This detailed and extensive study of his specific relationship with the Zionist movement affords us the opportunity to draw some general innovative conclusions about Einstein himself.

Einstein's organizational behavior vis-à-vis this specific movement has demonstrated that he needed considerable convincing before committing himself to a collective cause. It has also shown that to a large extent, he assumed a passive role in the organization and was completely comfortable with others determining agendas and itineraries. On the other hand, we have seen that he could be quite manipulative in his dealings with other members of the organization, whether in regard to the information he imparted to them or in his attempts to force other members of the organization to act as he wanted by exploiting his distinguished status and threatening to withdraw his support.

This study has also illustrated the palpable limits of Einstein's much-touted humanism. We have seen expressions of his extreme elitism, which, during World War I, even extended to matters of life and death when he differentiated between lives dedicated to science, which were worth living (and worth saving in warfare), and those that were not. Furthermore, we have seen how Einstein dehumanized his opponents by employing biological terminology to describe them in his correspondence. He conceived of those he disagreed with, whether it was the Brandeis faction in the United States or his rivals in the clash over the Hebrew University, as elements that had to be "purged" (at best) or as "downright vermin" that had to be eradicated (at worst).

We have also learned that, in contrast to accepted wisdom, Einstein had an amazing belief in his own competency in nonscientific matters: although totally lacking training in political affairs, he became embroiled in a cut-throat clash between two highly polished political machines, the European and the American Zionists. Despite having no prior experience in fundraising, he embarked on a major campaign in a country he had never previously visited. And with very little administrative experience (even though he functioned as director of the Kaiser Wilhelm Institute for Physics, the administration of which he executed from his Berlin apartment and whose sole staff member consisted of one part-time secretary), he firmly believed he was sufficiently competent in administrative matters to make executive decisions regarding a major institution such as the Hebrew University.

Nevertheless, a highly significant insight regarding Einstein that can be derived from this study is that he was far less naive than he is usually portrayed. He would carefully select the Zionist projects he agreed to support and only become active on behalf of a specific cause if its goal was in alignment with his own agenda. He could even manipulate such wily figures as Blumenfeld and Weizmann, he could use the Zionist movement to further his own personal goals, he could present concrete political proposals, he could adopt varying registers for various target audiences (most notably in the aftermath of the 1929 riots), and he could hold his own in the bitter struggle over the Hebrew University with Magnes and his supporters.

In the wider context of Einstein studies, this study has shown to what extent the scientific, the political, and the personal are intricately interrelated and interwoven in such a complex individual as Albert Einstein. Let two examples suffice to illustrate this point: we have seen what crucial impact Einstein's search for an organizational framework as an instrument for handing over the baton of science to the next generation played in his support for a Jewish enterprise, the Hebrew University. And this work has demonstrated how Einstein could utilize his Zionist trip to the United States to further the dissemination of his theories among his esteemed American colleagues. I believe this necessitates a far more holistic and interdisciplinary approach to Einstein studies than has been the case to date.

This study began with a reference to the new Israeli historians and the controversy over post-Zionism. One of the consequences of that debate has been the call to pay more attention to the voices of alternative groups in Israeli society: Jews from the Diaspora, Oriental Jews, and Palestinians. This development has also led to an increased interest in those intellectuals who wanted an alternative reality to unfold in Palestine, not a Jewish nation-state but rather some form of joint homeland for both Jews and Arabs. This has resulted in a flurry of publications on the positions toward Jewish nationalism of such figures as Ahad Ha'am, Gershom Scholem, Martin Buber, Hans Kohn, and Hannah Arendt.[10] One historian has termed this type of nationalism "non-nationalist nationalism." Others have called it "enlightened Zionism" or "proto-Zionism."[11] The proponents of this approach advocated an "attempt to redefine nationalism so as to eliminate what they considered its morally reprehensible aspects, notably the idea of sovereignty, and thus presumably deprive it of the potential to become an oppressive ideology."[12] In my opinion, Albert Einstein can definitely be added to this list of illustrious individuals. Like them, he would have preferred to refrain from engaging in the issues of nation, state, and power. I see a parallel here in Einstein's reluctance to interact with worldly affairs and his preference to pursue scientific research as a means to flee the "merely personal."[13]

Yet Einstein distinguished himself from the other non-nationalist nationalists in two highly significant ways: with the possible exception of Ahad Ha'am (who died in 1927), he was also the only one among them who became a major Zionist icon during the first third of the twentieth century. Moreover, unlike them, he did not develop succinct political or philosophical theories on Zionism and nationalism in an attempt to resolve the tension he experienced between the universalistic and particularistic strands of his ideological worldview. This is in keeping with Einstein's principal identity as a natural scientist.

However, this brings us to another noteworthy issue: unlike the Zionists (and other nationalists), Einstein did not supplant religion with the ideal of the nation; he replaced it with *Wissenschaft* (science). And this played a major role in his aversion to embrace Zionism without reservations. One Jewish historian views the Jews "as the ultimate diasporic population beyond and above the nation-state"[14] Does this then mean that Einstein was more Jewish than the Zionists in his outlook because he was opposed to the creation of a Jewish nation-state in the British Mandate of Palestine?

If they choose to, mainstream Zionism and Israeli society have much to learn from the alternative voices they have marginalized, proto-Zionists and post-Zionists alike. Perhaps Einstein's most important lesson to the Zionist movement is a twofold message: it can definitely view his association with Jewish nationalism as a badge of honor, yet it should also be ever mindful of his warnings of the pitfalls of nationalistic fervor. Moreover, his prescient conclusion that psychological barriers were the most significant impediments in the interethnic conflict in Palestine/Israel still rings true today. Ironically, it would seem that as far as Einstein's views on Zionism are concerned, we need to strive much harder to hear the forgotten voice of this most prominent of individuals.

EPILOGUE

I should much rather see reasonable agreement with the Arabs on the basis of living together in peace than the creation of a Jewish state. Apart from practical consideration[s], my awareness of the essential nature of Judaism resists the ideal of a Jewish state with borders, an army and a measure of temporal power no matter how modest. I am afraid of the inner damage Judaism will sustain—especially from the development of a narrow nationalism within our own ranks, against which we have already to fight strongly, even without a Jewish state.[1]

This was how Einstein expressed his opposition to the establishment of a Jewish state at the "Third Seder" celebration of the National Labor Committee for Palestine in New York in April 1938. By then five years had passed since he had chosen not to return to Germany following the Nazi rise to power in early 1933. Later that year, Einstein emigrated to the United States, took up residence in Princeton, New Jersey, and joined Princeton's Institute for Advanced Study.

This book closes with a dramatically incisive upheaval in Einstein's life: his years in Berlin, where he had been inducted into the Zionist movement, came to an abrupt (and undesired) conclusion. In the spring of 1933, Einstein suddenly became a refugee, and a completely new period in his life began. Needless to say, Jewish nationalism and the Zionist movement continued to preoccupy him, perhaps even more than before, in consequence of the Nazi persecution of the Jews, the continued interethnic strife in Palestine, and eventually the Holocaust and the establishment of the state of Israel. Various publications have dealt with Einstein's affiliation with the Zionist movement dur-

ing the period between 1933 and his death in 1955. However, for the
most part they have been based on limited source materials and their
conclusions have been quite generalized. Detailed and meticulous research on these years remains to be carried out. As with the period
covered by this book, new insights are to be expected from a study of
those years too.

Future research on Einstein and Zionism during the years 1933–
55 would have to deal with the effect the rise of Nazism had on Einstein's views on nationalism in general and on Jewish nationalism in
particular: Did it render him more or less of a nationalist? Did his
self-definition as a non-Zionist and a non-nationalist undergo any
major change as a consequence?

Furthermore, the effect of the increasing turmoil in Europe and
the Middle East on Einstein's positions could also be analyzed. Did it
lead to a transformation in his advocacy of peace and reconciliation
between Jews and Arabs? How did the rise of radical nationalism
among the Jewish and Arab populations of Palestine affect his views?
What relationships did he foster with Arab leaders and intellectuals?
Did he reach out to any representatives of more radical Arab groups?

Prospective research would also examine Einstein's position vis-à-
vis the various Zionist factions in the Yishuv. It would monitor whether
his sympathy for Labor Zionism persisted or whether he became disillusioned with its leaders. It would consider whether he continued to
idealize the pioneers on the kibbutzim. It would also study his reaction
to the rise of Revisionist Zionism and to that movement's far less tolerant attitude toward the Arab population. Furthermore, it would analyze his position on British policy in Palestine: How did he respond to
the increasingly restrictive measures against Jewish immigration,
which became a hotly contested issue in the interethnic conflict, particularly in light of Nazi persecution? Against the background of the
increasingly desperate plight of Jewish refugees from Europe, did he
continue to look to other options for Jewish settlement? Did he persist
in considering the establishment of other academic institutions to absorb Jewish scholars besides the Hebrew University?

The solutions to the Arab-Jewish conflict advocated by Einstein would also be examined. Did he support a compromise with the Arabs? What were his positions on partition and on binationalism? How did he view the role of the United Nations? Did he continue to oppose Jewish statehood, even after the Holocaust and the creation of Israel?

Further study could consider what effect the establishment of Israel had on Einstein's opinions. How did he react to the Jewish "return to history"? What was his response to the use of force and violence by his ethnic brethren? Was he prepared to compromise his ethical values for the sake of national goals? Additional research could also deliberate whether he continued to believe that Zionism was a positive force for Jews in the Diaspora and that it could raise their self-esteem. Did he view the nascent state as a rallying point for world Jewry or did he think they were now in a more precarious situation? How did Einstein feel about Israel's domestic and foreign policies—did he acquiesce to its government's actions or did he criticize its leaders? What did he think about the multitude of political parties in Israel?

Einstein's vision for Israeli society would also be examined. Did he support the cause of social justice in Israeli society? Did he continue to believe that the Zionist enterprise could be a "light unto the nations"? How did he view Israel's treatment of its Arab minority? Did he think the Jews had learned their lesson from being a persecuted minority and were doing a better job at being in the majority than other nations had in the past or present?

Einstein's views on the relations between the various sectors in Israeli society would also be analyzed, including his opinions on the interactions between Jews and Arabs, "old-timers" and immigrants, the religious and secular, and Jews of European and Middle Eastern descent. In addition, the study would consider Einstein's position on Israel's political, military, and economic alliance with the United States. Did he advocate neutrality for Israel vis-à-vis the United States and the Soviet Union? How did the dawn of the nuclear era affect his position on Israel's role in the Middle East? And what role did he envisage for academics in pre-state Palestine and in Israeli society? Did the Hebrew University continue to play a dominant role in his vision for the future of the Middle East?

A study focusing on this later period would also have to deliberate why Einstein did not immigrate to Palestine once it was certain that he would not return to Germany. Why did he not take up residence in Israel once the state was established? Why did he not accept a professorship at the Hebrew University, even though he was repeatedly invited to do so? And last, why did he not accept the presidency of Israel on the death of Chaim Weizmann in 1952, even though at the time he declared that his relationship with the Jewish people had become his "strongest human bond"?[2]

Ultimately, the crucial question for such a study would be, did any of these pivotal events—his upheaval from Europe and immigration to the United States, the rise of Nazism, mounting Arab-Jewish violence in Palestine, the systematic annihilation of European Jewry, and the establishment of the state of Israel—bring about the Zionization of Albert Einstein?

NOTES

INTRODUCTION

1. See Einstein to Paul Ehrenfest, 18 May 1922 (AEA, 10 053).
2. See Neil Caplan, "Talking Zionism, Doing Zionism, Studying Zionism," *Historical Journal* 44, no. 4 (December 2001): 1089.
3. See Neil Caplan, "Zionism and the Arabs: Another Look at the 'New' Historiography," *Journal of Contemporary History* 36, no. 2 (April 2001): 346.
4. See Stanley Meisler, *United Nations: The First Fifty Years* (New York: Atlantic Monthly Press, 1995), 216.
5. See Caplan, "Zionism and the Arabs," 348.
6. See Andreas Gotzmann, "Historiography as Cultural Identity: Toward a Jewish History beyond National History," in *Modern Judaism and Historical Consciousness: Identities, Encounters, Perspectives*, ed. Andreas Gotzmann and Christian Wiese (Leiden: Brill, 2007), 519.
7. See Walter Laqueur, *A History of Zionism* (London: Weidenfeld and Nicolson, 1972), 156.
8. See Melvin I. Urofsky, *American Zionism from Herzl to the Holocaust* (Garden City, NY: Doubleday, 1975), 287.
9. See Howard M. Sachar, *A History of Israel: From the Rise of Zionism to Our Time* (New York: Alfred A. Knopf, 1976), 66, 161.
10. See Alfred M. Lilienthal, *The Zionist Connection: What Price Peace?* (New York: Dodd, Mead, 1978), 340–43, 746.
11. See Stephen M. Poppel, *Zionism in Germany 1897–1933: The Shaping of a Jewish Identity* (Philadelphia: Jewish Publication Society, 1977).
12. See Hagit Lavsky, *Before Catastrophe: The Distinctive Path of German Zionism* (Jerusalem: Magnes Press, 1996), 98, 100, 197.
13. See Albert Einstein, *About Zionism: Speeches and Lectures by Professor Albert Einstein*, trans. and edited with an introduction by Leon Simon (London: Soncino Press, 1930), and Fred Jerome, *Einstein on Israel and Zionism: His Provocative Ideas about the Middle East* (New York: St. Martin's Press, 2009). A selection of Einstein's writings and speeches on Zionism was assembled and translated into English, Hebrew, and Arabic by the Israeli physicist Gerald Tauber on the occasion of Einstein's centenary and the signing of the peace treaty between Egypt and Israel, yet was never published (see Gerald Tauber, "Einstein on Zionism, Arabs and Palestine: A Collection of Papers, Letters, and Speeches" [typescript, Tel Aviv University, Tel Aviv, 1979]).
14. See David E. Rowe, and Robert Schulmann, eds., *Einstein on Politics: His*

Private Thoughts and Public Stands on Nationalism, Zionism, War, Peace, and the Bomb (Princeton, NJ: Princeton University Press, 2007).

15. See Alexander Moszkowski, *Einstein: Einblicke in seine Gedankenwelt; gemeinverständliche Betrachtungen über die Relativitätstheorie und ein neues Weltsystem, entwickelt aus Gesprächen mit Einstein von Alexander Moszkowski* (Berlin: F. Fontane, 1922), 219–20, 236.

16. See Anton Reiser [Rudolf Kayser], *Albert Einstein: A Biographical Portrait* (New York: A. & C. Boni, 1930), 131–34, 138; and Philipp Frank, *Einstein: His Life and Times*, 2nd ed. (New York: A. A. Knopf, 1953), 84, 149, 152.

17. See Frank, *Einstein*, 2nd ed., 150; Fritz Stern, "Together and Apart: Fritz Haber and Albert Einstein," in *Einstein's German World* (Princeton, NJ: Princeton University Press, 1999), 92, 138; John Stachel, "Einstein's Jewish Identity," in *Einstein from "B" to "Z"* (Boston: Birkhäuser, 2002), 63–64, 72; Rowe and Schulmann, *Einstein on Politics*, 1–3; and Walter Isaacson, *Einstein: His Life and Universe* (New York: Simon and Schuster, 2007), 291.

18. See Albrecht Fölsing, *Albert Einstein: A Biography* (New York: Penguin Books, 1997), 489–90, and Kurt Blumenfeld, "Einsteins Beziehungen zum Zionismus und zu Israel," in *Helle Zeit—Dunkle Zeit: In Memoriam Albert Einstein*, ed. Carl Seelig (Braunschweig/Wiesbaden: Friedr. Vieweg & Sohn, 1956), 75, 79. The original remark on "undignified mimicry" is from a letter by Einstein to former German minister Willy Hellpach, 8 October 1929 (AEA, 46 657).

19. See Stern, "Together and Apart," 92, 138.

20. See Jürgen Neffe, *Einstein: Eine Biographie* (Reinbek bei Hamburg: Rowohlt, 2005), 289.

21. See Reiser, *Albert Einstein*, 133, 135; David Reichinstein, *Albert Einstein: Sein Lebensbild und seine Weltanschauung* (Prague: Selbstverlag des Verfassers, 1935), 159; Frank, *Einstein*, 2nd ed., 149; Banesh Hoffmann, "Albert Einstein," *Leo Baeck Institute Yearbook* 21 (1976): 286; and Rowe and Schulmann, *Einstein on Politics*, 6.

22. See Reichinstein, *Albert Einstein*, 133.

23. See Frank, *Einstein*, 2nd ed., 183.

24. See Michael Berkowitz, *Western Jewry and the Zionist Project, 1914–1933* (Cambridge: Cambridge University Press, 1997), 53–55.

25. See Fritz Stern, "Einstein's Germany," in *Albert Einstein: Historical and Cultural Perspectives. The Centennial Symposium in Jerusalem*, ed. Gerald Holton and Yehuda Elkana (Princeton, NJ: Princeton University Press, 1982), 334.

26. See Blumenfeld, "Einsteins Beziehungen zum Zionismus," 85, and idem, *Erlebte Judenfrage: Ein Viertel Jahrhundert deutscher Zionismus* (Stuttgart: Deutsche Verlags-Anstalt, 1962), 133.

27. See Isaacson, *Einstein*, 281, 300.

28. See Yosef Gotlieb, "Einstein the Zionist," *Midstream, A Monthly Jewish Review* 25, no. 6 (1979): 44, 46; Isaiah Berlin, "Einstein and Israel," in *Albert Einstein: Historical and Cultural Perspectives. The Centennial Symposium in Jerusalem*, ed. Gerald Holton and Yehuda Elkana (Princeton, NJ: Princeton University Press, 1982), 285; Stachel, "Einstein's Jewish Identity," 57; and Rowe and Schulmann, *Einstein on Politics*, xxv.

29. See Detlev Claussen, "Das Genie als Autorität: Über den Non-Jewish

Jew Albert Einstein," in *Jüdische Geschichte als allgemeine Geschichte: Festschrift für Dan Diner zum 60. Geburtstag*, ed. Raphael Gross and Yfaat Weiss (Göttingen: Vandenhoeck & Ruprecht, 2006), 90.

30. See Jerome, *Einstein on Israel and Zionism*, 4.

31. See Ronald W. Clark, *Einstein: The Life and Times. An Illustrated Biography* (New York: Abrams, 1984), 229.

32. See Neffe, *Einstein*, 348.

33. See Reiser, *Albert Einstein*, 137.

34. See Peter Honigmann,"Albert Einsteins jüdische Haltung," *Tribüne* 98 (1986): 104.

CHAPTER 1
"A VIVID SENSE OF STRANGENESS"

1. See Einstein to [Paul Nathan], 3 April 1920, in *The Collected Papers of Albert Einstein*, vol. 9, *The Berlin Years: Correspondence, January 1919–April 1920*, ed. Diana Kormos Buchwald et al. (Princeton, NJ: Princeton University Press, 2004), doc. 366, p. 492.

2. On the anti-Semitic disturbances in Einstein's lectures, see Albert Einstein, "Uproar in the Lecture Hall," in *The Collected Papers of Albert Einstein*, vol. 7, *The Berlin Years: Writings, 1918–1921*, ed. Michael Janssen et al. (Princeton, NJ: Princeton University Press, 2002), doc. 33, pp. 284–88. For Einstein's article, see Albert Einstein, "Assimilation und Antisemitismus," 3 April 1920, in Janssen et al., *The Collected Papers of Albert Einstein*, vol. 7, doc. 34, pp. 289–90.

3. See Maja Winteler-Einstein, "Albert Einstein—Beitrag für sein Lebensbild," excerpt, in *The Collected Papers of Albert Einstein*, vol. 1, *The Early Years: 1879–1902*, ed. John Stachel et al. (Princeton, NJ: Princeton University Press, 1987), xlviii. In many autobiographies of Holocaust survivors who were former rural Jews, the authors often attempt to prove their rootedness in Germany (see Monika Richarz, "Ländliches Judentum als Problem der Forschung," in *Jüdisches Leben auf dem Lande: Studien zur deutsch-jüdischen Geschichte*, ed. Monika Richarz and Reinhard Rürup [Tübingen: Mohr Siebeck], 2).

4. On the Jewish inhabitants in the rural regions of Germany, see Richarz, "Ländliches Judentum als Problem der Forschung," 1–8; Steven M. Lowenstein, "Jüdisches religiöses Leben in deutschen Dörfern: Regionale Unterschiede im 19. und frühen 20. Jahrhundert," in Richarz and Rürup, *Jüdisches Leben auf dem Lande*, 219–29; Werner J. Cahnman, "Village and Small-Town Jews in Germany: A Typological Study," *Leo Baeck Institute Yearbook* 19 (1974): 107–30; Steven M. Lowenstein, "The Rural Community and the Urbanization of German Jewry," *Leo Baeck Institute Yearbook* 25 (1980): 218–36; Paula E. Hyman, "Village Jews and Jewish Modernity: The Case of Alsace in the Nineteenth Century," in *Jewish Settlement and Community in the Modern Western World*, ed. Ronald Dotterer, Deborah Dash Moore, and Steven M. Cohen (Selinsgrove, PA: Susquehanna University Press, 1991), 13–26; and idem, "Jüdische Familie und kulturelle Kontinuität im Elsaß des 19. Jahrhunderts," in Richarz and Rürup, *Jüdisches Leben auf dem Lande*, 249–67.

5. Winteler-Einstein, "Albert Einstein," xlix, and Arnold Tänzer, "Der Stammbaum Prof. Albert Einsteins," in *Jüdische Familienforschung*, Sonderdruck aus dem Heft 28 vom Dezember 1931, 420.

6. Tänzer, "Der Stammbaum Prof. Albert Einsteins," 420.

7. See Winteler-Einstein, "Albert Einstein," xlix, and Tänzer, "Der Stammbaum Prof. Albert Einsteins," 420. There is no record of the Hebrew names of Einstein's maternal grandparents.

8. See Paula E. Hyman, "The Social Contexts of Assimilation: Village Jews and City Jews in Alsace," in *Assimilation and Community: The Jews in Nineteenth-Century Europe*, ed. Jonathan Frankel and Steven J. Zipperstein (Cambridge: Cambridge University Press: 1992), 124.

9. See "Auszug aus dem Familienregister," Buchau a. F., Band IC, Bl. 145 230 (AEA, 29 180).

10. See Winteler-Einstein, "Albert Einstein," l.

11. Ibid., l–liv.

12. See Evyatar Friesel, "The German-Jewish Encounter as a Historical Problem: A Reconsideration," *Leo Baeck Institute Yearbook* 41 (1996): 273.

13. See Frank, *Einstein*, 2nd ed., 7.

14. See Max Talmey, *The Relativity Theory Simplified and the Formative Period of Its Inventor* (New York: Falcon Press, 1932), 159, and Frank, *Einstein*, 2nd ed., 7.

15. See Frank, *Einstein*, 2nd ed., xx, 6–7.

16. See Marion Kaplan, "*Unter uns*: Jews Socialising with other Jews in Imperial Germany," *Leo Baeck Institute Yearbook* 48 (2003): 52, and Till van Rahden, *Juden und andere Breslauer: Die Beziehung zwischen Juden, Protestanten und Katholiken in einer deutschen Grossstadt von 1860 bis 1925* (Göttingen: Vandenhoeck & Ruprecht, 2000), 19–21.

17. See Samuel Moyn, "German Jewry and the Question of Identity: Historiography and Theory," *Leo Baeck Institute Yearbook* 41 (1996): 295, and Hyman, "Jüdische Familie und kulturelle Kontinuität," 250.

18. It seems it was Shulamit Volkov who was the first to point this out: Shulamit Volkov, "Juden als wissenschaftliche 'Mandarine' im Kaiserreich und in der Weimarer Republik: Neue Überlegungen zu sozialen Ursachen des Erfolgs jüdischer Naturwissenschaftler," *Archiv für Sozialgeschichte* 37 (1997): 9. In her study, 75 percent of the Jewish scientists she surveyed belonged to the *Besitzbürgertum* and 25 percent belonged to the *Bildungsbürgertum*.

19. See Kaplan, "*Unter uns*," 51. George Mosse has defined *Bildung* as the specifically German bourgeois "process of self-cultivation based upon classical learning and the development of esthetic sensibilities." He defines *Sittlichkeit* as the bourgeois "belief in a certain moral order expressed through the concept of respectability" (see George L. Mosse, "Jewish Emancipation. Between *Bildung* and Respectability," in *The Jewish Response to German Culture: From the Enlightenment to the Second World War*, ed. Jehuda Reinharz and Walter Schatzberg [Hanover, NH: University Press of New England, 1985], 2).

20. See Frank, *Einstein*, 2nd ed., 7, and Winteler-Einstein, "Albert Einstein," lvi.

21. Bernal and Knight define enculturation as "the cultural teaching that parents, families, peers, and the rest of the ethnic community provide to children

during the childhood years" (see David O. Sears and Sheri Levy, "Childhood and Adult Political Development," in *Oxford Handbook of Political Psychology*, ed. David O.. Sears et al. [Oxford: Oxford University Press, 2003], 68.).

22. See Siegfried Wagner, "Wie aus der Einsteinschen Fabrik Münchens Endzeitsynagoge wurde," *Tribüne* 28 (1989): 168.

23. See Chronology, entry for 1885, ca. 1 October, in Stachel et al., *The Collected Papers of Albert Einstein*, vol. 1, 370; Winteler-Einstein, "Albert Einstein," lvi. For a detailed study in which Einstein's violin playing is seen as emblematic of the cultural aspirations of the Jewish *Bildungsbürgertum*, see Sander L. Gilman, "Einstein's Violin: Jews and the Performance of Identity," *Modern Judaism* 25, no. 3 (2005): 219–36; Fölsing, *Albert Einstein*, 15; Chronology, entry for 1886, ca. 1 October, in Stachel et al., *The Collected Papers of Albert Einstein*, vol. 1, 371; and Winteler-Einstein, "Albert Einstein," lx, fn. 44.

24. See Wintleler-Einstein, "Albert Einstein," lviii; Moszkowski, *Einstein*, 221; and Frank, *Einstein*, 2nd ed., 24–25.

25. See, e.g., Albert Einstein, *Autobiographical Notes* (La Salle, IL: Open Court, 1979), 5, and Clark, *Einstein, Illustrated*, 8.

26. See Einstein to [Paul Nathan], 3 April 1920, in Buchwald et al., *The Collected Papers of Albert Einstein*, vol. 9, doc. 306, p. 492.

27. Einstein, "Assimilation und Antisemitismus," 289–90.

28. See Stephen M. Quintana, "Children's Developmental Understanding of Ethnicity and Race," *Applied and Preventive Psychology* 7 (1998): 27–45.

29. See Einstein to [Paul Nathan], 3 April 1920, in Buchwald et al., *The Collected Papers of Albert Einstein*. Vol. 9, doc. 366, p. 492.

30. See, e.g., Erik H. Erikson, *Childhood and Society*, 2nd rev. ed. (New York: W. W. Norton, 1963), 260.

31. For descriptions of this incident, see Reiser, *Albert Einstein*, 30, and Max Jammer, *Einstein and Religion: Physics and Theology* (Princeton, NJ: Princeton University Press, 1999), 21.

32. See Winteler-Einstein, "Albert Einstein," lix, fn. 42.

33. See Moszkowski, *Einstein*, 221. For a description of school life for Jewish pupils in Munich, see Gerhard Pischel, "1833–1933: 100 Jahre jüdische Schüler am Wilhelmsgymnasium in München," in *Jüdisches Leben in München: Geschichtswettbewerb 1993/94*, ed. Landeshauptstadt München (Munich: Buchendorfer Verlag, 1995), 77–83. For details of the Jewish religious instruction Einstein received, see "Luitpold-Gymnasium, Curriculum," appendix B, in Stachel et al., *The Collected Papers of Albert Einstein*, vol. 1, 346–55.

34. See Einstein, "Assimilation und Antisemitismus, 289–90.

35. On the development of stereotypes pertaining to in-groups and out-groups, see *Introduction to Political Psychology*, ed. Martha Cottam et al. (Mahwah, NJ: Lawrence Erlbaum, 2004), 9–11, 42–45.

36. On the internalization of majority group stereotypes by ethnic minority members, see Sears et al., *Oxford Handbook of Political Psychology*, 66.

37. See Heinrich Friedmann to Albert Einstein, 12 March 1929 (AEA, 30 404).

38. See Reiser, *Albert Einstein*, 30; Moszkowski, *Einstein*, 219; Winteler-Einstein, "Albert Einstein," lx; and Frank, *Einstein*, 2nd ed., 9.

39. See Einstein, *Autobiographical Notes*, 3, 5.

40. Jammer, *Einstein and Religion*, 20.

41. See Einstein, *Autobiographical Notes*, 2; Talmey, *The Relativity Theory Simplified*, 159; and Frank, *Einstein*, 2nd ed., 13–14.

42. See Einstein, *Autobiographical Notes*, 5.

43. See Heinrich Friedmann to Albert Einstein, 12 March 1929 (AEA, 30 404), in which Friedmann states that he was Einstein's instructor in his Bar Mitzvah preparation lessons.

44. See Einstein, *Autobiographical Notes*, 5.

45. Einstein's paternal uncle, Jakob Einstein, also played a crucial role in introducing him to algebra and to the world of technology (see Stachel, *Einstein's Miraculous Year*, xliv–xlv).

46. Talmey's elder brother Bernhard actually preceded him at the Einsteins' dinner table. He was the family's weekly guest for two years before Max assumed that role (see Talmey, *The Relativity Theory Simplified*, 161).

47. See Avraham Barkai, "Sozialgeschichtliche Aspekte der deutschen Judenheit in der Zeit der Industrialisierung," *Jahrbuch des Instituts für Deutsche Geschichte*, 1982: 249.

48. See Richarz, "Ländliches Judentum als Problem der Forschung," 1–2.

49. See Fölsing, *Albert Einstein*, 21.

50. On the perceived authenticity of Eastern European Jewry among some sectors of German Jewry, see, e.g., Moshe Zimmermann, *Die Deutschen Juden, 1914–1945* (Munich: R. Oldenbourg Verlag, 1997), 8.

51. See Winteler-Einstein, "Albert Einstein," liii, lxiii.

52. See Frank, *Einstein*, 2nd ed., 16, and Winteler-Einstein "Albert Einstein," lxiii, fn. 57.

53. See "Release from Württemberg Citizenship," in Stachel et al., *The Collected Papers of Albert Einstein*, vol. 1, doc. 16, p. 20.

54. See Einstein to Julius Katzenstein, 27 December 1931 (EPPA, 78 936).

55. On the formation of oppositional identities, see Sears et al., *Oxford Handbook of Political Psychology*, 69.

56. See Albert Einstein "Autobiographische Skizze," in Seelig, *Helle Zeit—Dunkle Zeit*, 145.

57. See "Biography" for Gustav Maier, in Stachel et al., *The Collected Papers of Albert Einstein*, vol. 1, 384, and Silvan S. Schweber, "Einstein and Oppenheimer: Interactions and Intersections," *Science in Context* 19, no. 4 (2006): 540–41.

58. See Winteler-Einstein, "Albert Einstein," lxv; and "Biography" for Jost and Pauline Winteler, in Stachel et al., *The Collected Papers of Albert Einstein*, vol. 1, 388.

59. See Schweber, "Einstein and Oppenheimer," 540–41. I am indebted to Don Howard for drawing my attention to the impact of Gustav Maier and the Society for Ethical Culture on Einstein.

60. See Lewis Pyenson, *The Young Einstein: The Advent of Relativity* (Boston: Adam Hilger, 1985), 9; Georg Kreis, "Judenfeindschaft in der Schweiz," Schweizerischer Israelitischer Gemeindebund—Fédération suisse des communautés israélites (ed.): Jüdische Lebenswelt Schweiz—Vie et culture juives en Suisse. 100 Jahre Schweizerischer Israelitischer Gemeindebund—Cent ans

Fédération suisse des communautés israélites (Zurich: Chronos Verlag, 2004), 425; and Aram Mattioli, "Antisemitismus in der Geschichte der modernen Schweiz: Begriffserklärungen und Thesen," in *Antisemitismus in der Schweiz 1848–1960*, ed. Aram Mattioli (Zurich: Orell Füssli Verlag, 1998), 10.

61. See editorial note, "The Swiss Federal Polytechnical School (ETH)," in Stachel et al., *The Collected Papers of Albert Einstein*, vol. 1, 43.

62. See Winteler-Einstein, "Albert Einstein—Beitrag für sein Lebensbild," typescript, 1924 (EPPA, 73 880), 16, and Einstein to Maja Einstein, 1898, in Stachel et al., *The Collected Papers of Albert Einstein*, vol. 1, doc. 38, p. 211.

63. See Winteler-Einstein, "Albert Einstein," typescript, 17; Einstein to Conrad Habicht, between 2 and 5 April, 1911, in Klein et al., *The Collected Papers of Albert Einstein*, vol. 5, doc. 30, pp. 34–35, and vol. 11, Errata to vol. 5, p. 614; and Municipal Police Detective's Report, 4 July 1900, in Stachel et al., *The Collected Papers of Albert Einstein*, vol. 1, doc. 66, p. 246.

64. See, e.g., Walter Gross, "The Zionist Students' Movement," *Leo Baeck Institute Yearbook* 4 (1959): 143–64; and Moshe Zimmermann, "Jewish Nationalism and Zionism in German-Jewish Students' Organizations," *Leo Baeck Institute Yearbook* 27 (1982): 129–53.

65. See Winteler-Einstein, "Albert Einstein," liii–lv.

66. See Einstein to Pauline Winteler, May? 1897, in Stachel et al., *The Collected Papers of Albert Einstein*, vol. 1, doc. 34, pp. 55–56.

67. See, e.g., Pauline Einstein to Marie Winteler, 13 December 1896, in Stachel et al., The *Collected Papers of Albert Einstein*, vol. 1, doc. 31, p. 53.

68. See Einstein to Mileva Marić, 29? July 1900, in Stachel et al., *The Collected Papers of Albert Einstein*, vol. 1, doc. 68, p. 248.

69. See Einstein to Mileva Marić, 14? August 1900 (*Collected Papers of Albert Einstein*, vol. 1, doc. 72, p. 254). The phrase is taken from Ludwig Uhland's poem "Schwäbische Kunde."

70. See "Introduction," in Stachel et al., *The Collected Papers of Albert Einstein*, vol. 1, xxxvi–xxxvii; Albert Einstein, *Mein Weltbild* (Amsterdam: Querido Verlag, 1934), 33; and editorial note, "Einstein as a student of physics and his notes on H. F. Weber's course," in Stachel et al., *The Collected Papers of Albert Einstein*, vol. 1, 61.

71. See "Introduction," in Stachel et al., *The Collected Papers of Albert Einstein*, vol. 1, xxxvii, and Einstein to Mileva Marić, 23 March 1901, ibid., doc. 93, pp. 279, 281.

72. See Einstein to Mileva Marić, 27 March 1901, in Stachel et al., *The Collected Papers of Albert Einstein*, vol. 1, doc. 94, p. 282. On anti-Semitism in the physics departments of German universities, see Christa Jungnickel and Russell McCormmach, *Intellectual Mastery of Nature: Theoretical Physics from Ohm to Einstein*, vol. 2, *The Now Mighty Theoretical Physics 1870–1925* (Chicago: University of Chicago Press, 1986), 286–87.

73. See Mileva Marić to Helene Savić, ca. 23 November—mid-December 1901, in Stachel et al., *The Collected Papers of Albert Einstein*, vol. 1, doc. 125, pp. 319–20.

74. See Hermann Einstein to Wilhelm Ostwald, 13 April 1901, in Stachel et al., *The Collected Papers of Albert Einstein*, vol. 1, doc. 99, p. 289.

75. See Einstein to Marcel Grossmann, 6? September 1901, in Stachel et al., *The Collected Papers of Albert Einstein*, vol. 1, doc. 122, p. 315.

76. See Einstein to Mileva Marić, 12 December 1901, in Stachel et al., *The Collected Papers of Albert Einstein*, vol. 1, doc. 127, pp. 322–24.

77. See Einstein to Mileva Marić, 28 December 1901, in Stachel et al., *The Collected Papers of Albert Einstein*, vol. 1, doc. 131, p. 330.

78. See Einstein to Swiss Patent Office, 18 December 1901, in Stachel et al., *The Collected Papers of Albert Einstein*, vol. 1, doc. 129, p. 327.

79. See Einstein to Mileva Marić, 4 February 1902, in Stachel et al., *The Collected Papers of Albert Einstein*, vol. 1, doc. 134, p. 332, and "Introduction," ibid., p. xxxvii. Owing to the lack of any documentation of their daughter's existence outside of their correspondence, it is not known whether the child was actually named Lieserl or whether this was merely a nickname.

80. See "Receipt for the Return of Doctoral Fees," 1 February 1902, in Stachel et al., *The Collected Papers of Albert Einstein*, vol. 1, doc. 132, p. 331; "Expert Opinion by Alfred Kleiner and Heinrich Burkhardt on Einstein's Dissertation," 22–23 July 1905, in Klein et al., *The Collected Papers of Albert Einstein*, vol. 5, doc. 31, pp. 35–37; and Winteler-Einstein, "Albert Einstein," typescript, 23–24.

81. See Einstein to Marcel Grossmann, 3 January 1908, in Klein et al., *The Collected Papers of Albert Einstein*, vol. 5, doc. 71, p. 84; Georg Kreis, "Judenfeindschaft in der Schweiz," in *Jüdische Lebenswelt Schweiz: Vie et culture juives en Suisse. 100 Jahre Schweizerischer Israelitischer Gemeindebund—Cent ans Fédération Suisse des communautés israélites*, ed. Schweizerischer Israelitischer Gemeindebund—Fédération Suisse des communautés israélites 423–45 (Zürich: Chronos Verlag, 2004), 423–45; and Mattioli, "Antisemitismus in der Geschichte der modernen Schweiz."

82. See Winteler-Einstein, "Albert Einstein," typescript, 25–26. On the influx of Eastern European Jews into Switzerland from 1890 on and the fear of an "inundation" by "caftan Jews," see Kreis, "Judenfeindschaft in der Schweiz," 427, and Mattioli, "Antisemitismus in der Geschichte der modernen Schweiz," 11.

83. See Winteler-Einstein, "Albert Einstein," typescript, 26.

84. See Einstein to Jakob Laub, 27 August 1910, in Klein et al., *The Collected Papers of Albert Einstein*, vol.5, doc. 224, p. 253.

85. See ibid., p. 254, n. 3, and Paul Ehrenfest to Tatiana Ehrenfest, 25 February 1912 (Museum Boerhaave, Ehrenfest Archive, Personal Correspondence, EPC:3, section 6).

86. See Einstein to Alfred Kleiner, 3 April 1912, in Klein et al., *The Collected Papers of Albert Einstein*, vol. 5, doc. 381, p. 446.

87. Einstein to Hedwig Born, 8 September 1916, in *The Collected Papers of Albert Einstein*, vol. 8, *The Berlin Years: Correspondence*, ed. Robert Schulmann et al. (Princeton, NJ: Princeton University Press, 1998), doc. 257, p. 336. The book by Brod was probably *Tycho Brahes Weg zu Gott* (Berlin: Deutsche Buch-Gemeinschaft, K. Wolff, 1915), in which the character of Kepler was allegedly inspired by Einstein.

88. See Max Brod, *Streitbares Leben 1884–1968* (Munich: Herbig, 1969), 170–71; and Hugo Bergman[n], "Personal Remembrance of Albert Einstein," in *Logical*

and Epistemological Studies in Contemporary Physics, ed. Robert S. Cohen and Marx W. Wartofsky (Dordrecht: Reidel, 1974), 390.

89. See Philipp Frank, *Einstein: His Life and Times*, 1st ed. (New York: Alfred A. 1947), 82. Israeli physicist Gerald Tauber also believed that his sojourn in Prague had a notable impact on stimulating Einstein's interest in the affairs of the Jewish people (see Gerald E. Tauber, "Einstein and Zionism," in *Einstein: A Centenary Volume*, ed. A. P. French [Cambridge, MA: Harvard University Press, 1979], 199.) On the Jews in Prague, see, e.g., Gary Cohen, "Jews in German Society: Prague, 1860–1914," in *Jews and Germans: The Problematic Symbiosis*, ed. David Bronsen (Heidelberg: Winter, 1979), 306–37. See also Einstein to Heinrich Zangger, 7 April 1911, in Klein et al., *The Collected Papers of Albert Einstein*, vol. 5, doc. 263, p. 289.

90. See Einstein to Lucien Chavan, 5–6 July 1911, in Klein et al., *The Collected Papers of Albert Einstein*, vol. 5, doc. 271, p. 304; Einstein to Heinrich Zangger, 20 September 1911, ibid., doc. 286, p. 325; Chronology, entry for 30 January 1912, ibid., p. 630; and Einstein to Conrad and Paul Habicht, 9 February 1912, ibid., doc. 356, p. 408.

91. See Einstein to Ludwig Hopf, 12 June 1912 in Klein et al., *The Collected Papers of Albert Einstein*, vol. 5, doc. 408, p. 483.

92. See Einstein to Helene Savić, after 17 December 1912, in Klein et al., *The Collected Papers of Albert Einstein*, vol. 5, doc. 424, p. 508.

93. See Einstein to Jakob Laub, 22 July 1913, in Klein et al., *The Collected Papers of Albert Einstein*, vol. 5, doc. 455, p. 538, and Einstein to Elsa Einstein, before 2 December 1913, ibid., doc. 488, p. 573.

94. See Einstein to Pëtr Petrovich Lazarev, 16 May 1914, in Schulmann et al., *The Collected Papers of Albert Einstein*, vol. 8, doc. 7, pp. 18–19. Einstein may have read a report on the trial which appeared in the *Neue Zürcher Zeitung*, 11 November 1913, 1st Evening Supplement, 1.

95. It is intriguing that Einstein repeatedly used the term "*Stammesgenossen*" (ethnic comrades) or "*Stammesbrüder*" (ethnic brethren) to refer to his fellow Jews, yet never the term "*Volksgenossen*" (national comrades), which may have had a pejorative connotation for Einstein, even before the term subsequently became imbued with racial overtones by Nazi usage. One wonders whether he was influenced in any way by the Zionist sociologist Franz Oppenheimer's distinction between Western Jewry who are merely imbued with "ethnic consciousness" (*Stammesbewusstsein*) as opposed to Eastern Jewry, who are imbued with "national consciousness" (*Volksbewusstsein*) as well (see Franz Oppenheimer, "Stammesbewusstsein und Volksbewusstsein," *Jüdische Rundschau*, 25 February 1910, 86–89).

96. See Einstein to Elsa Einstein, 26 July 1914, in Schulmann et al., *The Collected Papers of Albert Einstein*, vol. 8, doc. 26, p. 47.

97. See draft of letter from Michele Besso to Einstein, 17 January 1928 (AEA, 7 100). It is not certain whether the letter was actually sent or not.

98. See Einstein to Paul Ehrenfest, 19 August 1914, in Schulmann et al., *The Collected Papers of Albert Einstein*, vol. 8, doc. 34, p. 56.

99. For recent literature on German Jewry during World War I, see Ulrich

Sieg, "'Nothing More German Than the German Jews'? On the Integration of a Minority in a Society at War," in *Towards Normality? Acculturation and Modern German Jewry*, ed. Rainer Liedtke and David Rechter (Tübingen: Mohr Siebeck, 2003), 201–16; Zimmermann, *Die deutschen Juden 1914–1945*; and Peter G. J. Pulzer, "The First World War," in *German-Jewish History in Modern Times*, vol. 3, *Integration in Dispute 1871–1918*, ed. Michael A. Meyer (New York: Columbia University Press, 1998), 360–84.

100. On their opposition to the war, see Pulzer "The First World War," 364, and Rivka Horwitz, "Voices of Opposition to the First World War among Jewish Thinkers," *Leo Baeck Institute Yearbook* 33 (1988): 233–59. Horwitz's article includes a short section on Einstein's views on the war in which she explains his opposition as being based on his internationalism (see 237–40).

101. See Einstein to Paul Ehrenfest, beginning of December 1914, in Schulmann et al., *The Collected Papers of Albert Einstein*, vol. 8, doc. 39, p. 63.

102. See Einstein to Paolo Straneo, 7 January 1915, in Schulmann et al., *The Collected Papers of Albert Einstein*, vol. 8, doc. 45, p. 77.

103. See Albert Einstein, "Manifesto to the Europeans," mid-October 1914, in *The Collected Papers of Albert Einstein*, vol. 6, *The Berlin Years: Writings*, ed. Anne J. Kox et al. (Princeton, NJ: Princeton University Press, 1996), doc. 8, especially n. 2, pp. 69–71.

104. See Einstein to Heinrich Zangger, 11 January 1915, in *The Collected Papers of Albert Einstein*, vol. 10, *The Berlin Years: Correspondence, May–December 1920. Supplementary Correspondence, 1909–1920*, ed. Diana Kormos Buchwald et al. (Princeton, NJ: Princeton University Press, 2006), doc. 8, 45a, p. 28, and Einstein to Władisław Natanson, 29 December 1915, ibid., doc. 175, p. 231.

105. See Sears, *Oxford Handbook of Political Psychology*, 90–91.

106. He had migrated from Germany to Italy, then to Switzerland, then to the Austro-Hungarian Empire, from there back to Switzerland, and then finally back to Germany.

107. See Einstein to Heinrich Zangger, 10 April 1915, in Schulmann et al., *The Collected Papers of Albert Einstein*, vol. 8, doc. 73, pp. 116–17.

108. See Einstein to Heinrich Zangger, 17 May 1915, in Schulmann et al., *The Collected Papers of Albert Einstein*, vol. 8, doc. 84, p. 129.

109. See Einstein to Hendrik A. Lorentz, 21 July 1915, in Schulmann et al., *The Collected Papers of Albert Einstein*, vol. 8, Doc. 98, n. 2, p. 151, and Calendar, entry for 27 July 1915, ibid., p. 1029.

110. See Einstein to Hendrik A. Lorentz, 2 August 1915, in Schulmann et al., *The Collected Papers of Albert Einstein*, vol. 8, doc. 103, pp. 155–56.

111. Einstein to Walther Schücking, 22 October 1915, in Schulmann et al., *The Collected Papers of Albert Einstein*, vol. 8, doc. 131, pp. 186–87.

112. See Sapiro, "Political Socialization during Adulthood," 204, and Sears, *Oxford Handbook to Political Psychology*, 88. The simultaneous coexistence in Einstein's political socialization of seemingly contradictory factors should not perturb us greatly; they are merely indicative of the limitations of theoretical models in the social sciences.

113. See Einstein to Walther Schücking, 22 October 1915, in Schulmann et al., *The Collected Papers of Albert Einstein*, vol. 8, doc. 131, pp. 186–87.

114. See Albert Einstein, "Meine Meinung über den Krieg," 23 October–11 November 1915, in Kox et al., *The Collected Papers of Albert Einstein*, vol. 6, doc. 20. This passage was deleted as the Goethebund deemed it too controversial for publication (see Einstein to Berliner Goethebund, 11 November 1915, in Schulmann et al., *The Collected Papers of Albert Einstein*, vol. 8, doc. 138, pp. 193–94).

115. See Einstein to Heinrich Zangger, before 4 December 1915, in Buchwald et al., *The Collected Papers of Albert Einstein*, vol. 10, doc. 8, 159a, p. 36, and Einstein to Mileva Marić, 1 December 1915, in Schulmann et al., *The Collected Papers of Albert Einstein*, vol. 8, doc. 159, p. 213.

116. See Einstein to Heinrich Zangger, 16 January 1917, in Buchwald et al., *The Collected Papers of Albert Einstein*, vol. 10, doc. 8, 287b, p. 67.

117. See Sieg, "Nothing More German," 208–11, and Peter G. J. Pulzer, "The First World War," in *German-Jewish History in Modern Times*, vol. 3, *Integration in Dispute 1871–1918*, ed. Michael A. Meyer (New York: Columbia University Press, 1997), 372.

118. See Einstein to Heinrich Zangger, 13 February 1917, in Buchwald et al., *The Collected Papers of Albert Einstein*, vol. 10, doc. 8, 297a, p. 70.

119. See Einstein to Heinrich Zangger, 6 December 1917, in Schulmann et al., *The Collected Papers of Albert Einstein*, vol. 8, doc. 403, p. 561.

120. Einstein to Heinrich Zangger, 8 August 1917, in Buchwald et al., *The Collected Papers of Albert Einstein*, vol. 10, doc. 8, 370a, p. 117.

121. See, e.g., Einstein to Elsa Einstein, 16 July 1917, in Buchwald et al., *The Collected Papers of Albert Einstein*, vol. 10, doc. 8, 361c, p. 102, and Einstein to Heinrich Zangger, 16 February 1917, ibid., doc. 8, 299a, p. 72.

122. See Einstein to Heinrich Zangger, 16 February 1917, in Buchwald et al., *The Collected Papers of Albert Einstein*, vol. 10, doc. 8, 299a, p. 72.

123. See Einstein to Hendrik A. Lorentz, 3 April 1917, in Schulmann et al., *The Collected Papers of Albert Einstein*, vol. 8, doc. 322, pp. 429–30. Planck and Waldeyer-Hartz had written to him regarding the feasibility and timing of establishing a war-crimes commission (see Hendrik A. Lorentz to Einstein, 23 March 1917, ibid., n. 5, p. 420).

124. See Einstein to Hendrik A. Lorentz, 18 December 1917, in Schulmann et al., *The Collected Papers of Albert Einstein*, vol. 8, doc. 413, p. 575.

125. See Einstein to Heinrich Zangger, 15 September 1917, in Buchwald et al., *The Collected Papers of Albert Einstein*, vol. 10, doc. 8, 380a, p. 133.

126. See Einstein to Michele Besso, 15 October 1917, in Buchwald et al., *The Collected Papers of Albert Einstein*, vol. 10, doc. 8, 390a, p. 136.

127. See Einstein to Otto Heinrich Warburg, 23 March 1918, in Schulmann et al., *The Collected Papers of Albert Einstein*, vol. 8, doc. 491, p. 696.

128. See, e.g., Reiser, *Albert Einstein*, 134, and Clark, *Einstein, Illustrated*, 8.

129. See, e.g., Pulzer, "The First World War," 378–81, and Zimmermann, *Die Deutschen Juden*, 7–8.

130. See Arthur Hantke and Otto Warburg, Zionistische Vereinigung für Deutschland, to Einstein, 23 May 1918, in Schulmann et al., *The Collected Papers of Albert Einstein*, vol. 8, doc. 547, pp. 772–73.

131. See Einstein to Kurt Hiller, 9 September 1918, in Schulmann et al., *The Collected Papers of Albert Einstein*, vol. 8, doc. 614, pp. 871–72.

132. See Einstein to Pauline Einstein, 11 November 1918, in Schulmann et al., *The Collected Papers of Albert Einstein*, vol. 8, doc. 651, p. 944.

CHAPTER 2
A DIFFERENT KIND OF NATIONALISM

1. See Albert Einstein to Hedwig Born, 8 September 1916, in Schulmann et al., *The Collected Papers of Albert Einstein*, vol. 8, doc. 257, p. 336.

2. See, e.g., Eberhard Kolb, *Die Weimarer Republik* (Munich: Oldenbourg, 2000).

3. See "Divorce Decree," 14 February 1919, in Buchwald et al., *The Collected Papers of Albert Einstein*, vol. 9, doc. 6, pp. 8–11; "Marriage Certificate," 2 June 1919, ibid, doc. 54, p. 82; and "Introduction," ibid., pp. xxx–xxxi.

4. See Einstein to Auguste Hochberger, 20 August 1919, (ibid, doc. 94, p. 139).

5. Einstein to Hendrik A. Lorentz, 21 Sept. 1919, in Buchwald et al., *The Collected Papers of Albert Einstein*, vol. 9, doc. 108, p. 163.

6. Einstein to Hendrik A. Lorentz, 1 August 1919, in Buchwald et al., *The Collected Papers of Albert Einstein*, vol. 9, doc. 80, p. 121.

7. See Einstein to Heinrich Zangger, 24 December 1919, in Buchwald et al., *The Collected Papers of Albert Einstein*, vol. 9, doc. 233, p. 326, and Einstein to Paul Ehrenfest, 7 April 1920, ibid., doc. 371, p. 498.

8. See "Introduction," in Buchwald et al., *The Collected Papers of Albert Einstein*, vol. 9, pp. xliii–xliv; Einstein to Adriaan D. Fokker, 30 July 1919, ibid., doc. 78, p. 117; Moritz Schlick to Einstein, 19 December 1919, ibid., doc. 222, p. 314; Einstein to Robert Holtzmann, 17 August 1919, ibid., doc. 91, p. 134; and Einstein to Robert W. Lawson, 26 December 1919, ibid., doc. 234, p. 328.

9. See Einstein to Adriaan D. Fokker, after 1 December 1919, in Buchwald et al., *The Collected Papers of Albert Einstein*, vol. 9, doc. 187, p. 264; Einstein to Heinrich Zangger, 24 December 1919, ibid., doc. 233, p. 326; and Einstein to Heinrich Zangger, 26 March 1920, ibid., doc. 361, p. 487.

10. See Einstein to Pauline Einstein and Maja Winteler-Einstein, 4 April 1919, in Buchwald et al., *The Collected Papers of Albert Einstein*, vol. 9, doc. 17, p. 29; Einstein to Ida Hurwitz, 22 November 1919, ibid., doc. 172, p. 242; and Einstein to Heinrich Zangger, 1 June 1919, ibid., doc. 52, p. 80.

11. Einstein to Max Born, 3 March 1920, in Buchwald et al., *The Collected Papers of Albert Einstein*, vol. 9, doc. 337, p. 460.

12. See Felix Rosenblüth to Einstein, 9 December 1918, in Schulmann et al., *The Collected Papers of Albert Einstein*, vol. 8, doc. 666; p. 963, "Entwurf einer Notiz für die Jüdische Rundschau" (AEA, 45 333); and Lavsky. *Before Catastrophe*, 42–43. For a contemporary Zionist article on the planned German Jewish Congress, see Julius Berger, "Der jüdische Kongreß und die deutschen Juden," *Jüdische Rundschau*, 17 December 1918, 425–26; "Entwurf eines Einladungsschreibens" (AEA, 45 332); and Felix Rosenblüth to Albert Einstein, 12 December 1918, in Schulmann et al., *The Collected Papers of Albert Einstein*, vol. 8, doc. 671, p. 970.

13. See Jehuda Reinharz, "East European Jews in the *Weltanschauung* of

German Zionists, 1882–1914," *Studies in Contemporary Jewry* 1 (1984): 59; idem, "Three Generations of German Zionism," *Jerusalem Quarterly* 9 (Fall 1978): 103; and idem, "Advocacy and History: The Case of the Centralverein and the Zionists," *Leo Baeck Institute Yearbook* 33 (1988): 121.

14. See Einstein to Paul Epstein, 5 October 1919, in Buchwald et al., *The Collected Papers of Albert Einstein*, vol. 9, doc. 122, pp. 180–181.

15. See Blumenfeld, *Erlebte Judenfrage*, 126–27; and idem, "Einsteins Beziehungen zum Zionismus," 75.

16. Blumenfeld, *Erlebte Judenfrage*, 126–27.

17. If Blumenfeld was correct and the lecture took place in February 1919, it must have been held during the last week of that month, as Einstein only returned from Switzerland around 23 February (see Buchwald et al., *The Collected Papers of Albert Einstein*, vol. 9, Calendar 1919, p. 553). Lectures were held at numerous locations in Berlin on 27 February by various *Bezirksgruppen* of the Berliner Zionistische Vereinigung (see *Jüdische Rundschau*, 25 February 1919, 115). A more likely candidate is a lecture delivered by both Blumenfeld and Rosenblüth on 3 March entitled "Unsere Forderungen an den Jüdischen Kongreß in Deutschland" (*Jüdische Rundschau*, 25 February 1919, 115).

18. Blumenfeld, "Einsteins Beziehungen zum Zionismus," 76, and idem, *Erlebte Judenfrage*, 127–28.

19. See Blumenfeld, *Erlebte Judenfrage*, 128, and Einstein to Kurt Blumenfeld, 25 March 1955 (AEA, 59 274).

20. See Blumenfeld, "Einsteins Beziehungen zum Zionismus," 74–75, 77; and Hans Tramer; "Kurt Blumenfeld. Seine Lehre und seine Leistung," in Blumenfeld, *Erlebte Judenfrage*, 13.

21. See Hans Mühsam to Albert Einstein, 24 October 1918 (AEA, 38 321; also in Schulmann et al., *The Collected Papers of Albert Einstein*, vol. 8, doc. 639, pp. 918–30); and Paul Mühsam, *Ich bin ein Mensch gewesen: Lebenserinnerungen*, ed. Ernst Kretzschmar (Gerlingen: Bleicher, 1989), 148.

22. See *Korrespondenzblatt des Vereins zur Gründung und Erhaltung einer Akademie für die Wissenschaft des Judentums*, vol. 1 (Berlin: Baruch, 1919), 2–4; and David N. Myers, "The Fall and Rise of Jewish Historicism: The Evolution of the Akademie für die Wissenschaft des Judentums (1919–1934)," *Hebrew Union College Annual* 63 (1992): 140.

23. Albert Einstein to Paul Ehrenfest, 22 March 1919, Buchwald et al., *The Collected Papers of Albert Einstein*, vol. 9, doc. 10, p. 16.

24. "President Gives Hope to Zionists," *New York Times*, 3 March, 1919; see also "Mr. Wilson and Zionists," *Times*, 4 March 1919, and "Wilson für den Zionismus," *Jüdische Rundschau*, 11 March 1919.

25. See Einstein to Heinrich Zangger, 8 August 1917, in Buchwald et al., *The Collected Papers of Albert Einstein*, vol. 10, doc. 8, 370a, p. 117; and Einstein to Elsa Einstein, 7 August 1917, in Schulmann et al., *The Collected Papers of Albert Einstein*, vol. 8, doc. 369b, in vol. 10, p. 115.

26. For an example of the Zionist view of the Jews as non-Europeans, see Kurt Blumenfeld, "Der Zionismus," *Preussische Jahrbücher* 161/1 (July 1915): 110, in which he described the Jews as an "old Oriental people."

27. See Lavsky, *Before Catastrophe*, 142–43.

28. See, e.g., Einstein to Elsa Einstein, 4 August 1919, in Buchwald et al., *The Collected Papers of Albert Einstein*, vol. 9, doc. 84a, in vol. 10, pp. 212–13.

29. Paul Epstein to Albert Einstein, 11 September 1919, in Buchwald et al., *The Collected Papers of Albert Einstein*, vol. 9, doc. 102, pp. 151–52. Attempts to locate the letter by the unnamed Zionist have not been successful.

30. Hagit Lavsky, "Beyn Hanachat Even ha-Pina li-F'ticha: Yesud ha-Universita ha-Ivrit, 1918–1925," in *Toldot ha-Universita ha-Ivrit bi-Yerushalyim: Shorashim ve-Hatchalot*, ed. Shaul Katz and Michael Heyd (Jerusalem: Magnes Press, 1997), 127.

31. Albert Einstein to Paul Epstein, 5 October 1919, in Buchwald et al., *The Collected Papers of Albert Einstein*, vol. 9, doc. 122, pp. 180–81.

32. See, e.g., Michael Brenner, *The Renaissance of Jewish Culture in Weimar Germany* (New Haven, CT: Yale University Press, 1998), 23.

33. Albert Einstein to Paul Epstein, 5 October 1919, in Buchwald et al., *The Collected Papers of Albert Einstein*, vol. 9, doc. 122, p. 181.

34. See Julius Berger to Albert Einstein, 13 October 1919 (CZA/F4/18 and CZA/L12/102/1). For a typed draft of the appeal with textual variants, see CZA/F4/18.

35. See Julius Berger to Hugo Bergmann, 15 October 1919 (CZA/Z3/690).

36. See Paul Epstein to Albert Einstein, 15 October 1919, in Buchwald et al., *The Collected Papers of Albert Einstein*, vol. 9, doc. 136, p. 197.

37. Hugo Bergmann to Albert Einstein, 22 October 1919, in Buchwald et al., *The Collected Papers of Albert Einstein*, vol. 9, doc. 147, p. 212.

38. Hugo Bergman[n], *Tagebücher und Briefe*, ed. Miriam Sambursky (Königstein a. T.: Jüdischer Verlag bei Athenäum, 1985), 131, citing a letter from Bergmann to Martin Buber, 10 November 1919.

39. See *Jüdische Pressezentrale Zürich*, 28 November 1919.

40. See Albert Einstein to Hugo Bergmann, 5 November 1919, in Buchwald et al., *The Collected Papers of Albert Einstein*, vol. 9, doc. 155, p. 222.

41. See Hugo Bergmann to Chaim Weizmann, 16 November 1919 (Weizmann Archives).

42. See the articles "Revolution in Science," which had a report on the session of the Royal Society, and "The Fabric of the Universe," which was a lead article on general relativity, *Times*, 7 November 1919.

43. CZA, L12/102/1.

44. Albert Einstein to Paul Ehrenfest, 8 November 1919, in Buchwald et al., *The Collected Papers of Albert Einstein*, vol. 9, doc. 160, p. 227.

45. See Reinharz, "East European Jews," 59.

46. See *Berliner Illustrirte Zeitung*, 14 November 1919.

47. See Hugo Bergmann to Julius Berger, 17 November, 1919 (CZA/Z3/1673).

48. Arthur Ruppin to the Culture Department of the Zionist Organization, London, 18 November 1919 (CZA/Z3/690).

49. Hugo Bergmann to Albert Einstein, 21 November 1919, in Buchwald et al., *The Collected Papers of Albert Einstein*, vol. 9, doc. 171, p. 240.

50. Paul Ehrenfest to Albert Einstein, 24 November 1919, in Buchwald et al., *The Collected Papers of Albert Einstein*, vol. 9, doc. 175, p. 248.

51. See Shmarya Levin to Albert Einstein, 27 November 1919, in Buchwald et al., *The Collected Papers of Albert Einstein*, vol. 9, doc. 178, p. 254, and AEA, 36–827.1.

52. See Selig Brodetsky to Solomon Ginsberg, 30 November 1919 (CZA/L12/65); Berlin Zionist Bureau, 19 December, 1919 (CZA/Z3/1673); and Education Department, Zionist Organization, London to Julius Berger, 3 December, 1919 (CZA/Z3/1673).

53. Einstein to Paul Ehrenfest, 4 December 1919, in Buchwald et al., *The Collected Papers of Albert Einstein*, vol. 9, doc. 189, p. 267. Carathéodory was actually involved in plans for establishing the University of Smyrna, not Saloniki.

54. See Einstein to Michele Besso, 12 December 1919, in Buchwald et al., *The Collected Papers of Albert Einstein*, vol. 9, doc. 207, p. 293 and n. 5, p. 294.

55. See Elsa Einstein to Paul Ehrenfest, 10 December 1919, in Buchwald et al., *The Collected Papers of Albert Einstein*, vol. 9, Calendar, p. 587.

56. Paul Ehrenfest to Einstein, 9 December 1919 in Buchwald et al., *The Collected Papers of Albert Einstein*, vol. 9, doc. 203, p. 287 and n. 15, p. 289.

57. See interview of 18 December in *Neues Wiener Journal*, 25 December 1919. This is possibly another instance of Paul Epstein influencing Einstein's opinion, as Epstein had voiced a similar opinion in his letter to Einstein of 15 October 1919 (see Buchwald et al., *The Collected Papers of Albert Einstein*, vol. 9, doc. 136, p. 197). For the general context of the discussions among the planners of the university on both the issue of whether teaching or research should be prioritized during its initial phases and of which institutes should be established first, see Lavsky, "Beyn Hanachat Even ha-Pina li-F'ticha," 123–24.

58. Einstein to Felix Ehrenhaft, 14 December 1919, in Buchwald et al., *The Collected Papers of Albert Einstein*, vol. 9, doc. 211, p. 298.

59. Einstein to Heinrich Zangger, 15 or 22 December 1919, in Buchwald et al., *The Collected Papers of Albert Einstein*, vol. 9, doc. 217, pp. 306–7.

60. Einstein to Heinrich Zangger, 24 December 1919, in Buchwald et al., *The Collected Papers of Albert Einstein*, vol. 9, doc. 233, p. 326.

61. See Education Department, Zionist Organization, London to Central Zionist Bureau, Berlin, 24 December, 1919 (CZA/Z3/1673). See also *Jüdische Pressezentrale Zürich*, 31 December 1919. For the invitees' replies, see CZA/L12/65 and CZA/Z3/1673, and Lavsky, "Beyn Hanachat Even ha-Pina li-F'ticha," 129. See also letters by Otto Warburg and Theodor Zloscisti to Education Dept. of the Zionist Organization, 12 December 1919 (CZA/Z3/1673), and Lavsky, "Beyn Hanachat Even ha-Pina li-F'ticha," 129.

62. Einstein to Michele Besso, 6 January 1920, in Buchwald et al., *The Collected Papers of Albert Einstein*, vol. 9, doc. 245, p. 342.

63. See Otto Warburg to Zionist Organization Central Office London, 12 December 1919 (CZA/Z3/1673).

64. Einstein, "Die Zuwanderung aus dem Osten," *Berliner Tageblatt*, 30 December 1919 (Morgen-Ausgabe), 2.

65. See Ibald. "En Revolution i Videnskaben: Professor Einsteins epokeg

ørende Teorier bekræftet; Newtons Tyngdelov omstødt," *Politiken*, 18 November 1919; Charlotte Weigert to Einstein, 10 January 1920, in Buchwald et al., *The Collected Papers of Albert Einstein*, vol. 9, doc. 253, pp. 350–351; and Einstein to Charlotte Weigert, 8 March 1920, in Buchwald et al., *The Collected Papers of Albert Einstein*, vol. 12, doc. 9, 343a, p. 11.

66. Albert Einstein, "In Support of Georg Nicolai," in Janssen et al., *The Collected Papers of Albert Einstein*, vol. 7, doc. 32, pp. 282–83.

67. See Einstein, "Uproar in the Lecture Hall," 287.

68. See Einstein and Leopold Landau to Konrad Haenisch, 19 February 1920, in Buchwald et al., *The Collected Papers of Albert Einstein*, vol. 9, doc. 317, p. 434, and Konrad Haenisch to Leopold Landau, 9 March 1920, ibid., doc. 344, p. 466.

69. See Zionist Organization, "Statement on the Present Position of the University Question," 21 January 1920 (AEA, 36 830), and Hugo Bergmann to Einstein, 19 January 1920, in Buchwald et al., *The Collected Papers of Albert Einstein*, vol. 9, doc. 266, p. 365.

70. Various versions of this draft are extant. See AEA, 36 828, and a transcription of the draft that is dated 18 February 1920 (AEA, 122 796). A more complete version is available in CZA/A185/31.

71. See Paul Ehrenfest to Einstein, 8 February 1920, in Buchwald et al., *The Collected Papers of Albert Einstein*, vol. 9, doc. 303, p. 415, and Einstein to Paul Ehrenfest, 1 March 1920, ibid., doc. 335, p. 457.

72. Julius Brodnitz, Centralverein to Einstein, 29 March 1920, in Buchwald et al., *The Collected Papers of Albert Einstein*, vol. 9, doc. 363, p. 490.

73. See Avraham Barkai, *"Wehr Dich!" Der Centralverein deutscher Staatsbürger jüdischen Glaubens (C.V.) 1893–1938* (Munich: Beck, 2002), 117.

74. Memo by I. Moos to the Centralverein, 29 December 1919 (see Politisches Archiv des Auswärtigen Amts, Berlin [PA AA], Pol III, Prof. Sobernheim Allgemeines).

75. Einstein, "Assimilation und Antisemitismus."

76. See Reinharz, "Three Generations of German Zionism," 102.

77. Einstein to the Centralverein, 5 April 1920, in Buchwald et al., *The Collected Papers of Albert Einstein*, vol. 9, doc. 368, pp. 494–95. The original handwritten version of this letter is dated 3 April 1920 (see AEA, 120 288).

78. Reinharz, "Three Generations of German Zionism," 97–99, 101.

79. Ibid., 100–101, 109.

80. See Keith A. Roberts, *Religion in Sociological Perspective* (Belmont, CA: Wadsworth/Thomson, 2004), 98, 100–103.

81. Einstein to Elsa Einstein, 19 October 1920, in Buchwald et al., *The Collected Papers of Albert Einstein*, vol. 10, doc. 179, p. 464.

82. See Roberts, *Religion in Sociological Perspective*, 105.

83. Ibid., 105, 107.

84. See Robert Wistrich, "Theodor Herzl: Zionist Icon, Myth-Maker, and Social Utopian," in *The Shaping of Israeli Identity: Myth, Memory and Trauma*, ed. Robert Wistrich and David Ohana (London: Frank Cass, 1995), 8–9, 15, 17; Peter J. Loewenberg, "Theodor Herzl: A Psychoanalytic Study in Charismatic Political Leadership," in *The Psychoanalytic Interpretation of History*, ed. Benjamin B. Wolman (New York: Basic Books, 1971), 169–171; Henry J. Cohn, "Theodor

Herzl's Conversion to Zionism," *Jewish Social Studies* 32 (1970): 104, 107; and Shlomo Avineri, "Herzls Weg zum Zionismus—eine Neubewertung," in *100 Jahre Zionismus: Von der Verwirklichung einer Vision*, ed. Ekkehard W. Stegemann (Stuttgart: Verlag W. Kohlhammer, 2000), 27–30. For Herzl's views on Jews and modernization, see Jacques Kornberg, "Theodor Herzl: The Zionist as Austrian Liberal," *Jerusalem Quarterly* 31 (Spring 1984), 111.

85. Sarah Schmidt, "The Zionist Conversion of Louis D. Brandeis," *Jewish Social Studies* 37 (1975): 20–25, 30, 32–33; and Allon Gal, "In Search of a New Zion: New Light on Brandeis' Road to Zionism," *American Jewish History* 68, no. 1 (1978): 19.

86. See Niels Bokhove, "'The Entrance to the More Important': Kafka's Personal Zionism," in *Kafka, Zionism, and Beyond*, ed. Mark H. Gelber (Tübingen: Max Niemeyer Verlag, 2004), 23–24; Vivian Liska. "Neighbors, Foes, and Other Communities: Kafka and Zionism," *Yale Journal of Criticism* 13, no. 2 (2000): 344; and Hans-Richard Eyl, "'Der letzte Zipfel': Kafka's State of Mind and the Making of the Jewish State," in Gelber, *Kafka*, 59–61.

87. Quoted in Bokhove, "'The Entrance to the More Important,'" 23.

88. A reference to those Jews who only frequented the synagogue on the High Holidays.

89. See Bokhove, "'The Entrance to the More Important,'" 25–26, 37, 42, 54–55, 57.

CHAPTER 3
THE "PRIZE-WINNING OX" IN "DOLLARIA"

1. See Julian Mack to Chaim Weizmann, 30 March 1921 (Weizmann Archives, Weizmann Institute).

2. On the uproar in Einstein's lecture hall, see "Introduction to Volume 9," in Buchwald et al., *The Collected Papers of Albert Einstein*, vol. 9, p. xlvii. On his clash with the antirelativists, see "Introduction to Volume 10," in Buchwald et al., *The Collected Papers of Albert Einstein*, vol. 10, pp. xxxviii–xlii.

3. On the rumors that Einstein planned to leave Germany, see Fritz Haber to Einstein, 30 August 1920, in Buchwald et al., *The Collected Papers of Albert Einstein*, vol. 10, doc. 119, pp. 395–97, and Max Planck to Einstein, 5 September 1920, ibid., doc. 133, p. 412. On the Bad Neuheim confrontation, see "Introduction to Volume 10," ibid., pp. xxxviii–xli.

4. See, e.g., Howard L. Adelson, "Ideology and Practice in American Zionism," in *Essays in American Zionism*, ed. Melvin I. Urofsky (New York: Herzl Press, 1978), 1–17; Harry Barnard, *The Forging of an American Jew: The Life and Times of Judge Julian W. Mack* (New York: Herzl Press, 1974), 269–83; George L. Berlin, "The Brandeis-Weizmann Dispute," *American Jewish Historical Quarterly* 60 (1970): 37–68; Ben Halpern, *A Clash of Heroes: Brandeis, Weizmann and American Zionism* (New York: Oxford University Press, 1987), 221–32; Esther L. Panitz, "'Washington versus Pinsk': The Brandeis-Weizmann Dispute," in Urofsky, *Essays in American Zionism*, 77–94; Jehuda Reinharz, *Chaim Weizmann: The Making of a Statesman* (New York: Oxford University Press, 1993), 346–48, 363–64; Norman Rose,

Chaim Weizmann: A Biography (New York: Elisabeth Sifton Books/Viking, 1986), 206–13; Chaim Weizmann, *Trial and Error: The Autobiography of Chaim Weizmann* (New York: Harper & Brothers, 1949), 265–69; and Melvin I. Urofsky, "Zionism: An American Experience," *American Jewish Historical Quarterly* 63 (1974): 211–21.

5. See, e.g., Clark, *Einstein, Illustrated*. 382–92; Fölsing, *Albert Einstein*, 494–509; Frank, *Einstein: His Life and Times*, 176–87; Armin Hermann, *Einstein: Der Weltweise und sein Jahrhundert: Eine Biographie* (Munich: Piper, 1994), 261–66; Isaacson, *Einstein*, 289–301; and Neffe, *Einstein*, 397–400. For an annotated collection of newspapers clippings on the tour, see József Illy, ed., *Albert Meets America: How Journalists Treated Genius during Einstein's 1921 Travels* (Baltimore: Johns Hopkins University Press, 2006).

6. See Luther P. Eisenhart to Einstein, 1 October 1920, in Buchwald et al., *The Collected Papers of Albert Einstein*, vol. 10, doc. 160, p. 441, and Albert G. Schmedeman to Einstein, 30 October 1920, ibid., Calendar, p. 604.

7. Einstein to Paul Ehrenfest, 26 November 1920, in Buchwald et al., *The Collected Papers of Albert Einstein*, vol. 10, doc. 209, p. 494.

8. See Paul Ehrenfest to Einstein, 7 November 1920, in Buchwald et al., *The Collected Papers of Albert Einstein*, vol. 10, doc. 191, p. 480; Einstein to Paul Ehrenfest, 26 November 1920, ibid., doc. 209, p. 494; and Paul Ehrenfest to Einstein, 7 November 1920, ibid., doc. 191, p. 480.

9. See John G. Hibben to Einstein, 24 December 1920, in Buchwald et al., *The Collected Papers of Albert Einstein*, vol. 10, doc.243, p. 539; and Charles S. Slichter to A. L. Barrows, 12 January 1921 (University of Wisconsin–Madison Archives, Dean Slichter Papers).

10. See Charles S. Slichter to Paul M. Warburg, 31 January 1921 (University of Wisconsin–Madison Archives, Dean Slichter Papers), and Max M. Warburg to Einstein, 8 February 1921, in Buchwald et al., *The Collected Papers of Albert Einstein*, vol. 12, doc. 43, p. 73.

11. Einstein to Paul Ehrenfest, 13 February 1921, in Buchwald et al., *The Collected Papers of Albert Einstein*, vol. 12, doc. 48, p. 84.

12. See Chaim Weizmann to Kurt Blumenfeld, 16 February 1921 (AEA, 33 345).

13. See Deborah E. Lipstadt, "Louis Lipsky and the Emergence of Opposition to Brandeis, 1917–1920," in *Essays in American Zionism 1917–1948*, ed. Melvin I. Urofsky (New York: Herzl Press, 1978), 55. On the London conference, see, e.g., Jacob de Haas, *Louis D. Brandeis: A Biographical Sketch* (New York: Bloch, 1929), 126–38; and A. Ulitzur, *Foundations: A Survey of 25 Years of Activity of the Palestine Foundation Fund Keren Hayesod. Facts and Figures 1921–1946* (Jerusalem: Jerusalem Press, 1946), 9–12.

14. Barnard, *The Forging of the American Jew*, 279–80.

15. See Chaim Weizmann to the Executive of the Zionist Organization of America, January 6, 1921, in *The Letters and Papers of Chaim Weizmann*, vol. 10, Series A, *July 1920–December 1921*, ed. Bernard Wasserstein and Joel S. Fishman (New Brunswick, NJ: Transaction Books; Jerusalem: Israel Universities Press, 1977), 133–34.

16. See Chaim Weizmann to Kurt Blumenfeld, 16 February 1921 (AEA, 33 345).

17. See Kurt Blumenfeld to Chaim Weizmann, 19 February 1921 (Weizmann Archives, Weizmann Institute).

18. The other three U.S. institutions were Columbia, Yale, and the National Research Council (see Einstein to Fritz Haber, 6 October 1920, in Buchwald et al., *The Collected Papers of Albert Einstein*, vol. 10, doc. 162, pp. 442–43; Einstein to John G. Hibben, 21 February 1921 (AEA, 36 244); Einstein to Max. M. Warburg, 8 December 1920, in Buchwald et al., *The Collected Papers of Albert Einstein*, vol. 10, doc. 223, p. 514.

19. See Kurt Blumenfeld to Chaim Weizmann, 20 February 1921 (Weizmann Archives, Weizmann Institute).

20. See Chaim Weizmann, "Speech to Fourth Session of the Plenary Session of the Action Committee of the Zionist Organization," Prague, 11 July 1921 (CZA/Z4/257/2).

21. See Einstein to John G. Hibben, 21 Feb. 1921, in Buchwald et al., *The Collected Papers of Albert Einstein*, vol. 12, doc. 53, p. 89.

22. For Lorentz's official invitation as president of the Congress's International Scientific Committee, see Hendrik A. Lorentz to Einstein, 9 June 1920, in Buchwald et al., *The Collected Papers of Albert Einstein*, vol. 10, doc. 49, pp. 302–3.

23. On the fascinating relationship between Einstein and Lorentz, see Anne J. Kox, "Einstein and Lorentz. More Than Just Good Colleagues," *Science in Context* 6, no. 1 (1993): 43–56.

24. See Emile Tassel to Michel Huisman, 23 March 1921 (translated from French) (Archives des Instituts Internationaux de Physique et de Chimie fondes par Ernest Solvay, Fonds de l'IIPC conserve aux Archives de l'Universite Libre).

25. See Ernest Rutherford to Bertram B. Boltwood, 28 February 1921, in: Lawrence Badash, *Rutherford and Boltwood: Letters on Radioactivity* (New Haven, CT: Yale University Press, 1969), 342. Even though Einstein himself repeatedly claimed that he only had Swiss citizenship during this period, the Prussian Kultusministerium and the Prussian Academy of Sciences later came to the conclusion that Einstein had become a Prussian citizen in 1913–14. For the published versions of the relevant documents, see Christa Kirsten and Hans-Jürgen Treder, eds., *Albert Einstein in Berlin 1913–1933*. Part 1, *Darstellung und Dokumente* (Berlin: Akademie-Verlag, 1979), 112–20.

26. See Einstein to Hendrik A. Lorentz, 22 February 1921, in Buchwald et al., *The Collected Papers of Albert Einstein*, vol. 12, doc. 57, p. 95.

27. See Einstein to Hendrik A. Lorentz, 22 February 1921, in Buchwald et al., *The Collected Papers of Albert Einstein*, vol. 12, doc. 57, p. 95.

28. See Chaim Weizmann to Einstein, 23 February 1921, in Buchwald et al., *The Collected Papers of Albert Einstein*, vol. 12, doc. 63, p. 101.

29. See Telegram from Nathan Ratnoff to Einstein, 27 February 1921, in Buchwald et al., *The Collected Papers of Albert Einstein*, vol. 12, doc. 67, pp. 104–5. However, there was a connection between Ratnoff and Weizmann, as they had been classmates in Russia (see Samuel B. Finkel, "American Jews and the Hebrew University," in *The American Jewish Year Book 5698* [Philadelphia: Jewish Publication Society of America, 1937], 196).

30. See David J. Kaliski, "The Physicians Committee: Its Efforts and Achievements," *New Palestine*, 27 March 1925, 345.

31. See Zionist Organization(?), ed., *The Proposed Hebrew University on Mount Scopus Jerusalem Palestine* (London(?): Zionist Organization(?), 1924(?), [4]; and Kaliski, "The Physicians Committee," 345.

32. See Paul Ehrenfest to Einstein, 28 February 1921, in Buchwald et al., *The Collected Papers of Albert Einstein*, vol. 12, doc. 68, p. 105. In his previous letter, Ehrenfest had expressed his ambivalent feelings in regard to Einstein's previous intention to travel to the United States on a private lecture tour (see Paul Ehrenfest to Einstein, 21 February 1921, ibid., doc. 55, p. 92).

33. See Einstein to Paul Ehrenfest, 8 March 1921, in Buchwald et al., *The Collected Papers of Albert Einstein*, vol. 12, doc. 83, p. 121.

34. See Jehuda Reinharz, *Chaim Weizmann: The Making of a Zionist Leader* (New York: Oxford University Press, 1985), 338, 506; Henry A. Miers, Manchester University, to Einstein, 20 April 1921, in Buchwald et al., *The Collected Papers of Albert Einstein*, vol. 12, doc. 125, p. 166, and Ernest Barker, King's College, London, to Einstein, 24 March 1921, ibid., Calendar, p. 438.

35. See Manfred Jonas, *The United States and Germany: A Diplomatic History* (Ithaca, NY: Cornell University Press, 1984), 158–59; and Erich Eyck, *A History of the Weimar Republic*, trans. Harlan P. Hanson and Robert G. L. Waite (Cambridge, MA: Harvard University Press, 1967), 177.

36. See Fritz Haber to Einstein, 9 March 1921, in Buchwald et al., *The Collected Papers of Albert Einstein*, vol. 12, doc. 87, pp. 124–27.

37. See Einstein to Fritz Haber, 9 March 1921, in Buchwald et al., *The Collected Papers of Albert Einstein*, vol. 12, doc. 88, pp. 127–30.

38. Einstein was probably displeased with Roethe's position in regard to the issue of whether the academy should take an open stance on Einstein's public dispute with the anti-relativity movement. Roethe had informed Planck that he was opposed to the academy publicly expressing its support for Einstein (see Gustav Roethe to Max Planck, 10 September 1920 [Archiv der Berlin-Brandenburgischen Akademie der Wissenschaften-Literaturarchiv, Hist. Abt. II. Tit. Personalia II–III, 38, Fo. 80]). See also Einstein to Ulrich Wilamowitz-Moellendorff, 19 April 1920, in Buchwald et al., *The Collected Papers of Albert Einstein*, vol. 9, doc. 379, p. 511, and "Introduction to Volume 10," in Buchwald et al., *The Collected Papers of Albert Einstein*, vol. 10, pp. xxxviii–xlii.

39. See Einstein to Fritz Haber, 9 March 1921, in Buchwald et al., *The Collected Papers of Albert Einstein*, vol. 12, doc. 88, pp. 127–30.

40. See Chaim Weizmann to Einstein, 28 February 1921, in Buchwald et al., *The Collected Papers of Albert Einstein*, vol. 12, doc. 70, p. 108.

41. See Einstein to Maurice Solovine, 8 March 1921, in Buchwald et al., *The Collected Papers of Albert Einstein*, vol. 12, doc. 85, pp. 122–23.

42. See S[olomon]. Ginossar (Ginzberg). "Early Days," in *The Hebrew University of Jerusalem: Semi-Jubilee Volume* (Jerusalem: Hebrew University of Jerusalem, 1950), 72.

43. See "Circular to all Committees of the Keren Hayesod," n.d. (received on 16 January 1921 by the Central Zionist Bureau in Berlin) (CZA/KH2/86).

44. See Solomon Ginzberg to Julian Mack, 11 March 1921 (CZA/A405/123/ß).

45. See Ginossar, "Early Days," 72.

46. See Erich Marx to Einstein, 2 March 1921, in Buchwald et al., *The Collected*

Papers of Albert Einstein, vol. 12, doc. 73, pp. 111–12, and Einstein to Erich Marx, 3 March 1921, draft, ibid., doc. 75, p. 113.

47. See "Minutes of the Eighth Meeting of the Provisional Executive for Palestine," 10 March 1921 (Weizmann Archives, Weizmann Institute), and "Prof. Einstein Here, Explains Relativity," *New York Times*, 3 April 1921.

48. See Chaim Weizmann to Julian Mack, 28 February 1921 (CZA/KH1/193), and "Minutes of the First Meeting of the Weizmann Committee," 23 March 1921 (Weizmann Archives).

49. See Louis D. Brandeis to Regina W. Goldmark, 1 March 1921, in *Letters of Louis D. Brandeis*, vol. 4 *(1916–1921): Mr. Justice Brandeis*, ed. Melvin I. Urovsky and David W. Levy (Albany: State University of New York Press, 1975), 536–37.

50. For illuminating articles on this issue, see, e.g., Adelson, "Ideology and Practice in American Zionism," 1–17, and Urofsky, "Zionism," 211–21.

51. See Kurt Blumenfeld to Chaim Weizmann, 15 March 1921, in Kurt Blumenfeld, *Im Kampf um den Zionismus: Briefe aus fünf Jahrzehnten*, ed. Miriam Sambursky and Jochanan Ginat (Stuttgart: Deutsche Verlags-Anstalt, 1976), 66–68. As far as we can tell, this is one of the first instances in which Einstein's holding divergent political opinions was perceived by a critic as "naive." In later years the accusation of naivety would eventually become a recurrent theme among those who opposed his political views.

52. See Solomon Ginzberg to Julian Mack, 11 March 1921 (CZA/A405/123/ß).

53. See "Professor Einstein über die Universität Jerusalem," *Jüdische Rundschau*, 30 March 1921.

54. See Kurt Blumenfeld to Chaim Weizmann, 16 March 1921 (Weizmann Archives, Weizmann Institute).

55. See Heinrich Löwe to Einstein, 20 March 1921 (Jewish National and University Library Archives). This was approximately equivalent to $32,800, according to the rate of exchange at the time (see *Vossische Zeitung*, 19 March 1921).

56. See *Jüdische Rundschau*, 23 March 1921.

57. See Chaim Weizmann, "Speech to Fourth Session of the Plenary Session of the Action Committee of the Zionist Organization," Prague, 11 July 1921 (CZA/Z4/257/2).

58. See Julian Mack and Felix Frankfurter to Chaim Weizmann, 2 March 1921 (Weizmann Archives, Weizmann Institute).

59. Some of the leading figures of the Zionist Organization of America had close ties to the Ivy League. For example, Brandeis had studied at Harvard Law School, and Felix Frankfurter was a professor of law at Harvard.

60. See Chaim Weizmann to Einstein, 9 March 1921, in Buchwald et al., *The Collected Papers of Albert Einstein*, vol. 12, doc. 91, pp. 132–33.

61. See See Chaim Weizmann to Julian Mack, 16 March 1921 (Weizmann Archives, Weizmann Institute), and Chaim Weizmann to Julian Mack, 18 March 1921 (CZA/KH1/193).

62. See Julian Mack, Bernard Flexner, and Felix Frankfurter to Chaim Weizmann, 19 March 1921 (Weizmann Archives, Weizmann Institute).

63. See Chaim Weizmann to Julian Mack, 20 March 1921 (Weizmann Archives, Weizmann Institute).

64. See Frederic A. Hall to Arthur Compton, 11 March 1921 (Washington University Archives, Compton Papers).

65. See Chaim Weizmann to Zionist Organization of America, 28 March 1921 (Weizmann Archives, Weizmann Institute), and Chaim Weizmann to Julian Mack, 30 March 1921, in *The Letters and Papers of Chaim Weizmann*, vol. 10, Series A, *July 1920–December 1921*, ed. Bernard Wasserstein (New Brunswick, NJ: Transaction Books; Jerusalem: Israel Universities Press, 1977), 177.

66. See cable from Julian Mack to Chaim Weizmann, 1 April 1921 (Weizmann Archives, Weizmann Institute).

67. See Levin Neumann to Chaim Weizmann, 1 April 1921 (CZA/Z4/303/1)].

68. See statement by Chaim Weizmann on board the *Rotterdam*, 1 April 1921 (Weizmann Archives, Weizmann Institute).

69. See *New York American*, 3 April 1921; *New York Times*, 2 and 3 April 1921; *Yidishes Tageblatt/Jewish Daily News*, 3 April 1921; and Weizmann, *Trial and Error*, 266.

70. See cable from Weizmann, Einstein, Ussishkin, and Mossinson to *Der Tog*, 2 April 1921 (Weizmann Archives, Weizmann Institute).

71. Einstein to Carl Beck, 8 April 1921, in Buchwald et al., *The Collected Papers of Albert Einstein*, vol. 12, doc. 115, pp. 158–60.

72. See, e.g., his refusal to debate relativity, as he claimed he had come to the United States only to promote the Hebrew University (see *New York American*, 12 April 1921).

73. Einstein to Carl Beck, 8 April 1921, in Buchwald et al., *The Collected Papers of Albert Einstein*, vol. 12, doc. 115, pp. 158–60.

74. *New York Times*, 11 April 1921; *New Palestine*, 15 April 1921; and *Yidishes Togeblat / The Jewish Daily News*, 12 April 1921.

75. See "Reception in honor of Dr. Weizmann and his associates," 13 April 1921 (Weizmann Archives, Weizmann Institute), and *New York Times*, 13 April 1921.

76. See *New York Call* and *New York Times*, both 6 April 1921; *New York Call*, *New York American*, and *New York Times*, all 9 April 1921; and "Resolution Granting the Freedom of the City of New York," 16 April 1921 (Weizmann Archives, Weizmann Institute).

77. See Melivin I. Urofsky and David E. Levy, eds., *"Half Brother, Half Son": The Letters of Louis D. Brandeis to Felix Frankfurter* (Norman: University of Oklahoma Press, 1991), 73.

78. Ibid.

79. See Einstein to Judah L. Magnes, 18 April 1921, in Buchwald et al., *The Collected Papers of Albert Einstein*, vol. 12, doc. 122, pp. 164–65; and Judah L. Magnes to Einstein, 19 April 1921, ibid., doc. 124, p. 166.

80. This is even more ironic in light of Einstein's future criticism of Magnes's yielding to the influence of American philanthropists after he became chancellor of the Hebrew University.

81. See Shmarya Levin to Berthold Feiwel, 21 April 1921 (CZA/KH1/193).

82. See *New York Times*, 26 and 27 April 1921, and *Washington Post*, 27 April 1921.

83. See Louis D. Brandeis to Alice Goldmark Brandeis, 27 April 1921, in

Urovsky and Levy, *Letters of Louis D. Brandeis*, vol. 4, 554; and Louis D. Brandeis to Julian Mack, Stephen Wise, Bernard Flexner, Jacob deHaas, Felix Franfurter, and Robert Szold, 26 April 1921 (American Jewish Historical Society, Archives, Stephen Wise Collection).

84. See Bernard Flexner to Felix Frankfurter, 2 May 1921 (University of Louisville, Law Library, Brandeis School of Law, Louis D. Brandeis Papers, reel 85).

85. Louis D. Brandeis to Einstein, 29 April 1921, in Buchwald et al., *The Collected Papers of Albert Einstein*, vol. 12, doc. 128, p. 170.

86. See Louis D. Brandeis to Felix Frankfurter, 5 May 1921, quoted in Urofsky and Levy, *"Half Brother, Half Son,"* 75.

87. See Julian Mack to Leonard Stein, 15 May 1921 (CZA/A185/61), and Julian Mack to Leonard Stein, 16 May 1921 (CZA/A185/61).

88. See Urofsky and Levy, *Letters of Louis D. Brandeis*, vol. 4, 556, n. 1.

89. See Leonard Stein to Julian Mack, 20 May 1921 (Weizmann Archives, Weizmann Institute).

90. See Julian W. Mack to Leonard J. Stein, 15 May 1921 (Weizmann Archives, Weizmann Institute).

91. See Leonard Stein to Julian Mack, 20 May 1921 (CZA/A405/123/β).

92. See Julian Mack to Leonard Stein, 2 June 1921 (CZA/A405/123/β).

93. See Chaim Weizmann, "Speech to Fourth Session of the Plenary Session of the Action Committee of the Zionist Organization," Prague, 11 July 1921 (CZA/Z4/257/2).

94. Einstein to Solomon Rosenbloom, 27 April 1921, in Buchwald et al., *The Collected Papers of Albert Einstein*, vol. 12, doc. 127, pp. 167–69.

95. Einstein to Jacques Loeb, 9 May 1921, in Buchwald et al., *The Collected Papers of Albert Einstein*, vol. 12, doc. 131, p. 172.

96. See Solomon Ginzberg to Stephen Wise, 13 May 1921 (American Jewish Historical Society, Archives, Stephen Wise Collection).

97. See Max M. Warburg to Einstein, 8 February 1921, in Buchwald et al., *The Collected Papers of Albert Einstein*, vol. 12, doc. 43, p. 73, and Erich Marx to Einstein, 2 March 1921, ibid., doc. 73, pp. 111–12.

98. See Paul M. Warburg to Einstein, 13 May 1921, in Buchwald et al., *The Collected Papers of Albert Einstein*, vol. 12, doc. 133, pp. 173–74. On Paul Warburg's lack of support for Zionism, see Ron Chernow, *The Warburgs: The Twentieth-Century Odyssey of a Remarkable Jewish Family* (New York: Random House, 1993), 249.

99. See Ginossar, "Early Days," 73.

100. See Solomon Rosenbloom to Einstein, 18 May 1921, in Buchwald et al., *The Collected Papers of Albert Einstein*, vol. 12, doc. 135, pp. 175–76.

101. See Ginossar, "Early Days," 73.

102. See Julian Mack to Leonard Stein, 15 May 1921 (Weizmann Archives, Weizmann Institute).

103. See Urofsky and Levy, *"Half Brother, Half Son,"* 78.

104. See Julian Mack to Einstein, 24 May 1921, in Buchwald et al., *The Collected Papers of Albert Einstein*, vol. 12, doc. 137, p. 178.

105. See *Boston Evening Transcript*, 18 and 19 May 1921; *Boston Herald* and

Boston Post, both 19 May 1921; and Felix Frankfurter to Einstein, 28 May 1921 (Library of Congress, Felix Frankfurter Papers).

106. Feilx Frankfurter to Einstein, 17 May 1921, in Buchwald et al., *The Collected Papers of Albert Einstein*, vol. 12, doc. 134, pp. 174–75.

107. See Julian Mack to Chaim Weizmann, 30 March 1921 (Weizmann Archives, Weizmann Institute). On the emergence of antisemitic attitudes among some of the administrators at the Ivy League universities (including Harvard) during this period, see Jerome Karabel, *The Chosen: The Hidden History of Admission and Exclusion at Harvard, Yale, and Princeton* (Boston: Houghton Mifflin, 2005), 86–89.

108. See cable from Norman E. Himes, Harvard Student Liberal Club, to Einstein, 30 April 1921 (AEA, 36 223).

109. See Einstein to Felix Frankfurter, 28 May 1921, in Buchwald et al., *The Collected Papers of Albert Einstein*, vol. 12, doc. 139, pp. 179–81.

110. See Einstein to Felix Frankfurter, 29 May 1921, in Buchwald et al., *The Collected Papers of Albert Einstein*, vol. 12, doc. 140, pp. 181–82.

111. See Felix Frankfurter to Einstein, 1 July 1921, in Buchwald et al., *The Collected Papers of Albert Einstein*, vol. 12, doc. 166, pp. 208–9.

112. See "Invitation to the reception and banquet" (Library of Congress, Papers of Jacques Loeb/mm 73030429, Container 4); *New York Times*, 22 May 1921; *Jüdische Rundschau*, 1 July 1921; and Nathan Ratnoff, "What We Have Accomplished," *New Palestine*, 30 December 1921, 7.

113. See "Preliminary Statutes of the American Jewish Physicians Foundation for the Establishment and Support of the Medical Department of the Hebrew University in Jerusalem," 21 May 1921 (CZA/L12/43/I).

114. See "Memorandum on the Present Position of the University Scheme," December 1921 (CZA/KH2/86).

115. See Einstein to Michele Besso, before 30 May 1921, in Buchwald et al., *The Collected Papers of Albert Einstein*, vol. 12, doc. 141, pp. 182–83.

116. See Shmarya Levin to Berthold Feiwel, 21 April 1921 (CZA/KH1/193), and cable from Solomon Ginzberg to Keren Hayesod, London, 9 May 1921 (CZA/KH1/193).

117. See Einstein to Michele Besso, before 30 May 1921, in Buchwald et al., *The Collected Papers of Albert Einstein*, vol. 12, doc. 141, pp. 182–83.

118. See "Jewish Nationalism and Anti-Semitism: Their Relativity," *Jewish Chronicle*, 17 June 1921, and [Otto] W[arburg?] to Jewish Correspondence Bureau, Vienna, 28 June 1921 (CZA/KH1/193).

119. See Einstein to Paul Ehrenfest, 18 June 1921, in Buchwald et al., *The Collected Papers of Albert Einstein*, vol. 12, doc. 152, pp. 194–95.

120. See Solomon Ginzberg to Patrick Geddes, 23 June 1921 (CZA/L12/66).

121. See "Der Aufbau Palästinas als Aufgabe der Judenheit: Eine jüdische Massenkundgebung in Berlin," *Jüdische Rundschau*, 1 July 1921; and "Einsteins Vortrag in Berlin," *Wiener Morgenzeitung*, 1 July 1921.

122. See "Einstein über die Friedensmission des Zionismus," *Vossische Zeitung*, 5 July 1921 (Morning Edition).

123. See "Een Interview met Prof. Albert Einstein," *Nieuwe Rotterdamsche Courant*, 4 July 1921.

124. "An Interview with Prof. Albert Einstein," Appendix D, in Jannsen et al., eds., *The Collected Papers of Albert Einstein*, vol. 7, p. 624.

125. See "Einstein Declares Women Rule Here," *New York Times*, 8 July 1921.

126. See Solomon Ginzberg to Einstein, 12 July 1921, in Buchwald et al., *The Collected Papers of Albert Einstein*, vol. 12, doc. 173, p. 216; and Solomon Ginzberg to Einstein, 14 July 1921, ibid., doc. 183, pp. 224–25.

127. See Einstein to Solomon Ginzberg, 14 July 1921, in Buchwald et al., *The Collected Papers of Albert Einstein*, vol. 12, doc. 180, p. 222.

128. See Chaim Weizmann to Einstein, 7 October 1921, in Buchwald et al., *The Collected Papers of Albert Einstein*, vol. 12, doc. 259, pp. 304–5.

129. On Einstein and Haber as emblematic figures, see Stern, "Together and Apart."

130. See, e.g., Panitz, "'Washington versus Pinsk.'"

131. "An Interview with Prof. Albert Einstein." An English translation of the interview published in the *Nieuwe Rotterdamsche Courant*, 4 July 1921, in Janssen et al., *The Collected Papers of Albert Einstein*, vol. 7, appendix D.2., p. 623.

CHAPTER 4
SECULAR PILGRIM OR ZIONIST TOURIST?

1. "The Opening of the Hebrew College," *Doar Hayom*, 9 February 1923, 1; "Prof. Einstein's lecture on Mt. Scopus," *Ha'aretz*, 11 February 1923, 3.

2. See Frank, *Einstein*, 2nd ed., 199–200; Clark, *Einstein, Illustrated*, 393–95; and Fölsing, *Albert Einstein*, 531.

3. See *Jüdische Rundschau*, 30 December 1921, 740.

4. See Einstein to Paul Ehrenfest, 18 May 1922 (AEA, 10 053).

5. See Einstein to Hantaro Nagaoka, 20 May 1922 (AEA, 36 433).

6. See Einstein to Hermann Anschütz-Kaempfe, 1 July 1922 (Schleswig-Holsteinsche Landesbibliothek, Kiel, Anschütz-Kaempfe Collection); Einstein to Marie Curie, 11 July 1922 (AEA, 34 776); Einstein to Hermann Anschütz-Kaempfe, 16 July 1922 (Deutsches Museum, Munich) and Elsa Einstein to Hermann Anschütz-Kaempfe, 16 July 1922 (Deutsches Museum, Munich).

7. Einstein to German ambassador Wilhelm Solf, Tokyo, 20 December 1922 (transcription by Helen Dukas from appendix to Anneliese Griese, *Relativitätstheorie und Weltanschauung* [Berlin: Deutscher Verlag der Wissenschaften, 1967], pp. 268ff.).

8. This invitation was a written one, yet it is not extent. For its existence, see Arthur Ruppin to Zionist Executive, 16 October 1922 (CZA/A126/542). For Blumenfeld's notes of 12 October 1922, see CZA/A222/165.

9. Einstein was to travel to Batavia (Java) to show his gratitude to the joint Dutch-German expedition that had observed a solar eclipse there in one of the attempts to prove his theories (see Einstein to Paul Ehrenfest, 18 May 1922 [AEA, 10 053]). The Java leg of Einstein's trip to the Far East was subsequently canceled (see Einstein's travel diary to the Far East, Palestine and Spain, 11 January 1923 [AEA, 29 129]) and he traveled to Palestine directly following his lecture tour in Japan.

10. Notes by Blumenfeld, 12 October 1922 (CZA/A222/165).

11. See "Prof. Einstein besucht Palästina," *Zionistische Korrespondenz* 44, 6 October 1922, and "Einstein to Visit Palestine," *Latest News and Wires through Jewish Correspondence Bureau News and Telegraphic Agency, Ltd.*, 10 October 1922.

12. See Weizmann to Einstein, 6 October 1922 (AEA, 33 363); Ilse Einstein to Weizmann, 20 October 1922 (AEA, 33 363); and Weizmann, *The Letters and Papers of Chaim Weizmann*, Series A, vol. 11, "Introduction," n. 15.

13. See Arthur Ruppin to Zionist Executive, Jerusalem, 16 October 1922 (Weizmann Archives).

14. See Arthur Ruppin to Chaim Weizmann, 16 October 1922 (Weizmann Archives).

15. Barbara Wolff, Albert Einstein Archives, personal communication to the author, 25 February 2008.

16. See Einstein, travel diary, entry for 2 November 1922, and S. R. Sassoon to Israel Cohen, 9 November 1922 (CZA/Z4/2685).

17. See telegram from C. Aurivillius, secretary of the Nobel Prize Committee, Royal Swedish Academy of Sciences, to Einstein, 10 November 1922 (AEA, 30 003); and letter from C. Aurivillius to Einstein, 10 November 1922, which confirms the prize and invites Einstein to the ceremony in December (AEA, 30 004).

18. On the Yishuv during this period, see Abraham Malamat, Haim Hillel Ben-Sasson, and Shmuel Ettinger, eds. *Toldot Am Yisrael*, vol. 3, *Toldot Am Yisrael B'Et HaHadasha*, ed. Shmuel Ettinger (Tel Aviv: Dvir, 1969), 272–88; Binyamin Eliav, ed., *ha-Yishuv beYamei ha-Bait ha-Leumi* (Jerusalem: Keter, 1976); Yehoshua Porat and Ya'akov Shavit, eds., *ha-Mandat ve'ha-Bayit ha-Leumi (1917–1947)* (Jerusalem: Keter, 1982); and Moshe Lissak, ed., *Toldot ha-Yishuv ha-Yehudi b'Eretz Yisrael, Me'az ha-Aliyah ha-Rishona: Tkufat ha-Mandat ha-Briti* (Jerusalem: Israel Academy of Sciences, 1993).

19. See Einstein, travel diary, 1 February 1923 and 13 October 1922.

20. Ibid., 1 February 1923.

21. There is no confirmation in the available sources that Ginzberg's wife Rosa, who Blumenfeld had suggested should accompany Elsa Einstein during the visit, played any role in the tour.

22. See Einstein, travel diary, 2 February 1923; *Ha'aretz*, 4 February 1923; *Palestine Weekly*, 9 February 1923, 93; Fölsing, *Albert Einstein*, 529; Edwin Samuel, *A Lifetime in Jerusalem: The Memoirs of the Second Viscount Samuel* (London: Vallentine, Mitchell, 1970), 54; and Herbert Samuel, *Memoirs* (London: Cresset Press, 1945), 22.

23. See Einstein, travel diary, 3 February 1923.

24. See Einstein, travel diary, 4 February 1923.

25. See Einstein, travel diary, 5 February 1923; *Ha'aretz*, 7 and 8 February 1923; *Doar Hayom*, 7 February 1923; and *New Palestine*, 16 February 1923, 117.

26. See Boris Schatz to Einstein, 21 February 1923 (AEA, 43 261), and Einstein to Boris Schatz, 14 June 1923 (CZA/L42/36).

27. See Frederick Hermann Kisch, *Palestine Dairy* (London: V. Gollancz, 1938), 30; "As to the meeting with Prof. Einstein," *Doar Hayom*, 5 February 1923, 3; "From the Life of Jerusalem: In Honor of Prof. Einstein," *Ha'aretz*, 11 February 1923; "Reception for Prof. Einstein at the Lämel School," *Doar Hayom*,

8 February 1923; "Reception at the Lemel School," *Palestine Weekly*, 9 February 1923, 88; *New Palestine*, 9 February 1923, 101.

28. Kisch, *Palestine Diary*.

29. See Einstein, travel diary, 7 February 1923 (erroneously dated 6 February by Einstein).

30. See "The Opening of the Hebrew College," *Doar Hayom*, 9 February 1923, 1; "Prof. Einstein's Lecture on Mt. Scopus," *Ha'aretz*, 11 February 1923, 3.

31. See *Palestine Weekly*, 9 February 1923, 93.

32. See Einstein, travel diary, 8 February 1923; *Ha'aretz*, 8 February 1923, *Palestine Weekly*, 9 February 1923, 93, and 16 February 1923, 102; "In Jaffa and Tel Aviv, Prof. Einstein in Tel Aviv: At the Gymnasia," *Ha'aretz*, 9 February 1923, 4; "In Jaffa and Tel Aviv, Prof. Einstein in Tel Aviv: Next to the City Hall," *Ha'aretz*, 9 February 1923, 4; "Einstein's Reception in Tel Aviv," *Doar Hayom*, 9 February 1923, 3; ibid, 9 February 1923; *New Palestine*, 9 February 1923, 100; A. Shachori, *A Dream That Turned into a City—Tel Aviv: Birth and Growth* (Tel Aviv: Avivim, 1990), 270; "Prof. Einstein in Tel Aviv: The Public Assembly." *Ha'aretz*, 11 February 1923; "Prof. Einstein in Tel Aviv: At the Experimental Station," *Ha'aretz*, 11 February 1923; "Prof. Einstein in Tel Aviv: At the Association for Continuing Education in Science," *Ha'aretz*, 11 February 1923; "Prof. Einstein in Tel Aviv: At the Engineers Association." *Ha'aretz*, 11 February 1923; "Reception in honor of Prof. Einstein," *Ha'aretz*, 11 February 1923; "Prof. Einstein's Visit in T.A. and Its Environs," *Doar Hayom*, 11 and 12 February 1923; "Prof. Einstein's Visit in Tel Aviv and Its Environs," *Doar Hayom*, 12 and 13 February 1923; "Dvarim Shel Ma Bechach," *Ha'aretz*, 14 February 1923.

33. See Einstein, travel diary, 9 February 1923; minutes of the second semi-annual session of the Histadrut, Adar 1923. Also "Prof. Einstein at the Workers' Conference"; "Einstein in Mikve-Israel," *Ha'aretz*, 11 February 1923; "Prof. Einstein in Mikve Israel," *Doar Hayom*, 11 February 1923, 3; "Prof. Einstein in Rishon LeZion," *Doar Hayom*, 11 and 12 February 1923; "Prof. Einstein in Haifa," Doar Hayom, 12 and 14 February 1923; *New Palestine*, 16 February 1923, 117; "Eröffnung der Arbeiterkonferenz," *Jüdische Rundschau*, 16 February 1923; *Palestine Weekly*, 16 February 1923, 103.

34. See Einstein, travel diary, 9 February 1923, and "Prof. Einstein in Haifa," *Doar Hayom*, 12 and 14 February 1923.

35. See Einstein, travel diary, 10 February 1923, and Asis Domet to Einstein, 24 September 1929 (AEA, 46 055). For more on Domet, see Y. M. Landau, "The Arab Author Domet and His Relationship to the Zionist Project," *Shivat Zion*, 4 (Tashtav/Tashtaz), 264–69. I am indebted to Rafi Weiser of the National Library in Jerusalem for the identification of Domet.

36. See Einstein, travel diary, 10 February 1923; "Prof. Einstein in Haifa," *Doar Hayom*, 12 February 1923, 3; "The Einstein Story," *Doar Hayom*, 14 February 1923, 2; "Prof. Einstein at the Reali School," *Doar Hayom*, 14 February 1923.

37. Einstein, travel diary, 11 Feb. 1923; "Prof. Einstein in Haifa," *Doar Hayom*, 12 and 14 February 1923; *Palestine Weekly*, 16 February 1923, 102; "Prof. Einstein in Nahalal," *Ha'aretz*, 20 February 1923, 3.

38. See Einstein, travel diary, 12 February 1923, and "Tiberias: Professor Einstein's Visit," *Palestine Weekly*, 2 March 1923, p. 141

39. See "Prof. Einstein's Lecture at the Lämel School," *Doar Hayom*, 15

February 1923, 3; Prof. Einstein's Second Lecture," *Ha'aretz*, 16 February 1923, 4; *Palestine Weekly*, 16 February 1923, 107; Deutsches Auslandsinstitut Stuttgart to Auswärtiges Amt, 13 February 1923 (Politisches Archiv des Auswärtigen Amts: Abt. VI/Kunst und Wissenschaft Nr. 518: Vorträge des Professor Einsteins im Auslande. Bd. 1. R 64677).

40. See Einstein, travel diary, 14 February 1923.

41. See Einstein, travel diary, 2 February 1923; and Stenographic notes of Einstein's remarks while staying with Hermann Struck in Haifa (Stargardt auction catalogue, June 1974; the claim has been made that the notes are not in Einstein's hand but rather in Struck's, see "'Einstein-Stenogramme' nicht von Einstein," *Berliner Tagesspiegel* 9078, n.d.].

42. See Einstein, travel diary, 3 and 12 February 1923.

43. See Einstein, travel diary, 3 February 1923.

44. See Einstein to Maurice Solovine, "Pfingsten 1923" (University of Texas at Austin, Solovine Collection).

45. See "Einstein über das Palästinawerk," *Jüdische Rundschau*, 26 September 1924, 555; "Prof. Einstein über seine Eindücke in Palästina," *Jüdische Presszentrale Zürich*, 27 April 1923, 5; and Einstein, travel diary, 12 February 1923.

46. See, e.g., his rejection of Kurt Hiller's initiative in 1918 in chapter 1.

47. Stenographic notes of Einstein's remarks while staying with Hermann Struck in Haifa (Stargardt auction catalog, June 1974).

48. See Einstein, "My Impressions of Palestine," *New Palestine*, 11 May 1923, 341 (also published in German in *Jüdische Rundschau*, 24 April 1923, 195–96, and in Hebrew in *HaOlam*, 20 April 1923, 1), and *New Palestine*, 29 June 1923, 459.

49. See Einstein, travel diary, 8 February 1923.

50. See, e.g., Einstein to Elsa Einstein, 14 September 1920, Buchwald et al., *The Collected Papers of Albert Einstein*, vol. 10, doc. 149, p. 431), in which Einstein described Berlin as being "nerve-racking."

51. See "Einstein über das Palästinawerk," *Jüdische Rundschau*, 26 September 1924, 555.

52. See Einstein, travel diary, 3 February 1923.

53. See, e.g., Menachem Friedman, *Society and Religion: The Non-Zionist Orthodox in Eretz-Israel 1918–1936* (Jerusalem: Yad Yitzhak Ben-Zvi Publications, 1977) (Hebrew).

54. See Wolf Kaiser, "The Zionist Project in the Palestine Travel Writings of German-Speaking Jews," *Leo Baeck Institute Yearbook* 37 (1992): 271.

55. See Einstein, Travel diary, 4 and 10 February 1923, and Einstein, "My Impressions of Palestine," 341.

56. See Stenographic notes of Einstein's remarks while staying with Hermann Struck in Haifa (Stargardt auction catalog, June 1974).

57. On this issue, see, e.g., Yaakov Goldstein, "Were the Arabs Overlooked by the Zionists?," *Forum on the Jewish People, Zionism and Israel* 39 (1980): 15–30.

58. See Einstein, "My Impressions of Palestine," 341.

59. See Stenographic notes of Einstein's remarks while staying with Hermann Struck in Haifa [Stargardt auction catalog, June 1974).

60. See Einstein to Weizmann, 27 October 1923 (Weizmann Archives).

61. Einstein, "My Impressions of Palestine," 341.

62. Einstein to Maurice Solovine, "Pfingsten 1923" (University of Texas at Austin, Solovine Collection); Einstein, "My Impressions of Palestine," 341; "Reception for Prof. Einstein at the Lämel School," *Doar Hayom*, 8 February 1923; "Reception at the Lemel School," *Palestine Weekly*, 9 February 1923, 88; *New Palestine*, 9 February 1923, 101; "From the life of Jerusalem: In honor of Prof. Einstein," *Ha'aretz*, 11 February 1923.

63. See "Einstein über das Palästinawerk," 555.

64. See Einstein, travel diary, 13 February 1923.

65. See Einstein to Weizmann, 11 February 1923 (Weizmann Archives).

66. See Elsa Einstein to Hannah Ruppin, 26? February 1923 (CZA/A107/571).

67. See "Die Universität in Jerusalem: Eine Besprechung bei Prof. Einstein," *Jüdische Rundschau*, 6 April 1923, 167, and Heinrich Loewe to Dr. Sackheim, 29 July 1923 (AEA, 36 862); Heinrich Loewe to Einstein, 30 July 1923 (AEA, 36 860), letterhead.

68. See Einstein to Menachem Ussishkin, 19 June 1923 (CZA/Z4/213/4), and "Minutes of the 147th Meeting of the Executive held on June 14th, 1923, at the Berlin Office of the Zionist Organisation" (Weizmann Archives).

69. See Einstein, travel diary, 2 February 1923, and Einstein, "My Impressions of Palestine."

70. See Solomon Ginzberg to Ben-Zion Mossinson, 5 February 1923 (CZA/A45/39).

71. See Weizmann to Einstein, 4 February 1923 (CZA/KH1/163) (the text in square brackets was added by hand to a different version of the cable at the Weizmann Archives).

72. See "The Reception for Prof. Einstein at the Lämel School," *Doar Hayom*, 8 February 1923, 3.

73. See "The Einstein Story," *Doar Hayom*, 14 February 1923, 2. The council member was (?)Lifshitz; Boris Schatz to Einstein, 21 February 1923 (AEA, 43 261).

74. See "Prof. Einstein's Lecture on Mt. Scopus," *Ha'aretz*, 11 February 1923, 3.

75. See "Einstein in Eretz Yisrael," *Aspeklarya*, 1 February 1923, 5.

76. "Prof. Einstein's lecture on Mt. Scopus," *Ha'aretz*, 11 February 1923; "Einstein in Eretz Yisrael," *Aspeklarya*, 1 February 1923, 5, and *Doar Hayom*, 9 February 1923.

77. See "Prof. Einstein's Biography," *Doar Hayom*, 6 February 1923, 2.

78. See "Visit of a Great Man," *Palestine Weekly*, 9 February 1923, 81.

79. See *Palestine Weekly*, 9 February 1923, 101; *Ha'aretz*, 11 February 1923; *Doar Hayom*, 9 February 1923; "Lecture at the Jewish University," *Palestine Weekly*, 9 February 1923, 88–89; "From Behind the Disguise," *Doar Hayom*, 11 February 1923, 2–3; *Jüdische Rundschau*, 16 February 1923, 75; A. T.,"Dvarim Shel Ma Bechach," *Ha'aretz*, 14 February 1923; and *Doar Hayom*, 13 February 1923.

80. For a collection of studies on the reception of relativity in various countries, see Thomas Glick, *The Comparative Reception of Relativity* (Dordrecht: D. Reidel, 1987).

81. See "Prof. Einstein's Lecture on Mt. Scopus," *Ha'aretz*, 11 February 1923, 3.

82. See "Reception in Honor of Prof. Einstein," *Ha'aretz*, 11 February 1923.

83. See Aharon Czerniawski, *Ha'aretz*, 9 February 1923, 2; Aharon Zeitlin, "On the Notion of Time in Art," *Hatekufa*, 19, 1923, 469–76; "The Einstein Theory," *Palestine Weekly*, 9 February 1923, 83–84.

84. "From Behind the Disguise," *Doar Hayom*, 11 February 1923, 2–3, and *Ha'aretz*, 11 February 1923.

85. See A. T., "Trifles. Clowns," *Ha'aretz*, 8 February 1923, 5 (quoting from *Egyptian Gazette*, 1 February 1923).

86. See Maria Luisa Mansfield, "Jerusalem in the 19th Century: From Pilgrims to Tourism," *ARAM Periodical* (2006–7): 706, 711, and Malcolm E. Yapp, "Some European Travelers in the Middle East," *Middle Eastern Studies* 39, no. 2 (2003): 214, 217.

87. See Kaiser, "The Zionist Project," 261–62, 265.

88. See Arieh B. Saposnik, "Europe and Its Orients in Zionist Culture before the First World War," *Historical Journal* 49, no. 4 (2006): 1106, 1111–12.

89. See Kisch, *Palestine Diary*, 29–31.

90. See Yehoshua Ben-Arieh, "Perceptions and Images of the Holy Land," in *The Land That Became Israel: Studies in Historical Geography*, ed. Ruth Kark (New Haven, CT: Yale University Press, 1989), 37–53.

91. See Wistrich, "Theodor Herzl," 33.

92. See Einstein to Paul Ehrenfest, 22 March 1919, in Buchwald et al., *The Collected Papers of Albert Einstein*, vol. 9, doc. 10, p. 16.

CHAPTER 5
THE "BOTCHED UNIVERSITY"

1. See Einstein to Andor Fodor, 17 May 1928 (CZA/L12/147III).

2. See Yaacov Iram, "Curricular and Structural Developments at the Hebrew University, 1928–1948," *History of Universities* 11 (1992): 213, and Arthur A. Goren, "The View from Scopus: Judah L. Magnes and the Early Years of the Hebrew University," *Modern Jewish Studies* 45 (1996): 207.

3. See, e.g., Fölsing, *Albert Einstein*, 594–95; Isaacson, *Einstein*, 412–14; and Clark, *Einstein, Illustrated*, 228–29, 233, 236.

4. See Chaim Weizmann to Einstein, 24 October 1923 (AEA, 33 365).

5. See Protokoll der Sitzung des Palästina-Hochschul-Komitees, 11 November 1923, Berlin (CZA/L12/116/II).

6. See Leo Kohn(?) to Otto Warburg, 14 November 1923 (CZA/L12/116/II).

7. See David Eder to Einstein, 9 January 1924 (AEA, 36 869), and David Eder to Einstein, 15 January 1924 (AEA, 36 870).

8. See Einstein to Chaim Weizmann, 16 July 1924 (Chaim Weizmann Archives).

9. See Lavsky, "Beyn Hanachat Even ha-Pina li-F'ticha," 158.

10. See "Professor Einstein's visit to Argentine," in: Extract from letter from the Federacion Zionista Argentina to ?, 21 November 1924 (CZA/KH1/193); "Hebrew University Opening Celebrated in Argentine: Einstein Present," *JCB*

Bulletin, 25 April 1925; "Keren Hayesod Campaign in Argentine," *JCB Bulletin*, 26 May 1925; Einstein, "Ein Wort auf den Weg," *Jüdische Rundschau*, 3 April 1925, 244. The article was previously published under the title "The Mission of Our University," *New Palestine*, 27 March 1925, 294.

11. Minutes of the 1st meeting of the Board of Governors of the Hebrew University, 12 and 21 April 1925 (AEA, 36 884), and Iram. "Curricular and Structural Developments," 212.

12. See Weizmann to Einstein, 30 June 1925 (AEA, 33 377), and Einstein to Weizmann, 2 July 1925 (AEA, 33 378).

13. See David Eder(?) to Alfred Mond, 13 July 1925 (CZA/L12/40/II). Einstein's communication does not seem to be extant.

14. See David Eder(?) to J. L. Magnes, 14 July 1925 (AEA, 36 886); David Eder(?) to Einstein, 15 July 1925 (CZA/L12/92/2); and Einstein to Leo Kohn, 23 July 1925 (CZA/L12/147I).

15. See Ahad Ha'am, Chaim Nachman Bialik, Magnes to David Eder, 28 July 1925 (AEA, 36 891).

16. See Einstein to Magnes, 9 September 1925 (36 895).

17. See Protokoll der zweiten Sitzung des Kuratoriums der Universität Jerusalem, 23 September 1925, Munich (CZA/L12/83/1/1).

18. See Leo Kohn to Einstein, 4 October 1925 (AEA, 36 898); David Kaliski to Weizmann, 18 November 1925 (Chaim Weizmann Archives); and Weizmann to David Kaliski, 3 December 1925 (CZA/L12/92/1).

19. See Einstein to Leo Kohn, 21 November 1925 (CZA/L12/147I).

20. See Weizmann to Magnes, 15 December 1925 (Chaim Weizmann Archives).

21. See Einstein to Magnes, 29 December 1925 (CZA/L12/147I) and Jacob Klatzkin, "Gespräch mit Einstein," unpublished, undated (CZA/A40/85/1).

22. See Einstein to Felix Warburg, 1 January 1926 (CZA/L12/147I).

23. Ibid.

24. See Magnes to Einstein, 20 January 1926 (CZA/L12/95/1).

25. See Einstein to Leo Kohn, 7 February 1926 (CZA/L12/147I).

26. See Einstein to Leo Kohn, 30 January 1926 (CZA/L12/147I), and Einstein to Leo Kohn, 14 February 1926 (CZA/L12/147I).

27. See Leo Kohn(?) to Nathan Ratnoff, 19 February 1926 (CZA/L12/85/1b).

28. See Einstein to Weizmann, 9 April 1926 (AEA, 33 382), and Einstein to Magnes, 6 March 1926 (CZA/L12/147I).

29. See Weizmann to Einstein, 9 April 1926 (AEA, 33 383).

30. See Einstein to Weizmann, 15 June 1926 (CZA/L12/147I) and Einstein to Leo Kohn, 29 May 1926 (CZA/L12/147I).

31. See Einstein to Weizmann, 6 July 1926 (CZA/L12/147I).

32. See Weizmann to Einstein, 9 July 1926 (CZA/L12/147/I).

33. See Goren. "The View from Scopus," p. 218.

34. See Einstein to the Board of Governors of the Hebrew University in Jerusalem, 23 July 1926 (CZA/L12/147I).

35. See Einstein to Weizmann, 28 July 1926 (CZA/L12/147I).

36. See Eder to Magnes, 23 November 1926 (CZA/L12/85/1a).

37. See Leo Kohn to Einstein, 13 May 1927 (AEA, 36 937).

38. See Einstein to Leo Kohn, 28 May 1927 (CZA/L12/147I).

39. See Einstein to Leo Kohn, 21 August 1927 (CZA/L12/147I).

40. See Leo Kohn to Einstein, 7 September 1927 (AEA, 36 960).

41. See Einstein to Leo Kohn, 20 September 1927 (AEA, 36 963).

42. For this description of Einstein's relationship to the University, see Solomon Ginzberg to Einstein, 28 June 1929 (AEA, 37 019).

43. See Felix Danziger to Einstein, 4 October 1927 (AEA, 36 969).

44. See Einstein to Andor Fodor, 19 October 1927 (AEA, 36 984), and Einstein to Andor Fodor, 26 November 1927 (AEA, 36 990). It is not clear why Einstein included the Academic Council in this threat, as he had allegedly already resigned from that body a year previously.

45. See Einstein to Weizmann, 8 January 1928 (CZA/L12/147II).

46. See Einstein and Leopold Landau to Konrad Haenisch, 19 February 1920, in Buchwald et al., *The Collected Papers of Albert Einstein*, vol. 9, pp. 433–34; and Esther Gurland-Eliaschoff to Einstein, 4 March 1925 (AEA, 44 053).

47. See Weizmann to Felix Warburg, 17 January 1928 (CZA/L12/28); Weizmann to Einstein, 22 January 1928 (AEA, 33 389); and Einstein to Weizmann, 25 January 1928 (AEA, 33 390).

48. See Magnes to Weizmann, 13 February 1928 (AEA, 36 1001), and Einstein to Leo Kohn, 17 March 1928 (CZA/L12/147II).

49. See Magnes to Weizmann, 24 April 1928 (AEA, 36 1027).

50. See Andor Fodor, "Memorandum über die gegenwärtigen Zustände an der Hebräischen Universität Jerusalem," April(?) 1928 (AEA, 36 1024).

51. See Einstein to Leo Kohn, 17 May 1928 (CZA/L12/147III).

52. See Einstein to Andor Fodor, 17 May 1928 (CZA/L12/147III), and Andor Fodor to Kurt Blumenfeld, 23 May 1928 (AEA, 36 1038).

53. See Einstein to Weizmann, 29 May 1928 (CZA/L12/147III).

54. Ibid.

55. See Einstein to the Academic Council and the Board of Governors of the Hebrew University, 29 May 1928 (CZA/L12/147III).

56. See Felix Warburg to Einstein, 1 June 1928 (AEA, 37 001).

57. See Weizmann to Einstein, 6 June 1928 (AEA, 33 394).

58. See Einstein to Weizmann, 14 June 1928 (Weizmann Archives).

59. See Einstein to Weizmann, 20 June 1928 (AEA, 33 397).

60. See Einstein to Andor Fodor, 28 June 1928 (AEA, 37 003).

61. See Emanuel Libman to Weizmann, 9 September 1928 (CWA), and Magnes to Weizmann, 22 December 1928, translation (AEA, 36 1004).

62. See Einstein to Selig Brodetsky, 8 April 1929 (AEA, 37 018.2).

63. See Einstein to Leo Kohn, 7 July 1929 (CZA/L12/147III).

64. See Einstein to Selig Brodetsky, 7 July 1929 (CZA/A124/19).

65. See Weizmann to Einstein, 11 July 1929 (CZA/L12/147/III); Einstein to Weizmann, 1 June 1929 (AEA, 33 402); and Weizmann to Einstein, 19 August 1929 (AEA, 33 409).

66. See "Große Tage in Zürich," *Jüdische Rundschau*, 13 August 1929, 405.

67. See "Ovationen für Einstein," *Jüdische Rundschau*, 16 August 1929, 411.

68. See Einstein to Joseph Blumenfeld, 11 August 1929 (Weizmann Archives).

CHAPTER 6
"A GENUINE SYMBIOSIS"

1. See Einstein to Robert Weltsch, 16 May 1930 (AEA, 48 750).

2. See *Doar Hayom*, 28 April 1930, quoted in "Ussishkin in einer Pressekonferenz des Keren Kajemeth in Jerusalem am 27. April 1930," *Jüdische Rundschau*, 9 May 1930, 250. Intriguingly, the German translation omits the word "thousand."

3. See William L. Cleveland, *A History of the Middle East* (Boulder, CO: Westview Press, 2000), 250.

4. See Albert Einstein, "Botschaft," *Jüdische Rundschau*, 17 February 1925, 129.

5. See Klatzkin, "Mit Einstein," typescript, 5 March 1926 (CZA/A40/85/1).

6. See Klatzkin, untitled typescript, undated (CZA/A40/85/1).

7. See Josef Walk, "Das 'Deutsche Komitee Pro Palästina' 1926–1933," *Bulletin des Leo Baeck Instituts* 52 (1976): 162, 178.

8. See Einstein, United Jewish Appeal message, January 1929 (AEA, 48 646).

9. See Martin Kolinsky, "Premeditation in the Palestine Disturbances of August 1929?," *Middle Eastern Studies* 26, no. 1 (January 1990): 18.

10. See "Einsteins Brief," dated 31 August 1929, published in *Jüdische Rundschau*, 3 September 1929, 447.

11. See Einstein, travel diary, 4 and 10 February 1923, and idem, "My Impressions of Palestine," 341.

12. Einstein to Michele Besso, 4 September 1929.

13. See Stenographic notes of Einstein's remarks while staying with Hermann Struck in Haifa (Stargardt auction catalog, June 1974).

14. See Jerome, *Einstein on Israel and Zionism*, 242–243.

15. See Einstein, "Letter to the Editor," *Manchester Guardian*, 12 October 1929.

16. Einstein to the Editor of "Falastin," 28 January 1930 (AEA, 46 150).

17. Einstein to Willy Hellpach, 8 October 1929 (AEA, 46 656).

18. On the movement's policies on the Arab-Jewish conflict, see Raluca Munteanu Eddon, "Gershom Scholem, Hannah Arendt and the Paradox of 'Non-Nationalist' Nationalism," *Journal of Jewish Thought and Philosophy* 12, no. 1 (2003): 55–68; Christian Wiese, "'Doppelgesichtigkeit des Nationalismus': Die Ambivalenz zionistischer Identität bei Robert Weltsch und Hans Kohn," in *Janusfiguren: "Jüdische Heimstätte," Exil und Nation im deutschen Zionismus*, ed. Andrea Schatz and Christian Wiese (Berlin: Metropol Verlag, 2006), 213–61; and Shalom Ratsabi, *Between Zionism and Judaism: The Radical Circle in Brith Shalom, 1925–1933* (Boston: Brill, 2002).

19. Einstein to Hugo Bergmann, 27 September 1929 (AEA, 45 553).

20. Hugo Bergmann to Einstein, 8 October 1929 (CZA/A187/18a).

21. Einstein to Hugo Bergmann, 18 October 1929 (AEA, 45 556), and Hugo Bergmann to Einstein, 29 October 1929 (CZA/A187/18a).

22. Einstein to Hugo Bergmann, 6 November 1929, and Hugo Bergmann to Einstein, 14 November 1929 (CZA/A187/18a).

23. See Robert Weltsch to Einstein, 21 November 1929 (AEA, 123 152), and Einstein to Robert Weltsch, 21 November 1929 (EPPA 92 066).

24. Chaim Weizmann to Einstein, 25 November 1929 (AEA, 33 411).

25. Einstein to Hugo Bergmann, 25 November 1929 (AEA, 45 561).

26. Hugo Bergmann to Einstein, 17 December 1929 (AEA, 45 563).

27. Kurt Blumenfeld to Hugo Bergmann, 19 December 1929 (Jewish National and University Library Archives, Arc 4 1502/683).

28. Hugo Bergmann to Einstein, 24 April 1930 (AEA, 45 569).

29. Robert Weltsch to Einstein, 6 May 1930 (AEA, 48 749).

30. Einstein to Robert Weltsch, 16 May 1930 (AEA, 48 750).

31. Hugo Bergmann to Einstein, 5 June 1930 (AEA, 45 570).

32. Einstein to Hugo Bergmann, 19 June 1930 (AEA, 45 571).

33. Hugo Bergmann to Einstein, 3 July 1930 (AEA, 45 573).

34. See M.R.A., "Relativity and Propaganda," *Falastin*, 19 October 1929.

35. See *Palestine Bulletin*, 13 October 1929.

36. In contrast, Hannah Arendt advocated a position that was far more critical of the potential for colonial attitudes in Zionist ideology (see, e.g., Munteanu Eddon, "Gershom Scholem," 58.).

37. See Einstein to Azmi El-Nashashibi, Editor of *Falastin*, incomplete 1st draft, 28 December 1929, and 2nd draft, 28 January 1930 (AEA, 72 450 and 46 150).

38. See Typescript of Jewish Telegraphic Agency report, 27 January 1930 (AEA, 46 152) and "Einstein on Palestine," *Falastin*, 15 March 1930.

39. See Azmi El-Nashashibi to Einstein, 12 February 1930 (AEA, 46 153).

40. See Einstein to Azmi El-Nashashibi, undated draft (AEA, 72 508)).

41. See Einstein to Robert Weltsch, 3 March 1930 (Robert Weltsch Collection, Leo Baeck Institute, New York).

42. See Einstein to Azmi El-Nashashibi, 25 February 1930, published as "Einstein on Palestine," *Falastin*, 15 March 1930, 1. Intriguingly, Einstein addressed his letter to "Azmi El-Nashashili."

43. See Robert Weltsch to Einstein, 30 December 1929 (AEA, 48 743). Ironically, the Islam Institute has been awarding a media and peace prize in Tschelebi's name since the 1990s.

44. See Einstein to Weizmann, 25 November 1929 (AEA, 33 411).

45. See Weizmann to Einstein, 2 December 1929 (AEA, 33 413).

46. See Einstein to Weizmann, after 2 December 1929 (33 414).

47. See "Einstein über die jüdisch-arabische Verständigung," *Jüdische Rundschau*, 17 October 1930, 528.

48. See Weizmann to Einstein, 21 October 1930 (AEA, 33 415). The Passfield White Paper of the British government was issued in October 1930 and aimed to "narrow the scope of Jewish colonization" by restricting Jewish land purchases and "linking future [Jewish] immigration to its potential impact on Arab employment" (see Paul Kelemen, "Zionism and the British Labor Party: 1917–1939," *Social History* 21, no. 1 [January 1996]: 76).

49. See Hugo Bergmann to Einstein, 30 November 1929 (AEA, 45 562).

50. See Einstein to Max Warburg, 7 July 1930 (CZA/K12/50/10).

51. See Arthur Ruppin, *Briefe, Tagebücher, Erinnerungen*, edited by Schlomo Krolik (Königstein/Ts.: Jüdischer Verlag Athenäum, 1985), diary entry for 6 September 1930, p. 427.

CHAPTER 7
THE "BUG-INFESTED HOUSE"

1. Einstein to Weizmann, 7 May 1933 (AEA, 33 426).

2. See Einstein to Abraham A. Fraenkel, 14 March 1929 (AEA, 37 018.5), and Abraham A. Fraenkel to Einstein, 13 July 1930 (AEA, 37 020).

3. See Einstein to Abraham A. Fraenkel, 10 November 1930 (AEA, 37 023).

4. See Abraham A. Fraenkel to Einstein, 13 November 1930 (AEA, 37 024); Einstein to Abraham A. Fraenkel, 28 November 1930 (AEA, 37 025); Abraham A. Fraenkel to Einstein, 19 December 1930 (AEA, 37 026) and Abraham A. Fraenkel to Einstein, 17 March 1931 (AEA, 37 030).

5. See Einstein to Abraham A. Fraenkel, 27 March 1931 (AEA, 37 031.1)

6. See Abraham A. Fraenkel to Einstein, 21 April 1931 (AEA, 37 032).

7. See Einstein to Weizmann, 20 May 1931 (Weizmann Archives); Weizmann to Einstein, 21 May 1931 (AEA, 33 417); Einstein to Weizmann, 24 May 1931 (AEA, 33 418); Otto Warburg to Einstein, 17 June 1931 (AEA, 37 042); Einstein to Felix Warburg, 20 June 1931 (AEA, 37 045); Felix Warburg to Einstein, 30 June 1931 (AEA, 37 046); and Einstein to Abraham A. Fraenkel, 16 July 1931 (AEA, 37 037).

8. See Abraham A. Fraenkel to Einstein, 6 September 1932 (AEA, 37 048); Einstein to Abraham A. Fraenkel, 22 September 1932 (AEA, 37 049.1); Abraham A. Fraenkel to Einstein, 6 October 1932 (AEA, 37 050); and Einstein to Abraham A. Fraenkel, 13 October 1932 (AEA, 37 051).

9. See Weizmann to Einstein, 8 November 1932 (AEA, 33 421).

10. See Einstein to Weizmann, 20 November 1932 (AEA, 33 423.1); Einstein to Fraenkel, 16 April 1933 (AEA, 37 058); Iram, "Curricular and Structural Developments," 216–17; *The Hebrew University of Jerusalem: Its History and Development* (Jerusalem: Hebrew University of Jerusalem, 1942), 7; and Einstein to Hugo Bergmann, 6 December 1935 (AEA, 37 219).

11. See Goren, "The View from Scopus," 20; Lavsky, "Beyn Hanachat Even ha-Pina li-F'ticha ," 122–23; and Michael Heyd, "Beyn Leumiut le-Universaliut, beyn Mechkar le-Hora'a: Kavim le-Toldot Reshit ha-Universita ha-Ivrit," in *Chinuch ve-Historia: Heksherim Tarbutiyim ve-Politiyim* (Education and History: Cultural and Political Contexts), ed. R. Feldhay and Emanuel Atkes (Jerusalem: Shazar Center, 1999), 363.

12. See Goren, "The View from Scopus," 212, and David N. Myers, "A New Scholarly Colony in Jerusalem: The Early History of Jewish Studies at the Hebrew University," *Judaism* 45 (1996): 149.

13. See Yaacov Iram "The Idea of a Hebrew University 1882–1928," *New Education* 13, no. 2 (1991): 50, and Myers, "A New Scholarly Colony," 153–54.

14. See Martin Schmeiser, *Akademischer Hasard: Das Berufsschicksal des Professors und das Schicksal der deutschen Universität 1870–1920. Eine verstehend soziologische Untersuchung* (Stuttgart: Klett-Cotta, 1994), 47.

15. See Rüdiger vom Bruch, "Professoren im Deutschen Kaiserreich," in *Gelehrtenpolitik, Sozialwissenschaften und akademische Diskurse in Deutschland im 19. und 20. Jahrhundert*, ed. Björn Hofmeister and Hans-Christoph Liess (Stuttgart: Franz Steiner Verlag, 2006), 12.

16. See Iram, "Curricular and Structural Developments," 212, and Charles E. McClelland, *State, Society and University in Germany 1700–1914* (Cambridge: Cambridge University Press, 1980), 285–91.

17. See Schmeiser, *Akademischer Hasard*, 26–27.

18. See, e.g., Einstein to Marcel Grossmann, 27 February 1920 (AEA, 11 457).

19. See, e.g., Abraham Pais, *"Subtle Is the Lord . . .": The Science and the Life of Albert Einstein* (Oxford: Oxford University Press, 1982), 483.

20. See Iram, "Curricular and Structural Developments," 212–13.

21. See Matthias Middell, "Das 'Spiel mit den Maßstäben' gestern und heute: Kompatibilität oder Diversität europäischer Wissenschaftssysteme," in *". . . immer im Forschen bleiben": Rüdiger vom Bruch zum 60. Geburtstag*, ed. Marc Schalenberg and Peter Th. Walther (Stuttgart: Franz Steiner Verlag, 2004), 210.

22. See Goren, "The View from Scopus," 212.

23. See Middell, "Das 'Spiel mit den Maßstäben,'" 199, 200–202.

24. See Goren, "The View from Scopus," 203.

25. See Heyd, "Beyn Leumiut le-Universaliut," 365.

26. See Goren, "The View from Scopus," 203.

27. See Myers, "A New Scholarly Colony," 149.

28. See Goren, "The View from Scopus," 210, 214.

29. See Einstein to Weizmann, 16 July 1924 (Weizmann Archives).

30. See Alice Gallin, *Midwives to Nazism: University Professors in Weimar Germany 1925–1933* (Macon, GA: Mercer University Press, 1986), 18.

31. See "Introduction to Volume 8," in Schulmann et al., *The Collected Papers of Albert Einstein*, vol. 8, xli–xliv (see also there for references to earlier works); Einstein, "Manifesto to the Europeans," in Kox et al., *The Collected Papers of Albert Einstein*, vol. 6, doc. 8, pp. 69–71; and Einstein to Leo Arons, 12 November 1918 or later, in Schulmann et al., *The Collected Papers of Albert Einstein*, vol. 8, doc. 653, pp. 945–46.

32. See Kurt Sontheimer, "Die deutschen Hochschullehrer in der Zeit der Weimarer Republik," in *Deutsche Hochschullehrer als Elite 1815–1945*, ed. Klaus Schwabe (Boppard am Rhein: Harald Boldt Verlag, 1983), 216–17.

33. See Goren, "The View from Scopus," 210, and Lavsky, "Beyn Hanachat Even ha-Pina li-F'ticha," 124.

34. See Sven-Eric Liedman, "Institutions and Ideas: Mandarins and Non-Mandarins in the German Academic Intelligentsia," *Comparative Studies in Society and History* 28, no. 1 (1986): 134.

35. See Anita Shapira, "The Zionist Labor Movement and the Hebrew University," *Modern Jewish Studies* 45 (1996): 188, 190.

36. See Myers, "A New Scholarly Colony," 155.

37. See Roy MacLeod, "Balfour's Mission to Palestine: Science, Strategy and Vision in the Inauguration of the Hebrew University," *Studies in Contemporary Jewry* 14 (1998): 223.

38. See Heyd, "Beyn Leumiut le-Universaliut," 374.

39. Ibid., 367.

40. See Goren, "The View from Scopus," 205, and Lavsky, "Beyn Hanachat Even ha-Pina li-F'ticha," 124.

41. On Weizmann's position on these issues, see Goren, "The View from Scopus," 206–7.

42. On Weizmann's position on this issue, see Iram, "Curricular and Structural Developments," 222.

43. Quoted in Schmeiser, *Akademischer Hasard*, 37.

44. See Marita Baumgarten, *Professoren und Universitäten im 19. Jahrhundert: Zur Sozialgeschichte deutscher Geistes- und Naturwissenschaftler* (Göttingen: Vandenhoeck & Ruprecht, 1997), 19.

45. See Goren, "The View from Scopus," 210.

46. See Myers, "A New Scholarly Colony," 148.

47. See Konrad H. Jarausch, ed., *The Transformation of Higher Learning, 1860–1930: Expansion, Diversification, Social Opening, and Professionalization in England, Germany, Russia, and the United States* (Chicago: University Press of Chicago, 1983), 22.

48. See McClelland, *State, Society and University*, 32, and Hartmut Titze, "Die zyklische Überproduktion von Akademikern im 19. und 20. Jahrhundert," in *Geschichte und Gesellschaft* 10 (1984): 92, 115.

49. See Clark, *Einstein, Illustrated*, 228–29, 233, 236, and Einstein's correspondence with Pierre Comert of July 1922 (AEA, 34 778–34 782).

50. See Einstein to Weizmann, 7 May 1933 (AEA, 33 426), and Einstein to Raymond Klibansky, 18 November 1933 (AEA, 37 117).

51. See Goren, "The View from Scopus," 220.

52. For a reference by Einstein to the sacred qualities of science (*Wissenschaft*), see, e.g. Einstein to Heinrich Zangger, 8 January 1917, in *The Collected Papers of Albert Einstein*, vol. 8, doc. 287a, in vol. 10, p. 65.

CONCLUSION

1. Einstein, "Botschaft," 129.

2. See Einstein to Charlotte Weigert, 8 March 1920, in Buchwald et al., *The Collected Papers of Albert Einstein*, vol. 12, doc. 9, 343a, p. 11, and Einstein to Robert Weltsch, 16 May 1930 (AEA, 48 750).

3. See Gideon Shimoni, *The Zionist Ideology* (Hanover, NH: University Press of New England for Brandeis University Press, 1995), 5.

4. Alain Dieckhoff, *The Invention of a Nation: Zionist Thought and the Making of Modern Israel* (New York: Columbia University Press, 2003), 9.

5. It is intriguing that the prominent German Jewish intellectual Hannah Arendt also credited Blumenfeld with being instrumental in her early attraction to Zionism. See Richard J. Bernstein, "Hannah Arendt's Zionism?," in *Hannah Arendt in Jerusalem*, ed Steven E. Aschheim (Berkeley and Los Angeles: University of California Press, 2001), 196.

6. See Reinharz, "Three Generations of German Zionism.," 103, and Ratsabi, *Between Zionism and Judaism*, xii.

7. Anthony D. Smith, *The Ethnic Origin of Nations* (Oxford: Blackwell, 1986), 193.

8. This positive view of the Ostjuden as a healthy sector among world Jewry is similar to that of early Zionist ideologue Nathan Birnbaum, who held that Eastern European Jewry represented "the best hope for the regeneration of the Jewish nation" (see Robert S. Wistrich, "The Clash of Ideologies in Jewish Vienna (1880–1918): The Strange Odyssey of Nathan Birnbaum," *Leo Baeck Institute Yearbook* 33 [1988]: 230).

9. Albert Einstein to Paul Epstein, 5 October 1919, in Buchwald et al., *The Collected Papers of Albert Einstein*, vol. 9, doc. 122, pp. 180–81.

10. See Andrea Schatz and Christian Wiese, eds., *Janusfiguren: "Jüdische Heimstätte," Exil und Nation im deutschen Zionismus* (Berlin: Metropol Verlag, 2006), 14–15.

11. "Enlightenend Zionism" is the term favored by David N. Myers (see Munteanu Eddon, "Gershom Scholem," 57, fn. 16). "Proto-Zionism" is the term preferred by Dan Diner (see Dan Diner, "Ambiguous Semantics: Reflections on Jewish Political Concepts," *Jewish Quarterly Review* 98, no. 1 [Winter 2008]: 91–92).

12. See Munteanu Eddon, "Gershom Scholem, Hannah Arendt," 57.

13. See Einstein, *Autobiographical Notes*, 5.

14. See Diner, "Ambiguous Semantics," 89.

EPILOGUE

1. See "Our Debt to Zionism," *New Palestine* 28 (29 April 1938), 2: 3–4.

2. Einstein to Abba Eban, 18 November 1952 (Israel Defense Forces Archive, Givatayim, Israel).

BIBLIOGRAPHY

ARCHIVAL SOURCES

Albert Einstein Archives (AEA), Hebrew University of Jerusalem, Israel.

Albert Einstein Collection, University of Texas at Austin Archives, Austin, Texas, United States.

Archief H. A. Lorentz, Rijksarchief Noord-Holland, Haarlem, The Netherlands.

Archiv der Berlin-Brandenburgischen Akademie der Wissenschaften-Literatur-archiv, Berlin, Germany.

Archives des Instituts Internationaux de Physique et de Chimie fondes par Ernest Solvay, Fonds de l'IIPC conserve aux Archives de l'Universite Libre, Brussels, Belgium.

Arthur H. Compton Papers, Washington University Archives, St. Louis, Missouri, United States.

Central Zionist Archives (CZA), Jerusalem, Israel. The following record groups were used:

A40	Jacob Klatzkin
A45	Ben Zion Mossinson
A124	Hermann Struck
A126	Leo Motzkin
A185	Leonard Stein
A222	Kurt Blumenfeld
A405	Julian Mack
F4	Zionistische Vereinigung für Deutschland
K12	Ahad Ha-am
KH1	Keren Hayesod, Head Office, London
KH2	Keren Hayesod, Abteilung für Central Europa, Berlin
L12	Zionist Organization, Committee for the Hebrew University, London
L42	Bezalel School and Museum
Z3	Zionistisches Zentralbureau, Berlin
Z4	The Zionist Organization/The Jewish Agency for Palestine/Israel—Central Office, London

Chaim Weizmann Archives (CWA), Weizmann Institute, Rehovot, Israel.

Charlotte Weigert Papers, Privatarkiv No. 3464, Danish National Archives (Rigsarkivet), Copenhagen, Denmark.

Dean Slichter Papers, University of Wisconsin–Madison Archives, Madison, Wisconsin, United States.

Einstein Papers Project Archives (EPPA), California Institute of Technology, Pasadena, California, United States.

Felix Frankfurter Papers, Library of Congress, Washington, D.C., United States.

Hermann Anschütz-Kaempfe Collection, Deutsches Museum, Munich, Germany.

Hermann Anschütz-Kaempfe Collection, Schleswig-Holsteinsche Landesbibliothek, Kiel, Germany.

Hugo Bergmann Papers, Dept. of Manuscripts and Archives, Arc 4° 1502, Jewish National and University Library, Jerusalem, Israel.

Jacques Loeb Papers., Library of Congress, Washington, D.C., United States.

Jewish National and University Library Archives, Jerusalem, Israel.

Judah L. Magnes Papers, Central Archives for the History of the Jewish People, Jerusalem, Israel.

Louis D. Brandeis Papers, University of Louisville, Law Library, Brandeis School of Law, Louisville, Kentucky, United States.

Maurice Solovine Collection, University of Texas at Austin Archives, Austin, Texas, United States.

Michele Besso Papers, Basel, Switzerland.

Paul Ehrenfest Archive, Museum Boerhaave, Leyden, The Netherlands.

Politisches Archiv des Auswärtigen Amts, Berlin, Germany.

Stephen Wise Collection, American Jewish Historical Society, Archives, New York, United States.

PUBLISHED SOURCES

Newspapers

Berliner Illustrirte Zeitung
Berliner Tageblatt
The Boston Evening Transcript
The Boston Herald
The Boston Post
Doar Hayom
Ha'aretz
Haolam
Hatekufa
The Jewish Chronicle
Jewish Correspondence Bureau Bulletin
Jüdische Pressezentrale Zürich
Jüdische Rundschau
Korrespondenzblatt des Vereins zur Gründung und Erhaltung einer Akademie für die Wissenschaft des Judentums
Latest News and Wires through Jewish Correspondence Bureau News and Telegraphic Agency, Ltd.
Neues Wiener Journal
The New Palestine

The New York American
The New York Call
The New York Times
Neue Zürcher Zeitung
Nieuwe Rotterdamsche Courant
The Palestine Weekly
The Times (London)
Vossische Zeitung
The Washington Post
Wiener Morgenzeitung
Yidishes Tageblatt—The Jewish Daily News
Zionistische Korrespondenz

Books and Articles

Note: Works listed include both those cited and those consulted for the study but not cited.

Adelson, Howard L. "Ideology and Practice in American Zionism." In *Essays in American Zionism*, ed. Melvin I. Urofsky, 1–17. New York: Herzl Press, 1978.

Anderson, Benedict. *Imagined Communities: Reflections on the Origin and Spread of Nationalism*. London: Verso, 1983.

Aschheim, Steven E. *Brothers and Strangers: The East European Jew in German and German Jewish Consciousness, 1800–1923*. Madison: University of Wisconsin Press, 1982.

———. "The East European Jew and German Jewish Identity." *Studies in Contemporary Jewry* 1 (1984): 3–25.

———. "German History and German Jewry: Boundaries, Junctions and Interdependence." *Leo Baeck Institute Yearbook* 43 (1998): 315–322.

———. *In Times of Crisis: Essays on European Culture, Germans, and Jews*. Madison: University of Wisconsin Press, 2000.

Avineri, Shlomo. "Herzls Weg zum Zionismus—eine Neubewertung." In *100 Jahre Zionismus: Von der Verwirklichung einer Vision*, ed. Ekkehard W. Stegemann, 19–31. Stuttgart: Verlag W. Kohlhammer, 2000.

Badash, Lawrence. *Rutherford and Boltwood: Letters on Radioactivity*. New Haven, CT: Yale University Press, 1969.

Barkai, Avraham. "Sozialgeschichtliche Aspekte der deutschen Judenheit in der Zeit der Industrialisierung." *Jahrbuch des Instituts für Deutsche Geschichte* (1982): 237–60.

———. *"Wehr Dich!" Der Centralverein deutscher Staatsbürger jüdischen Glaubens (C.V.) 1893–1938*. Munich: Beck, 2002.

Barnard, Harry. *The Forging of an American Jew: The Life and Times of Judge Julian W. Mack*. New York: Herzl Press, 1974.

Barshai, Bezalel. "ha-Hachanot lif'tichat ha-Universita bi-Yerushalayim ve'Shnoteiya ha-Rishonot." *Kathedra* 29 (1983): 65–78.

———. "ha-Universitah ha-Ivrit bi-Yerushalayim, 1925–1935." *Kathedra* 53 (1990): 107–25.

Baumgarten, Marita. *Professoren und Universitäten im 19. Jahrhundert: Zur Sozial-*

geschichte deutscher Geistes- und Naturwissenschaftler. Göttingen: Vandenhoeck & Ruprecht, 1997.

Bein, Alex. *Theodore Herzl: A Biography.* Cleveland: World Publishing Co., 1962.

Beller, Mara, R. S. Cohen, and Jürgen Renn, eds. *Einstein in Context: A Special Issue of Science in Context.* Cambridge: Cambridge University Press, 1993.

Ben-Arieh, Yehoshua. "Perceptions and Images of the Holy Land." In *The Land That Became Israel: Studies in Historical Geography,* ed. Ruth Kark, 37–53. New Haven, CT: Yale University Press, 1989.

Benz, Wolfgang, Arnold Paucker, and Peter Pulzer, eds. *Jüdisches Leben in der Weimarer Republik / Jews in the Weimar Republic.* Tübingen: Mohr Siebeck, 1998.

Berger, Julius. "Der jüdische Kongreß und die deutschen Juden." *Jüdische Rundschau,* 17 December 1918, 425–26.

Bergman[n], Hugo. "Personal Remembrance of Albert Einstein." In *Logical and Epistemological Studies in Contemporary Physics,* ed. Robert S. Cohen and Marx W. Wartofsky, 388–94. Dordrecht: Reidel, 1974.

———. *Tagebücher und Briefe.* Edited by Miriam Sambursky. Königstein a. T.: Jüdischer Verlag bei Athenäum, 1985.

Berkowitz, Michael, ed. *Nationalism, Zionism and Ethnic Mobilization of the Jews in 1900 and Beyond.* Leiden: Brill, 2004.

———. *Western Jewry and the Zionist Project, 1914–1933.* Cambridge: Cambridge University Press, 1997.

Berlin, George L. "The Brandeis-Weizmann Dispute." *American Jewish Historical Quarterly* 60 (1970): 37–68.

Berlin, Isaiah. "Einstein and Israel." In *Albert Einstein: Historical and Cultural Perspectives. The Centennial Symposium in Jerusalem,* ed. Gerald Holton and Yehuda Elkana, 281–92. Princeton, NJ: Princeton University Press, 1982.

Bernstein, Richard J. "Hannah Arendt's Zionism?" In *Hannah Arendt in Jerusalem,* ed. Steven E. Aschheim, 194–202. Berkeley and Los Angeles: University of California Press, 2001

Blömer, Ursula, and Detlef Garz. "Jüdische Kindheit in Deutschland am Ende des 19. und Anfang des 20. Jahrhunderts." In *Jüdisches Kinderleben im Spiegel jüdischer Kinderbücher,* ed. Ursula Blömer and Detlef Garz, 67–79. Oldenburg: Bibliotheks- und Informationssystem der Universität Oldenburg, 1998.

———, eds. *"Wir Kinder hatten ein herrliches Leben . . .": Jüdische Kindheit und Jugend im Kaiserreich 1871–1918.* Oldenburg: Bibliotheks- und Informationssystem der Universität Oldenburg, 2000.

Blumenfeld, Kurt. "Der Zionismus." *Preussische Jahrbücher* 161/1 (July 1915): 110.

———. "Einsteins Beziehungen zum Zionismus und zu Israel." In *Helle Zeit— Dunkle Zeit: In Memoriam Albert Einstein,* ed. Carl Seelig, 74–85. Braunschweig/ Wiesbaden: Friedr. Vieweg & Sohn, 1956.

———. *Erlebte Judenfrage: Ein Viertel Jahrhundert deutscher Zionismus.* Stuttgart: Deutsche Verlags-Anstalt, 1962.

———. *Im Kampf um den Zionismus: Briefe aus fünf Jahrzehnten,* ed. Miriam Sambursky and Jochanan Ginat. Stuttgart: Deutsche Verlags-Anstalt, 1976.

Boehm, Adolf. *Die zionistische Bewegung.* Berlin: Jüdischer Verlag, 1935.

Boehm, Gero von. *Who Was Albert Einstein?* New York: Assouline, 2005.

Bokhove, Niels. "'The Entrance to the More Important': Kafka's Personal Zionism." In *Kafka, Zionism, and Beyond*, ed. Mark H. Gelber, 23–58. Tübingen: Max Niemeyer Verlag, 2004.

Borut, Jacob. "'Verjudung des Judentums': Was there a Zionist Subculture in Weimar Germany?" In *In Search of Jewish Community: Jewish Identities in Germany and Austria, 1918–1933*, ed. Michael Brenner and Derek J. Penslar, 92–114. Bloomington: Indiana University Press, 1998.

Brandeis, Louis D. *Letters of Louis D. Brandeis.* Vol. 4, *(1916–1921): Mr. Justice Brandeis.* Edited by Melvin I. Urovsky and David W. Levy. Albany: State University of New York Press, 1975.

Brenner, Michael. *The Renaissance of Jewish Culture in Weimar Germany.* New Haven, CT: Yale University Press, 1996.

Brenner, Michael, and David N. Myers, eds. *Jüdische Geschichtsschreibung heute: Themen, Positionen, Kontroversen.* Munich: Beck, 2002.

Brenner, Michael, Rainer Liedtke, and David Rechter, eds. *Two Nations: British and German Jews in Comparative Perspective.* Tübingen: Mohr Siebeck, 1999.

Brenner, Michael, and Derek Jonathan Penslar. *In Search of Jewish Community: Jewish Identities in Germany and Austria, 1918–1933.* Bloomington: Indiana University Press, 1998.

Brod, Max. *Streitbares Leben 1884–1968.* Munich: Herbig, 1969.

Cahnman, Werner J. "Village and Small-Town Jews in Germany: A Typological Study." *Leo Baeck Institute Yearbook* 19 (1974): 107–30.

Caplan, Neil. "Talking Zionism, Doing Zionism, Studying Zionism." *Historical Journal* 44, no. 4 (2001): 1083–97.

———. "The Yishuv, Sir Herbert Samuel, and the Arab Question in Palestine, 1921–25." In *Zionism and Arabism in Palestine and Israel*, ed. Elie Kedourie and Sylvia G. Haim, 1–51. London: Frank Cass, 1982.

———. "Zionism and the Arabs: Another Look at the 'New' Historiography." *Journal of Contemporary History* 36, no. 2 (2001): 345–60.

Caprara, Gian Vittorio, et al. "Personality and Politics: Values, Traits, and Political Choice." *Political Psychology* 27, no.1 (2006): 1–28.

Carver, Terell. "Methodological Issues in Writing a Political Biography." *Journal of Political Science* 20 (1992): 3–13.

Charpa, Ulrich, and Ute Deichmann. "Jewish Scientists as Geniuses and Epigones: Scientific Practice and Attitudes towards Albert Einstein, Ferdinand Cohn, Richard Goldschmidt." *Studia Rosenthaliana* 40 (2007–8): 75–108.

———. "Problems, Phenomena, Explanatory Approaches." In *Jews and Sciences in German Contexts: Case Studies from the 19th and 20th Centuries*, ed. Ulrich Charpa and Ute Deichman, 3–36. Tübingen: Mohr-Siebeck, 2007.

Chernow, Ron. *The Warburgs: The Twentieth-Century Odyssey of a Remarkable Jewish Family.* New York: Random House, 1993.

Clark, Ronald W. *Einstein: The Life and Times.* London: Hodder and Stoughton, 1973.

———. *Einstein: The Life and Times. An Illustrated Biography.* New York: Abrams, 1984.

Claussen, Detlev. "Das Genie als Autorität Über den Non-Jewish Jew Albert Einstein." In *Jüdische Geschichte als allgemeine Geschichte: Festschrift für Dan Diner zum 60. Geburtstag*, ed. Raphael Gross and Yfaat Weiss, 76–96. Göttingen: Vandenhoeck & Ruprecht, 2006.

Cleveland, William L. *A History of the Middle East*. Boulder, CO: Westview Press, 2000

Cohen, Gary. "Jews in German Society: Prague, 1860–1914." In *Jews and Germans: The Problematic Symbiosis*, ed. David Bronsen, 306–37. Heidelberg: Winter, 1979.

Cohen, Mitchell. "A Preface to the Study of Jewish Nationalism." *Jewish Social Studies* 1, no.1 (1994): 73–93.

Cohn, Henry J. "Theodor Herzl's Conversion to Zionism." *Jewish Social Studies* 32 (1970): 101–10.

Cottam, Martha, et al. *Introduction to Political Psychology*. Mahwah, NJ: Lawrence Erlbaum Associates, 2004.

de Haas, Jacob. *Louis D. Brandeis: A Biographical Sketch*. New York: Bloch, 1929.

Degen, Peter A. "Albert Einstein: Ein deutsch-jüdischer Physiker zwischen Assimilation und Zionismus." In *Den Menschen zugewandt leben: Festschrift für Werner Licharz*, ed. Ulrich Lilienthal and Lothar Stiehm, 147–58. Osnabrück: Secolo, 1999.

Dieckhoff, Alain. *The Invention of a Nation: Zionist Thought and the Making of Modern Israel*. New York: Columbia University Press, 2003.

Diner, Dan. "Ambiguous Semantics: Reflections on Jewish Political Concepts." *The Jewish Quarterly Review* 98, no. 1 (Winter 2008): 89–102.

———. "Geschichte der Juden: Paradigma einer europäischen Geschichtsschreibung" In *Gedächtniszeiten: Über jüdische und andere Geschichten*, 246–262. Munich: C. H. Beck, 2003.

———, ed. *Historiographie im Umbruch*. Gerlingen, Bleicher, 1996.

———. "Ubitiquär in Zeit und Raum: Annotationen zum jüdischen Geschichtsbewusstsein" In *Synchrone Welten: Zeiträume jüdischer Geschichte*, ed. Dan Diner, 13–34. Göttingen: Vandenhoeck & Ruprecht, 2005.

———. "Zweierlei Emanzipation: Westliche Juden und Ostjuden gegenübergestellt." In *Gedächtniszeiten: Über jüdische und andere Geschichten*, 125–34. Munich: C. H. Beck, 2003.

Dirks, Christian. "Anwalt der Sündenböcke. Albert Einstein und sein Engagement für die Ostjuden." In *Albert Einstein: Ingenieur des Universums*, ed. Jürgen Renn. Weinheim: Wiley-VCH, 2005.

Doron, Joachim. "Rassenbewusstsein und naturwissenschaftliches Denken im deutschen Zionismus während der wilhelminischen Ära." *Jahrbuch des Instituts für deutsche Geschichte* 9 (1980): 389–427.

———. "Social Concepts Prevalent in German Zionism: 1883–1914." *Studies in Zionism* 5 (1982): 1–31.

Edel, Leon. "Transference." *Biography* 7, no. 4 (1984): 283–91.

Einstein, Albert. *About Zionism: Speeches and Lectures by Professor Albert Einstein*. Translated and edited with an introduction by Leon Simon. London: Soncino Press, 1930.

———. "Assimilation und Antisemitismus." In *The Collected Papers of Albert Ein-*

stein. Vol. 7, *The Berlin Years: Writings, 1918–1921*. Edited by Michel Janssen et al., doc. 34, 289–290. Princeton, NJ: Princeton University Press, 2002.

———. *Autobiographical Notes*. La Salle, IL: Open Court Publishing Co., 1979.

———. "Autobiographische Skizze." In *Helle Zeit—Dunkle Zeit: In Memoriam Albert Einstein*, ed. Carl Seelig, 9–17. Zurich: Europa Verlag, 1956.

———. "Botschaft." *Jüdische Rundschau*, 17 February 1925, 129.

———. *The Collected Papers of Albert Einstein*. Vol. 1, *The Early Years: 1879–1902*. Edited by John Stachel et al. Princeton, NJ: Princeton University Press, 1987.

———. *The Collected Papers of Albert Einstein*. Vol. 5, *The Swiss Years: Correspondence, 1902–1914*. Edited by Martin J. Klein et al. Princeton, NJ: Princeton University Press, 1993.

———. *The Collected Papers of Albert Einstein*. Vol. 6, *The Berlin Years: Writings*. Edited by Anne J. Kox et al. Princeton, NJ: Princeton University Press, 1996.

———. *The Collected Papers of Albert Einstein*. Vol. 7, *The Berlin Years: Writings, 1918–1921*. Edited by Michel Janssen et al. Princeton, NJ: Princeton University Press, 2002.

———. *The Collected Papers of Albert Einstein*. Vol. 8, *The Berlin Years: Correspondence, 1914–1918*. Edited by Robert Schulmann et al. Princeton, NJ: Princeton University Press, 1998.

———. *The Collected Papers of Albert Einstein*. Vol. 9, *The Berlin Years: Correspondence, January 1919–April 1920*. Edited by Diana Kormos Buchwald et al. Princeton, NJ: Princeton University Press, 2004.

———. *The Collected Papers of Albert Einstein*. Vol. 10, *The Berlin Years: Correspondence, May–December 1920. Supplementary Correspondence, 1909–1920*. Edited by Diana Kormos Buchwald et al. Princeton, NJ: Princeton University Press, 2006.

———. *The Collected Papers of Albert Einstein*. Vol. 11, *Cumulative Index, Bibliography, List of Correspondence, Chronology, and Errata to Volumes 1–10*. Edited by Anne J. Kox et al. Princeton, NJ: Princeton University Press, 2009.

———. *The Collected Papers of Albert Einstein*.Vol. 12, *The Berlin Years: Correspondence, January–December 1921*. Edited by Diana Kormos Buchwald et al. Princeton, NJ: Princeton University Press, 2009.

———. "Die Zuwanderung aus dem Osten." *Berliner Tageblatt*, 30 December 1919 (Morgen-Ausgabe), 2.

———. "Ein Wort auf den Weg." *Jüdische Rundschau*, 3 April 1925, 244.

———. *Einstein on Peace*. Edited by Otto Nathan and Heinz Norden. New York: Simon and Schuster, 1960.

———. *Ideas and Opinions*. New York: Crown, 1954.

———. "In Support of Georg Nicolai." In *The Collected Papers of Albert Einstein*. Vol. 7: *The Berlin Years: Writings, 1918–1921*. Edited by Michel Janssen et al., doc. 32, 282–83. Princeton, NJ: Princeton University Press, 2002.

———. "Manifesto to the Europeans." In *The Collected Papers of Albert Einstein*. Vol. 6, *The Berlin Years: Writings*. Edited by A. J. Kox et al., doc. 8, 69–71. Princeton, NJ: Princeton University Press, 1996.

———. *Mein Weltbild*. Amsterdam: Querido Verlag, 1934.

———. *Mein Weltbild*. Edited by Carl Seelig, new and expanded ed. Zurich: Europa Verlag, 1953.

Einstein, Albert. "Meine Meinung über den Krieg." In *The Collected Papers of Albert Einstein*, Vol. 6. *The Berlin Years: Writings*. Edited by A. J. Kox et al., doc. 20, 211–13. Princeton, NJ: Princeton University Press, 1996.

———. "My Impressions of Palestine." *New Palestine*, 11 May 1923, 341.

———. *Out of My Later Years*. New York: Philosophical Library, 1950.

———. "The Mission of Our University." *New Palestine*, 27 March 1925, 294.

———. "Uproar in the Lecture Hall." In *The Collected Papers of Albert Einstein*. Vol. 7, *The Berlin Years: Writings, 1918–1921*. Edited by Michel Janssen et al., doc. 33, 284–88. Princeton, NJ: Princeton University Press, 2002.

Eliav, Binyamin, ed. *ha-Yishuv beYamei ha-Bait ha-Leumi*. Jerusalem: Keter, 1976.

Ellwein, Thomas. *Die deutsche Universität: Vom Mittelalter bis zur Gegenwart*. Königstein/Ts.: Athenäum Verlag, 1985.

Elton, Lewis. "Einstein, General Relativity, and the German Press, 1919–1920." *Isis* 77 (1986): 95–103.

Epstein, Simcha. "Einstein on Antisemitism: Highlights from 1920, 1933, and 1938." *Antisemitism International* 3–4 (2006): 80–93.

Erikson, Erik H. *Childhood and Society*, 2nd rev. ed. New York: W. W. Norton, 1963.

———. "Psychoanalytical Reflections on Einstein's Century." In *Albert Einstein: Historical and Cultural Perspectives. The Centennial Symposium in Jerusalem*, ed. Gerald Holton and Yehuda Elkana, 151–73. Princeton, NJ: Princeton University Press, 1982.

Evans, Richard J. *The Coming of the Third Reich*. London: A. Lane, 2003.

Eyck, Erich. *A History of the Weimar Republic*, trans. Harlan P. Hanson and Robert G. L. Waite. Cambridge, MA: Harvard University Press, 1967.

Eyl, Hans-Richard. "'Der letzte Zipfel': Kafka's State of Mind and the Making of the Jewish State." In *Kafka, Zionism, and Beyond*, ed. Mark H. Gelber, 59–68. Tübingen: Max Niemeyer Verlag, 2004.

Ezrahi, Yaron. "Einstein and the Light of Reason." In *Albert Einstein: Historical and Cultural Perspectives. The Centennial Symposium in Jerusalem*, ed. Gerald Holton and Yehuda Elkana, 253–78. Princeton, NJ: Princeton University Press, 1982.

Feuer, Lewis S. "The Social Roots of Einstein's Theory of Relativity: Part I." *Annals of Science* 27, no. 3 (1971): 277–98.

Finkel, Samuel B. "American Jews and the Hebrew University." In *The American Jewish Year Book 5698*, 193–201. Philadelphia: Jewish Publication Society of America, 1937.

Fischer, Klaus. "Jüdische Wissenschaftler in Weimar: Marginalität, Identität und Innovation." In *Jüdisches Leben in der Weimarer Republik / Jews in the Weimar Republic*, ed. Wolgang Benz, Arnold Paucker, and Peter Pulzer, 89–116. Tübingen: Mohr Siebeck, 1998.

Fölsing, Albrecht. *Albert Einstein: A Biography*. New York: Penguin Books, 1997.

Forman, Paul. "Scientific Internationalism and the Weimar Physicists: The Ideology and Its Manipulation in Germany after World War I." *Isis* 64, no. 2 (June 1973): 151–80.

Frank, Philipp. *Einstein: His Life and Times*. New York: Alfred A. Knopf, 1947.

———. *Einstein: His Life and Times*, 2nd ed. New York: Alfred A. Knopf, 1953.

Friedman, Alan J., and Carol C. Donley. *Einstein as Myth and Muse*. Cambridge: Cambridge University Press, 1985.

Friedman, Menachem. *Society and Religion: The Non-Zionist Orthodox in Eretz-Israel 1918–1936*. Jerusalem: Yad Yitzhak Ben-Zvi Publications, 1977 (Hebrew).

Friesel, Evyatar. "Criteria and Conception in the Historiography of German and American Zionism." *Zionism* 1, no. 2 (1980): 285–302.

———. "The German-Jewish Encounter as a Historical Problem. A Reconsideration." *Leo Baeck Institute Yearbook* 41 (1996): 263–75.

Gal, Allon. "In Search of a New Zion: New Light on Brandeis' Road to Zionism." *American Jewish History* 68, no. 1 (1978): 19–31.

———. "Independence and Universal Mission in Modern Jewish Nationalism: A Comparative Analysis of European and American Zionism (1897–1948)." *Studies in Contemporary Jewry* 5 (1989): 242–74

Gallin, Alice. *Midwives to Nazism: University Professors in Weimar Germany 1925–1933*. Macon, GA: Mercer University Press, 1986.

Gardner, Howard. "Albert Einstein: The Perennial Child." In idem, *Creating Minds: An Anatomy of Creativity Seen Through the Lives of Freud, Einstein, Picasso, Stravinsky, Eliot, Graham, and Gandhi*, 87–131. New York: Basic Books, 1993.

Gay, Peter. *Freud, Jews and Other Germans: Masters and Victims in Modernist Culture*. Oxford: Oxford University Press, 1978.

Gebhardt, Miriam. *Das Familiengedächtnis: Erinnerung im deutsch-jüdischen Bürgertum 1890 bis 1932*. Stuttgart: Franz Steiner Verlag, 1999.

Gelber, Mark H. *Melancholy Pride: Nation, Race, and Gender in the German Literature of Cultural Zionism*. Tübingen: M. Niemeyer, 2000.

Gessner, Dieter. *Die Weimarer Republik*. Darmstadt: Wissenschaftliche Buchgesellschaft, 2002.

Gilbert, Felix. "Einstein's Europe." In *Some Strangeness in the Proportion: A Centennial Symposium to Celebrate the Achievements of Albert Einstein*, ed. Harry Woolf, 13–27. Reading, MA: Addison-Wesley Publishing Co., 1980.

Gillerman, Sharon Ilise. "Between Public and Private: Family, Community and Jewish Identity in Weimar Berlin." PhD dissertation, University of California–Los Angeles, 1996.

Gilman, Sander L. "Einstein's Violin: Jews and the Performance of Identity." *Modern Judaism* 25, no. 3 (2005): 219–36.

Ginossar (Ginzberg), S[olomon]. "Early Days." In *The Hebrew University of Jerusalem: Semi-Jubilee Volume*, 71–74. Jerusalem: Hebrew University of Jerusalem, 1950.

Glick, Thomas F. "Between Science and Zionism: Einstein in Brazil." *Episteme*, Porto Alegre 9 (July–December 1999): 101–20.

———. *Einstein in Spain: Relativity and the Recovery of Science*. Princeton, NJ: Princeton University Press, 1988.

———, ed. *The Comparative Reception of Relativity*. Dordrecht, Holland, and Norwell, MA: D. Reidel, 1987.

Goenner, Hubert. *Einstein in Berlin 1914–1933*. München: Beck, 2005.

———. "The Reaction to Relativity Theory I: The Anti-Einstein Campaign in Germany in 1920." *Science in Context* 6, no. 1 (1993): 107–33.

Goenner, Hubert, and Giuseppe Castagnetti. "Albert Einstein as Pacifist and Democrat during World War I." *Science in Context* 9, no. 4 (1996): 325–86.

Goldmann, Stefan. "Im Mittelpunkt der Bildung: Zur Bildungsreligion und ihrem Berliner Tempel." In *Wissenschaften in Berlin*. Vol. 3, *Gedanken*, ed. Tilmann Buddensieg et al., 8–12. Berlin: Mann, 1987.

Goldstein, Yaakov. "Were the Arabs Overlooked by the Zionists?" *Forum on the Jewish People, Zionism and Israel* 39 (1980): 15–30.

Gordon, Neve, and Gabriel Motzkin. "Between Universalism and Particularism: The Origins of the Philosophy Department at Hebrew University and the Zionist Project." *Jewish Social Studies* 9 (2003): 99–122.

Goren, Arthur A. "The View from Scopus: Judah L. Magnes and the Early Years of the Hebrew University." *Modern Jewish Studies* 45 (1996): 203–24.

Gotlieb, Yosef. "Einstein the Zionist." *Midstream, A Monthly Jewish Review* 25, no. 6 (1979): 43–48.

Gotzmann, Andreas. "Historiography as Cultural Identity: Toward a Jewish History beyond National History." In *Modern Judaism and Historical Consciousness: Identities, Encounters, Perspectives*, ed. Andreas Gotzmann and Christian Wiese, 494–528. Leiden: Brill, 2007.

Grab, Walter, and Julius H. Schoeps, eds. *Die Juden in der Weimarer Republik: Studien zur Geistesgeschichte*, Vol. 6. Edited by Julius H. Schoeps. Stuttgart: Burg Verlag, 1986.

Griese, Anneliese. *Relativitätstheorie und Weltanschauung*. Berlin: Deutscher Verlag der Wissenschaften, 1967.

Gross, Walter. "The Zionist Students' Movement." *Leo Baeck Institute Yearbook* 4 (1959): 143–64.

Grundmann, Siegfried. *Einsteins Akte: Einsteins Jahre in Deutschland aus der Sicht der deutschen Politik*. Berlin: Springer-Verlag, 1998.

Grypma, Sonja J. "Critical Issues in the Use of Biographical Methods in Nursing History." *Nursing History Review* 13 (2005): 171–87.

Gutfreund, Hanoch. "Albert Einstein und die Hebräische Universität." In *Albert Einstein: Ingenieur des Universums*, ed. Jürgen Renn, 314–18. Weinheim: Wiley-VCH, 2005.

———. "Einstein's Jewish Identity." In *Einstein for the 21st Century: His Legacy in Science, Art, and Modern Culture*, ed. Peter L. Galison, Gerald Holton, and Silvan S. Schweber, 27–34. Princeton, NJ: Princeton University Press, 2008.

Hackeschmidt, Jörg. *Von Kurt Blumenfeld zu Norbert Elias: Die Erfindung einer jüdischen Nation*. Hamburg: Europäische Verlagsanstalt/Rotbuch Verlag, 1997.

Halpern, Ben. *A Clash of Heroes: Brandeis, Weizmann, and American Zionism*. New York: Oxford University Press, 1987.

Halpern, Ben, and Jehuda Reinharz. *Zionism and the Creation of a New Society*. New York: Oxford University Press, 1998.

Hammerstein, Notker. "Professoren in Kaisereich und Weimarer Republik und der Antisemitismus." In *Die Konstruktion der Nation gegen die Juden*, ed. Peter Alter et al., 119–36. Munich: Wilhelm Fink Verlag, 1999.

Hebrew University of Jerusalem, ed. *The Hebrew University of Jerusalem: Its History and Development*. Jerusalem: Hebrew University of Jerusalem, 1942.

Heil, Johannes. "Deutsch-jüdische Geschichte, ihre Grenzen, und die Grenzen ihrer Synthesen: Anmerkungen zur neueren Literatur." *Historische Zeitschrift* 269 (1999): 653–80.

Heilbronner, Oded, ed. *Yehude Vaimar: Hevrah be-Mashber ha-Moderniyut, 1918–1933.* Jerusalem: Magnes Press, 1994.

Hermann, Armin. "Antisemitische Vorurteile gegen Einstein." In *Reiz und Fremde jüdischer Kultur: 150 Jahre jüdische Gemeinden im Kanton Bern*, ed. Georg Eisner and Rupert Moser, 87–102. Bern: Peter Lang, 2000.

———. *Einstein: Der Weltweise und sein Jahrhundert: Eine Biographie.* Munich: Piper, 1994.

Herzig, Arno. "Zur Problematik deutsch-jüdischer Geschichtsschreibung." *Menora: Jahrbuch für deutsch-jüdische Geschichte* 1 (1990): 209–234.

Heyd, Michael. "Beyn Leumiut le-Universaliut, Beyn Mechkar le-Hora'a: Kavim le-Toldot Reshit ha-Universitah ha-Ivrit." In *Chinuch ve-Historia: Heksherim Tarbutiyim ve-Politiyim*, ed. Rivka Feldhay and Emanuel Atkes, 355–75. Jerusalem: Zalman Shazar Center for Jewish History, 1999.

Hobsbawm, E. J. *Nations and Nationalism since 1780: Programme, Myth, Reality.* Cambridge: Cambridge University Press, 1990.

Hoffmann, Banesh. "Albert Einstein." *Leo Baeck Institute Yearbook* 21 (1976): 279–88.

———. "Einstein und der Zionismus." In *Albert Einstein: Sein Einfluss auf Physik, Philosophie und Politik*, ed. P. C. Aichelburg and R. U. Sexl, 177–84. Braunschweig: Vieweg, 1979.

——— (with the collaboration of Helen Dukas). *Albert Einstein: Creator and Rebel.* New York: Viking Press, 1972.

Hoffmann, Christhard. "Between Integration and Rejection: The Jewish Community in Germany, 1914–1918." In *State, Society, and Mobilization in Europe during the First World War*, ed. John Horne, 89–104. Cambridge: Cambridge University Press, 1997.

———. "The German-Jewish Encounter and German Historical Culture." *Leo Baeck Institute Yearbook* 41 (1996): 277–90.

Holbok, Sándor. "Jüdische Kindheit zwischen Tradition und Assimilation." In *Die jüdische Familie in Geschichte und Gegenwart*, ed. Sabine Hödl and Martha Keil, 123–40. Berlin: Philo Verlagsgesellschaft, 1999.

Holmes, Virginia Iris. "Was Einstein Really a Pacifist? Einstein's Forward-Thinking, Pragmatic, Persistent Pacifism." *Peace & Change* 33, no. 2 (April 2008): 274–307.

Honigmann, Peter. "Albert Einsteins jüdische Haltung." *Tribüne* 98 (1986): 95–116.

Horowitz, Dan, and Moshe Lissak. *Origins of the Israeli Polity: Palestine under the Mandate.* Chicago: University of Chicago Press, 1978.

Horwitz, Rivka. "Voices of Opposition to the First World War among Jewish Thinkers." *Leo Baeck Institute Yearbook* 33 (1988): 233–59.

Hosking, Geoffrey, and George Schöpflin, eds. *Myths and Nationhood.* New York: Routledge, 1997.

Hutchinson, John, and Anthony D. Smith, eds. *Nationalism.* Oxford: Oxford University Press, 1994.

Hyman, Paula E. "Jüdische Familie und kulturelle Kontinuität im Elsaß des 19.

Jahrhunderts." In *Jüdisches Leben auf dem Lande: Studien zur deutsch-jüdischen Geschichte*, ed. Monika Richarz and Reinhard Rürup, 249–67. Tübingen: Mohr Siebeck, 1997.

———. "The Social Contexts of Assimilation: Village Jews and City Jews in Alsace." In *Assimilation and Community: The Jews in Nineteenth-Century Europe*, ed. Jonathan Frankel and Steven J. Zipperstein, 110–29. Cambridge: Cambridge University Press, 1992.

———. "Village Jews and Jewish Modernity: The Case of Alsace in the Nineteenth Century" In *Jewish Settlement and Community in the Modern Western World*, ed. Ronald Dotterer, Deborah Dash Moore, and Steven M. Cohen, 13–26. Selinsgrove, PA: Susquehanna University Press, 1991.

Ibald. "En Revolution I Videnskaben: Professor Einsteins epokeg ørende Teorier bekræftet. Newtons Tyngdelov omstødt." *Politiken*, 18 November 1919.

Illy, József. "Albert Einstein in Prague." *Isis* 70, no. 1 (March 1979): 76–84.

———, ed. *Albert Meets America: How Journalists Treated Genius during Einstein's 1921 Travels*. Baltimore: Johns Hopkins University Press, 2006.

Iram, Yaacov. "Curricular and Structural Developments at the Hebrew University, 1928–1948." *History of Universities* 11 (1992): 205–41.

———. "Higher Education Traditions of Germany, England, the U.S.A., and Israel: A Historical Perspective." *Paedagogica Historica* 22 (1982): 93–118.

———. "The Idea of a Hebrew University 1882–1928." *New Education* 13, no. 2 (1991): 47–52.

Isaacson, Walter. *Einstein: His Life and Universe*. New York: Simon & Schuster, 2007.

Jammer, Max. *Einstein and Religion: Physics and Theology*. Princeton, NJ: Princeton University Press, 1999.

Jarausch, Konrad H. "Die unfreien Professionen: Überlegungen zu den Wandlungsprozessen im deutschen Bildungsbürgertum 1900–1955." In *Bürgertum im 19. Jahrhundert*. Vol. 2, *Wirtschaftsbürger und Bildungsbürger*, ed. Jürgen Kocka, 124–46. Göttingen: Vandenhoeck & Ruprecht, 1995.

———. *Students, Society and Politics in Imperial Germany: The Rise of Academic Illiberalism*. Princeton, NJ: Princeton University Press, 1982.

———, ed. *The Transformation of Higher Learning, 1860–1930: Expansion, Diversification, Social Opening, and Professionalization in England, Germany, Russia and the United States*. Chicago: University of Chicago Press, 1983.

Jerome, Fred. *Einstein on Israel and Zionism: His Provocative Ideas about the Middle East*. New York: St. Martin's Press, 2009.

Jonas, Manfred. *The United States and Germany: A Diplomatic History*. Ithaca, NY: Cornell University Press, 1984.

Jungnickel, Christa, and Russell McCormmach. *Intellectual Mastery of Nature: Theoretical Physics from Ohm to Einstein*. Vol. 2, *The Now Mighty Theoretical Physics 1870–1925*. Chicago: University of Chicago Press, 1986.

Kaiser, Wolf. "The Zionist Project in the Palestine Travel Writings of German-Speaking Jews." *Leo Baeck Institute Yearbook* 37 (1992): 261–86.

Kaliski, David J. "The Physicians Committee: Its Efforts and Achievements." *New Palestine* 8, no. 13 (27 March 1925): 345.

Kampe, Norbert. "Jews and Antisemites at Universities in Imperial Germany (I).

Jewish Students: Social History and Social Conflict." *Leo Baeck Institute Yearbook* 30 (1985): 357–394.

Kaplan, Marion A. *Jewish Daily Life in Germany, 1618–1945.* Oxford: Oxford University Press, 2005.

———. "Priestess and Hausfrau: Women and Tradition in the German-Jewish Family." In *The Jewish Family. Myths and Reality,* ed. Steven M. Cohen and Paula E. Hyman, 62–81. New York: Holmes & Meier, 1986.

———. "Revealing and Concealing: Using Memoirs to Write German-Jewish History." In *Text and Context: Essays in Modern Jewish History and Historiography in Honor of Ismar Schorsch,* ed. Eli Lederhendler and Jack Wertheimer, 383–410. New York: Jewish Theological Seminary, 2005.

———. "Tradition and Transition: The Acculturaltion, Assimilation and Integration of Jews in Imperial Germany. A Gender Analysis." *Leo Baeck Institute Yearbook* 27 (1982): 3–35.

———. "*Unter uns*: Jews Socialising with Other Jews in Imperial Germany." *Leo Baeck Institute Yearbook* 48 (2003): 41–65.

Karabel, Jerome. *The Chosen: The Hidden History of Admission and Exclusion at Harvard, Yale, and Princeton.* Boston: Houghton Mifflin, 2005.

Kather, Regine. "'Die Wissenschaft ist und bleibt international': Das kosmopolitische Weltbild Albert Einsteins." *Menora: Jahrbuch für deutsch-jüdische Geschichte* 6 (1995): 65–91.

Katz, Shaul and Michael Heyd, eds. *Toldot ha-Universita ha-Ivrit biYerushalayim. Shorashim vehatchalot.* Jerusalem: Magnes Press, 1997.

Kelemen, Paul. "Zionism and the British Labor Party: 1917–1939." *Social History* 21(1) Jan. 1996: 71–87.

Kirsten, Christa, and Hans-Jürgen Treder, eds. *Albert Einstein in Berlin, 1913–1933.* Part I. *Darstellung und Dokumente.* Berlin: Akademie-Verlag, 1979.

Kisch, Frederick Hermann. *Palestine Dairy.* London: V. Gollancz, 1938.

Kocka, Jürgen, ed. *Bürger und Bürgerlichkeit im 19. Jahrhundert.* Göttingen: Vandenhoeck & Ruprecht, 1987.

———. "Losses, Gains and Opportunities: Social History Today." *Journal of Social History* 37 (Fall 2003): 21–28.

Kolb, Eberhard. *Die Weimarer Republik.* Munich: Oldenbourg, 2000.

Kolinsky, Martin. "Premeditation in the Palestine Disturbances of August 1929?" *Middle Eastern Studies* 26, no.1 (January 1990): 18–34.

Kohn, Hans. "Nationalism" In *International Encyclopedia of the Social Sciences,* ed. David L. Sills, Vol. 11, 63–70. New York: Macmillan/Free Press, 1968.

Kornberg, Jacques. "Theodor Herzl: The Zionist as Austrian Liberal." *Jerusalem Quarterly* 31 (Spring 1984): 107–17.

Kotzin, Daniel P. *An American Jewish Radical: Judah L. Magnes, American Jewish Identity, and Jewish Nationalism in America and Mandatory Palestine.* PhD dissertation, New York University, 1998.

Kox, Anne J. "Einstein and Lorentz: More Than Just Good Colleagues." *Science in Context* 6, no. 1 (1993): 43–56.

Kraul, Margret. "Bildung und Bürgerlichkeit." In *Bürgertum im 19. Jahrhundert. Deutschland im europäischen Vergleich,* ed. Jürgen Kocka, 45–73. Munich: Deutscher Taschenbuch Verlag, 1988.

Kreis, Georg. "Judenfeindschaft in der Schweiz." In *Jüdische Lebenswelt Schweiz: Vie et culture juives en Suisse. 100 Jahre Schweizerischer Israelitischer Gemeindebund—Cent ans Fédération Suisse des communautés israélites*, ed. Schweizerischer Israelitischer Gemeindebund—Fédération Suisse des communautés israélites, 423–45. Zurich: Chronos Verlag, 2004.

Lamberti, Marjorie. "From Coexistence to Conflict: Zionism and the Jewish Community in Germany, 1897–1914." *Leo Baeck Institute Yearbook* 27 (1982): 53–86.

Landau, Y. M. "The Arab author Domet and His Relationship to the Zionist Project." *Shivat Zion* 4 (Tashtav/Tashtaz): 264–69.

Laqueur, Walter. *A History of Zionism*. London: Weidenfeld and Nicolson, 1972.

Large, David Clay, and Ismar Schorsch. *The Perils of Prominence: Jews in Weimar Berlin*. New York: Leo Baeck Institute, 2001.

Lavsky, Hagit. *Before Catastrophe: The Distinctive Path of German Zionism*. Jerusalem: Magnes Press, 1996.

———. "Beyn Hanachat Even ha-Pina li-F'ticha: Yesud ha-Universita ha-Ivrit, 1918–1925." In *Toldot ha-Universita ha-Ivrit bi-Yesrushalayim. Shorashim ve-Hatchalot*, ed. Shaul Katz and Michael Heyd, 120–59. Jerusalem: Magnes Press, 1997.

———. "The Distinctive Path of German Zionism." *Studies in Contemporary Jewry* 6 (1990): 254–271.

———, ed. *Toldot ha-Universita ha-Ivrit bi-Yerushalayim: Hitbasesut Ve-Zmicha*. Part A. Jerusalem: Magnes Press, 2004/2005.

Levenson, Thomas. *Einstein in Berlin*. New York: Bantam Books, 2003.

Levinson, Daniel J. *The Seasons of a Man's Life*. New York: Alfred A. Knopf, 1988.

Liedman, Sven-Eric. "Institutions and Ideas. Mandarins and Non-Mandarins in the German Academic Intelligentsia." *Comparative Studies in Society and History* 28, no. 1 (1986): 119–44.

Liesenfeld, Cornelia. "Einsteins Zionismus—Einsteins Spinozismus." In *Jüdische Fragen als Themata der Philosophie*, ed. Sabine S. Gelhaar, 41–71. Cuxhaven/Dartford: Junghaus, 1996.

Lilienthal, Alfred M. *The Zionist Connection. What Price Peace?* New York: Dodd, Mead, 1978.

Lipstadt, Deborah E. "Louis Lipsky and the Emergence of Opposition to Brandeis, 1917–1920." In *Essays in American Zionism 1917–1948*, ed. Melvin I. Urofsky, 37–60. New York: Herzl Press, 1978.

Liska, Vivian. "Neighbors, Foes, and Other Communities: Kafka and Zionism." *Yale Journal of Criticism* 13, no. 2 (2000): 343–60.

Lissak, Moshe, ed. *Toldot ha-Yishuv ha-Yehudi b'Eretz Yisrael, Me'az ha-Aliyah ha-Rishona: Tkufat ha-Mandat ha-Briti*. Jerusalem: Israel Academy of Sciences, 1993.

Loewenberg, Peter J. "Theodor Herzl: A Psychoanalytic Study in Charismatic Political Leadership" In *The Psychoanalytic Interpretation of History*, ed. Benjamin B. Wolman, 150–91. New York: Basic Books, 1971.

Longerich, Peter. *Deutschland 1918–1933: Die Weimarer Republik: Handbuch zur Geschichte*. Hannover: Fackelträger, 1995.

Lowenstein, Steven M. "Jüdisches religiöses Leben in deutschen Dörfern: Regionale Unterschiede im 19. und frühen 20. Jahrhundert." In *Jüdisches Leben auf dem Lande: Studien zur deutsch-jüdischen Geschichte*, ed. Monika Richarz and Reinhard Rürup, 219–29. Tübingen: Mohr Siebeck, 1997.

———. "The Rural Community and the Urbanization of German Jewry." *Leo Baeck Institute Yearbook* 25 (1980): 218–236.

MacLeod, Roy. "Balfour's Mission to Palestine: Science, Strategy and Vision in the Inauguration of the Hebrew University." *Studies in Contemporary Jewry* 14 (1998): 214–31.

Malamat, Abraham, Haim Hillel Ben-Sasson, and Shmuel Ettinger, eds. *Toldot Am Yisrael*. Vol. 3, *Toldot Am Yisrael b'Et ha-Hadasha*, ed. Shmuel Ettinger. Tel Aviv: Dvir, 1969.

Mansfield, Maria Luisa. "Jerusalem in the 19th Century: From Pilgrims to Tourism." *ARAM Periodical* (2006–7): 705–14.

Mattioli, Aram. "Antisemitismus in der Geschichte der modernen Schweiz: Begriffserklärungen und Thesen." In *Antisemitismus in der Schweiz 1848–1960*, ed. Aram Mattioli, 3–22. Zurich: Orell Füssli Verlag, 1998.

Maurer, Trude. "Plädoyer für eine vergleichende Erforschung der jüdischen Geschichte Deutschlands und Osteuropas." *Geschichte und Gesellschaft* (2001): 308–26.

———. "Reife Bürger der Republik und bewusste Juden: Die jüdische Minderheit in Deutschland 1918–1933." In *Judenemanzipation—Antisemitismus—Verfolgung in Deutschland, Österreich-Ungarn, den böhmischen Ländern und in der Slowakei*, ed. Jörg K. Hoensch, 101–16. Essen: Klartext, 1999.

McClelland, Charles E. *State, Society and University in Germany 1700–1914*. Cambridge: Cambridge University Press, 1980.

Medoff, Rafael, and Chaim I. Waxman, eds. *Historical Dictionary of Zionism*. Lanham, MD: Scarecrow Press, 2000.

Meisler, Stanley. *United Nations: The First Fifty Years*. New York: Atlantic Monthly Press, 1995

Meyer, Michael A., ed. *German-Jewish History in Modern Times*. Vol. 3, *Integration in Dispute*. New York: Columbia University Press, 1997.

———, ed. *German-Jewish History in Modern Times*. Vol. 4, *Renewal and Destruction*. New York: Columbia University Press, 1998.

———. "German Jewry's Path to Normality and Assimilation: Complexities, Ironies, Paradoxes." In *Towards Normality? Acculturation and Modern German Jewry*, ed. Rainer Liedtke and David Rechter, 13–25. Tübingen: Mohr Siebeck, 2003.

———. "Jews as Jews versus Jews as Germans: Two Historical Perspectives. Introduction to Year Book XXXVI." *Leo Baeck Institute Yearbook* 36 (1991): xv–xxii.

Middell, Matthias. "Das 'Spiel mit den Maßstäben' gestern und heute: Kompatibilität oder Diversität europäischer Wissenschaftssysteme." In: ". . . *immer im Forschen bleiben.*" *Rüdiger vom Bruch zum 60. Geburtstag*, ed. Marc Schalenberg and Peter Th. Walther, 199–212. Stuttgart: Franz Steiner Verlag, 2004.

Missner, Marshall. "Why Einstein Became Famous in America." *Social Studies of Science* 15 (1985): 267–91.

Mosse, George L. "Central European Intellectuals in Palestine." *Judaism* 45, no. 2 (1996): 134–42.

Mosse, George L. *Confronting the Nation: Jewish and Western Nationalism.* Hanover, NH: University Press of New England, 1993.

———. *German Jews beyond Judaism.* Cincinnati, 1985.

———. *Germans and Jews.* New York: Howard Fertig, 1970

———. "Jewish Emancipation. Between *Bildung* and Respectibility" In *The Jewish Response to German Culture. From the Enlightenment to the Second World War*, ed. Jehuda Reinharz and Walter Schatzberg, 1–16. Hanover, NH: University Press of New England, 1985.

———. "The Jews and the Civic Religion of Nationalism." In *Confronting the Nation: Jewish and Western Nationalism*, 121–30. Hanover, NH: Brandeis University Press, 1993.

———. *The Jews and the German War Experience 1914–1918.* Leo Baeck Memorial Lecture 21. New York: Leo Baeck Institute, 1977.

Mosse, George L., and Klaus L. Berghahn, eds. *The German-Jewish Dialogue Reconsidered: A Symposium in Honor of George L. Mosse, German Life and Civilization*; vol. 20. New York: Peter Lang, 1996.

Mosse, Werner E., and Arnold Pauker, ed. *Deutsches Judentum in Krieg und Revolution, 1916–1923.* Tübingen: J.C.B. Mohr, 1971.

Moszkowski, Alexander. *Einstein: Einblicke in seine Gedankenwelt; gemeinverständliche Betrachtungen über die Relativitätstheorie und ein neues Weltsystem, entwickelt aus Gesprächen mit Einstein von Alexander Moszkowski.* Berlin: F. Fontane, 1922.

Moyn, Samuel. "German Jewry and the Question of Identity. Historiography and Theory." *Leo Baeck Institute Yearbook* 41 (1996): 291–308.

Mühsam, Paul. *Ich bin ein Mensch gewesen: Lebenserinnerungen*, ed. Ernst Kretzschmar. Gerlingen: Bleicher, 1989.

Munro, Doug. "On the Relationship between Biographer and Subject." *History Now* 8, no. 3 (2002): 12–16.

Munslow, Alun, and Robert A. Rosenstone. *Experiments in Rethinking History.* New York: Routledge, 2004.

Munteanu Eddon, Raluca. "Gershom Scholem, Hannah Arendt and the Paradox of 'Non-Nationalist' Nationalism." *Journal of Jewish Thought and Philosophy* 12, no. (2003): 55–68.

Myers, David N. "The Fall and Rise of Jewish Historicism: The Evolution of the Akademie für die Wissenschaft des Judentums (1919–1934)." *Hebrew Union College Annual* 63 (1992): 107–48.

———. "A New Scholarly Colony in Jerusalem: The Early History of Jewish Studies at the Hebrew University." *Judaism* 45 (1996): 142–59.

Nagi, Saad Z. "Nationalism." In *Encyclopedia of Sociology*, ed. Edgar F. Borgatta and Rhonda J.V. Montgomery, 1333–42. London: Macmillan Reference Books, 2000.

Neffe, Jürgen. *Einstein: Eine Biographie.* Reinbek: Rowohlt, 2005.

Nicosia, Francis R. "Resistance and Self-Defence: Zionism and Antisemitism in Inter-War Germany." *Leo Baeck Institute Yearbook* 42 (1997): 123–34.

Niewyk, Donald L. *The Jews in Weimar Germany*, 2nd ed. New Brunswick, NJ: Transaction Publishers, 2001.

Oppenheimer, Franz. "Stammesbewusstsein und Volksbewusstsein." *Jüdische Rundschau*, 25 February 1910, 86–89.

Pais, Abraham. *"Subtle Is the Lord . . .": The Science and the Life of Albert Einstein.* New York: Oxford University Press, 1982.

Panitz, Esther L. "'Washington versus Pinsk': The Brandeis-Weizmann Dispute." In *Essays in American Zionism*, ed. Melvin I. Urofsky, 77–94. New York: Herzl Press, 1978.

Parzen, Herbert. *The Hebrew University, 1925–1935.* New York: Ktav Publishing House, 1974.

———. "The Magnes-Weizmann-Einstein Controversy." *Jewish Social Studies* 32, no. 3 (1970): 187–213.

Passerini, Luisa. "Transforming Biography: From the Claim of Objectivity to Intersubjective Plurality." *Rethinking History* 4, no. 3 (2000): 413–16.

Pfister, Gertrud. "We Love Them and We Hate Them: On the Emotional Involvement and Other Challenges during Biographical Research." In *Writing Lives in Sport: Biographies, Life Histories and Methods*, ed. John Bale, Mette Krogh Christensen and Gertrud Pfister, 131–56. Aarhus: Aarhus University Press, 2004.

Picht, Clemens. "Zwischen Vaterland und Volk. Das deutsche Judentum im Ersten Weltkrieg." In *Der Erste Weltkrieg: Wirkung, Wahrnehmung, Analyse*, ed. Wolfgang Michalka, 736–55. Munich: Piper, 1994.

Pischel, Gerhard. "1833–1933: 100 Jahre jüdische Schüler am Wilhelmsgymnasium in München." In *Jüdisches Leben in München: Geschichtswettbewerb 1993/94*, ed. Landeshauptstadt München, 77–83. Munich: Buchendorfer Verlag, 1995.

Poppel, Stephen M. *Zionism in Germany 1897–1933: The Shaping of a Jewish Identity.* Philadelphia: Jewish Publication Society, 1977.

Porat, Yehoshua, and Ya'akov Shavit, eds. *HaMandat Ve'HaBayit HaLeumi (1917–1947).* Jerusalem: Keter, 1982.

Possing, Brigitte. "Biography: Historical." In *International Encyclopedia of the Social & Behavioral Sciences*, ed. Neil J. Smelser and Paul B. Baltes, 1213–17. Amsterdam: Elsevier, 2001.

———. "The Historical Biography: Genre, History and Methodology." In *Writing Lives in Sport: Biographies, Life-Histories and Methods*, ed. John Bale, Mette Krogh Christensen, and Gertrud Pfister, 17–24. Aarhus: Aarhus University Press, 2004.

Pulzer, Peter G. J. *Jews and the German State: The Political History of a Minority, 1848–1933.* Oxford: Blackwell, 1992.

———. "The First World War." In *German-Jewish History in Modern Times.* Vol. 3, *Integration in Dispute 1871–1918*, ed. Michael A. Meyer, 360–84. New York: Columbia University Press, 1997.

Pyenson, Lewis. *The Young Einstein: The Advent of Relativity.* Boston: Adam Hilger, 1985.

Quintana, Stephen M. "Children's Developmental Understanding of Ethnicity and Race." *Applied and Preventive Psychology* 7 (1998): 27–45.

Rahden, Till van. *Juden und andere Breslauer: Die Beziehung zwischen Juden, Protestanten und Katholiken in einer deutschen Grossstadt von 1860 bis 1925.* Göttingen: Vandenhoeck & Ruprecht, 2000.

Ratnoff, Nathan. "What We Have Accomplished." *New Palestine*, 30 December 1921, 7.

Ratsabi, Shalom. *Between Zionism and Judaism: The Radical Circle in Brith Shalom, 1925–1933*. Boston: Brill, 2002

Ravin, James G. "Albert Einstein and His Mentor Max Talmey." *Documenta Ophthalmologica* 94, (1997): 1–17.

Reichinstein, David. *Albert Einstein: Sein Lebensbild und seine Weltanschauung*. Prague: Selbstverlag des Verfassers, 1935

Reinharz, Jehuda. "Achad Haam und der deutsche Zionismus." *Bulletin of the Leo Baeck Institute* 61 (1982): 3–27.

———. "Advocacy and History: The Case of the Centralverein and the Zionists." *Leo Baeck Institute Yearbook* 33 (1988): 113–22.

———. *Chaim Weizmann: The Making of a Statesman*. New York: Oxford University Press, 1993.

———. *Chaim Weizmann: The Making of a Zionist Leader*. New York: Oxford University Press, 1985.

———. "Chaim Weizmann and German Jewry." *Leo Baeck Institute Yearbook* 35 (1990): 189–218.

———. *Dokumente zur Geschichte des deutschen Zionismus 1882–1933*. Tübingen, 1981.

———. "East European Jews in the *Weltanschauung* of German Zionists, 1882–1914." *Studies in Contemporary Jewry* 1 (1984): 55–95.

———. *Fatherland or Promised Land: The Dilemma of the German Jew, 1893–1914*. Ann Arbor: University of Michigan Press, 1975.

———. "Ideology and Structure in German Zionism, 1882–1933." *Jewish Social Studies* 42, no. 2 (1980): 119–46.

———. "Jewish Nationalism and Jewish Identity in Central Europe." *Leo Baeck Institute Yearbook* 37 (1992): 147–67.

———. "Three Generations of German Zionism." *Jerusalem Quarterly* 9, no. Fall (1978): 95–110.

———. "The Zionist Response to Antisemitism in Germany." *Leo Baeck Institute Yearbook* 30 (1985): 105–40.

Reinharz, Jehuda, and Walter Schatzberg, eds. *The Jewish Response to German Culture: From the Enlightenment to the Second World War*. Hanover, NH: University Press of New England, 1985.

Reinharz, Jehuda, and Anita Shapira, eds. *Essential Papers on Zionism*. New York: New York University Press, 1996.

Reiser, Anton [Rudolf Kayser]. *Albert Einstein: A Biographical Portrait*. New York: A. & C. Boni, 1930.

Renn, Jürgen, Giuseppe Castagnetti, and Peter Damerow. "Albert Einstein: Alte und neue Kontexte in Berlin." In *Die Königlich Preußische Akademie der Wissenschaften zu Berlin im Kaiserreich*, ed. Jürgen Kocka, 333–54. Berlin: Akademie Verlag, 1999.

Richarz, Monika. *Jüdisches Leben in Deutschland: Selbstzeugnisse zur Sozialgeschichte*. Vol. 3, *1918–1945*. Stuttgart: Deutsche Verlags-Anstalt, 1982.

———. "Ländliches Judentum als Problem der Forschung." In *Jüdisches Leben auf dem Lande: Studien zur deutsch-jüdischen Geschichte*, ed. Monika Richarz and Reinhard Rürup, 1–8. Tübingen: Mohr Siebeck, 1997.

Ringer, Fritz. *The Decline of the German Mandarins: The German Academic Community, 1890–1933*. Cambridge, MA: Harvard University Press, 1969.

Roberts, Brian. *Biographical Research*. Buckingham/Philadelphia: Open University Press, 2002.

Roberts, Keith A. *Religion in Sociological Perspective*. Belmont, CA: Wadsworth/Thomson, 2004.

Rogger, Franziska. *Einsteins Schwester: Maja Einstein—Ihr Leben und ihr Bruder Albert*. Zürich: Verlag Neue Zürcher Zeitung, 2005.

Rose, Norman. *Chaim Weizmann: A Biography*. New York: Viking, 1986.

Rosenkranz, Ze'ev. "Albert Einstein's Travel Diary to Palestine, February 1923." *Arkhiyyon: Reader in Archives Studies and Documentation* 10–11 (1999): 184–207 (in Hebrew).

———."Einstein and the German Zionist Movement." In *Albert Einstein: Ingenieur des Universums*. Catalog to the Exhibition in Berlin, 302–7. Berlin: Max-Planck-Institut für Wissenschaftsgeschichte, 2005.

———. "Das Genie als nationale Symbolfigur: Albert Einstein und die zionistische Bewegung in der Weimarer Republik." In *Relativ jüdisch: Albert Einstein Jude, Zionist, Nonkonformist*, ed. Christian Dirks and Hermann Simon, 98–111. Berlin: Stiftung Neue Synagoge Berlin-Centrum Judaicum, 2005.

———. "The Genius as National Icon: Albert Einstein, Zionism and Social Responsibility." *Physics & Society Newsletter*, October 2005, 27–37.

———."The German Zionists' Perception of Albert Einstein, 1919–1921." In: *Weimar Jewry and the Crisis of Modernization, 1918–1933*, ed. Oded Heilbronner, 108–21. Jerusalem: Magnes Press, 1994 (in Hebrew).

———." 'Lofty Spiritual Centre' or 'Bug-Infested House?' Albert Einstein's Involvement in the Affairs of the Hebrew University, 1919–1935." In: *The History of the Hebrew University of Jerusalem: Origins and Beginnings*, ed. Shaul Katz and Michael Heyd, 386–94. Jerusalem: Magnes Press, 1997 (in Hebrew).

Rowe, David E., and Robert Schulmann, eds. *Einstein on Politics: His Private Thoughts and Public Stands on Nationalism, Zionism, War, Peace, and the Bomb*. Princeton, NJ: Princeton University Press, 2007.

Rürup, Reinhard. "An Appraisal of German-Jewish Historiography." *Leo Baeck Institute Yearbook* 35 (1990): xv–xxiv.

Ruppin, Arthur. *Briefe, Tagebücher, Erinnerungen*. Edited by Schlomo Krolik. Königstein/Ts.: Jüdischer Verlag Athenäum, 1985.

Sachar, Howard M. *A History of Israel. From the Rise of Zionism to Our Time*. New York: Alfred A. Knopf, 1976.

Samuel, Edwin. *A Lifetime in Jerusalem. The Memoirs of the Second Viscount Samuel*. London: Vallentine, Mitchell, 1970.

Samuel, Herbert. *Memoirs*. London: Cresset Press, 1945.

Sapiro, Virginia. "Political Socialization during Adulthood: Clarifying the Political Time of Our Lives." *Research in Micropolitics* 4 (1994): 197–223.

Saposnik, Arieh B. "Europe and Its Orients in Zionist Culture before the First World War." *Historical Journal* 49, no. 4 (2006): 1105–23.

Schatz, Andrea, and Christian Wiese, eds. *Janusfiguren: "Jüdische Heimstätte," Exil und Nation im deutschen Zionismus*. Berlin: Metropol Verlag, 2006.

Scheideler, Britta. "The Scientist as Moral Authority: Albert Einstein between Elitism and Democracy, 1914–1933." *Historical Studies in the Physical and Biological Sciences* 32, no. 2 (2002): 319–46.

Schmeiser, Martin. *Akademischer Hasard: Das Berufsschicksal des Professors und das Schicksal der deutschen Universität 1870–1920. Eine verstehend soziologische Untersuchung.* Stuttgart: Klett-Cotta, 1994.

Schmidt, Sarah. "The Zionist Conversion of Louis D. Brandeis." *Jewish Social Studies* 37 (1975): 18–34.

Schulmann, Robert. "Albert Einstein—ein politischer Jude." In *Relativ jüdisch: Albert Einstein. Jude, Zionist, Nonkonformist,* ed. Christian Dirks and Hermann Simon, 83–97. Berlin: text.verlag, 2005.

———. "Einstein at the Patent Office: Exile, Salvation, or Tactical Retreat?" *Science in Context* 6, no. 1 (1993): 17–24.

———. "Einstein's Swiss Years." In *Albert Einstein—Chief Engineer of the Universe: One Hundred Authors for Einstein,* ed. Jürgen Renn, 156–60. Berlin: Wiley-VCH, 2005.

———. "From Periphery to Center: Einstein's Path from Bern to Berlin (1902–1914)." In *No Truth Except in the Details,* ed. A. J. Kox and D. M. Siegel, 259–71. Dordrecht: Kluwer, 1995.

Schwarzenbach, Alexis. *Das verschmähte Genie: Albert Einstein und die Schweiz.* München: Deutsche Verlags-Anstalt, 2005.

Schweber, Silvan S. "Einstein and Oppenheimer: Interactions and Intersections." *Science in Context* 19, no. 4 (2006): 513–59.

Schweid, Eliezer. "The Construction and Deconstruction of Jewish Zionist Identity." In *Ideology and Jewish Identity in Israeli and American Literature,* ed. Emily Miller Budick, 23–43. Albany: State University of New York Press, 2001.

Sears, David O., and Sheri Levy. "Childhood and Adult Political Development." In *Oxford Handbook of Political Psychology,* ed. David O. Sears et al. Oxford: Oxford University Press, 2003.

Sears, David O., et al. *Oxford Handbook of Political Psychology.* Oxford: Oxford University Press, 2003.

Shachori, Ilan. *A Dream That Turned into a City—Tel Aviv: Birth and Growth.* Tel Aviv: Avivim, 1990.

Shapira, Anita. "The Zionist Labor Movement and the Hebrew University." *Modern Jewish Studies* 45 (1996): 183–98.

Shimoni, Gideon. *The Zionist Ideology.* Hanover, NH: University Press of New England for Brandeis University Press, 1995.

Sieg, Ulrich. "'Nothing more German than the German Jews'? On the Integration of a Minority in a Society at War." In *Towards Normality? Acculturation and Modern German Jewry,* ed. Rainer Liedtke and David Rechter, 201–16. Tübingen: Mohr Siebeck, 2003.

Silbermann, Alphons. "Deutsche Juden oder jüdische Deutsche? Zur Identität der Juden in der Weimarer Republik." In *Juden in der Weimarer Republik,* ed. Walter Grab and Julius H. Schoeps, 347–55. Stuttgart: Burg Verlag, 1986.

Smith, Anthony D. *The Ethnic Origin of Nations.* Oxford: Blackwell, 1986

———. *Myths and Memories of the Nation.* Oxford: Oxford University Press, 1999.

———. *Nationalism and Modernism: A Critical Survey of Recent Theories of Nations and Nationalism*. London: Routledge, 1998.

Sontheimer, Kurt. "Die deutschen Hochschullehrer in der Zeit der Weimarer Republik." In *Deutsche Hochschullehrer als Elite 1815–1945*, ed. Klaus Schwabe, 215–25. Boppard am Rhein: Harald Boldt Verlag, 1983.

Sorkin, David. "Emancipation and Assimilation: Two Concepts and their Application to German-Jewish History." *Leo Baeck Institute Yearbook* 35 (1990): 17–33.

———. *The Transformation of German Jewry, 1780–1840*. New York: Oxford University Press, 1987.

Spector, Scott. "Forget Assimilation: Introducing Subjectivity to German-Jewish History." *Jewish History* 20 (2006): 349–61.

Stachel, John. "Einstein's Jewish Identity." In *Einstein from "B" to "Z,"* 57–86. Boston: Birkhäuser, 2002.

Stern, Fritz. "Einstein's Germany." In *Albert Einstein: Historical and Cultural Perspectives. The Centennial Symposium in Jerusalem*, ed. Gerald Holton and Yehuda Elkana, 319–43. Princeton, NJ: Princeton University Press, 1982.

———. "Together and Apart: Fritz Haber and Albert Einstein." In *Einstein's German World*, 59–164. Princeton, NJ: Princeton University Press, 1999.

Stone, Lilo. "German Zionists in Palestine before 1933." *Journal of Contemporary History* 32, no. 2 (1997): 171–86.

Strikovski, Arie, ed. *Albert Einstein VeYahaduto*. Jerusalem: Ministry of Education, Culture and Sport, Department of Torah Education, 2005–6.

Switzer-Rakos, Kennee. "Albert Einstein's Concept of the Jewish State." *Midstream* 31 (1985): 19–22.

Tal, Uriel. "Jewish and Universal Social Ethics in the Life and Thought of Albert Einstein." In *Albert Einstein: Historical and Cultural Perspectives. The Centennial Symposium in Jerusalem*, ed. Gerald Holton and Yehuda Elkana, 297–318. Princeton, NJ: Princeton University Press, 1982.

Talmey, Max. *The Relativity Theory Simplified and the Formative Period of Its Inventor*. New York: Falcon Press, 1932.

Tänzer, Arnold. "Der Stammbaum Prof. Albert Einsteins." In *Jüdische Familienforschung*. Sonderdruck aus dem Heft 28 vom Dezember 1931, 419–21.

Tauber, Gerald E. "Einstein and Zionism." In *Einstein: A Centenary Volume*, ed. A. P. French, 199–207. Cambridge, MA: Harvard University Press, 1979.

———. "Einstein on Zionism, Arabs and Palestine: A Collection of Papers, Letters, and Speeches." Typescript. Tel Aviv University, Tel Aviv, 1979.

Titze, Hartmut. "Die zyklische Überproduktion von Akademikern im 19. und 20. Jahrhundert." *Geschichte und Gesellschaft* 10 (1984): 92–121.

Tolmasquim, Alfredo Tiomno. "Einstein's Visit to the Emerging Jewish Community of Rio De Janeiro in 1925." *Judaica latinoamericana: Estudios histâorico-sociales* 4 (2001): 115–34.

Tramer; Hans. "Kurt Blumenfeld. Seine Lehre und seine Leistung" In Blumenfeld, *Erlebte Judenfrage*, 13.

Ulitzur, A. *Foundations: A Survey of 25 Years of activity of the Palestine Foundation Fund Keren Hayesod. Facts and Figures 1921–1946*. Jerusalem: Jerusalem Press, 1946.

Urofsky, Melvin I. *American Zionism from Herzl to the Holocaust*. Garden City, NY: Doubleday, 1975.

———. "Zionism: An American Experience."*American Jewish Historical Quarterly* 63 (1974): 211–21.

Urofsky, Melvin I., and David W. Levy, eds. *"Half Brother, Half Son": The Letters of Louis D. Brandeis to Felix Frankfurter*. Norman: University of Oklahoma Press, 1991.

———, eds. *Letters of Louis D. Brandeis*. Vol. 4 (1916–1921), *Mr. Justice Brandeis*. Albany: State University of New York Press, 1975.

Volkov, Shulamit. "The Ambivalence of *Bildung*: Jews and Other Germans." In *The German-Jewish Dialogue Reconsidered: A Symposium in Honor of George L. Mosse*, ed. Klaus L. Berghahn, 81–97. New York: Peter Lang, 1996.

———. "Antisemitism as a Cultural Code: Reflections on the History and Historiography of Antisemitism in Imperial Germany." *Leo Baeck Institute Yearbook* 23 (1978): 25–46.

———. *Das jüdische Projekt der Moderne: Zehn Essays*. München: C. H. Beck, 2001.

———. *Die Juden in Deutschland 1780–1918*. Munich: R. Oldenbourg Verlag, 1994.

———. "German Jewish History: Back to *Bildung* and Culture?" In *What History Tells Us: George L. Mosse and the Culture of Modern Europe*, ed. Stanley G. Payne, David J. Dorkin, and John S. Tortorice. Madison: University of Wisconsin Press, 2004.

———. "How German and How Jewish Were the German Jews? Reflections on the Problem of Identity." In *Text and Context: Essays in Modern Jewish History and Historiography in Honor of Ismar Schorsch*, ed. Eli Lederhendler and Jack Wertheimer, 411–31. New York: Jewish Theological Seminary, 2005.

———. "Juden als wissenschaftliche "Mandarine" im Kaiserreich und in der Weimarer Republik: Neue Überlegungen zu sozialen Ursachen des Erfolgs jüdischer Naturwissenschaftler." *Archiv für Sozialgeschichte* 37 (1997): 1–18.

———. "Jüdische Assimilation und jüdische Eigenart im Deutschen Reich: Ein Versuch." *Geschichte und Gesellschaft* 9 (1983): 331–48.

———. "Reflections on German-Jewish Historiography: A Dead End or a New Beginning?" *Leo Baeck Institute Yearbook* 41 (1996): 309–20.

———. "Reflexionen zum 'Modernen' und zum 'Uralten' jüdischen Nationalismus." In *Deutschlands Weg in die Moderne; Politik, Gesellschaft und Kultur im 19. Jahrhundert*, ed. Wolfgang and Harm-Hinrich Brandt Hardtwig. München: C. H. Beck, 1993.

———. "Soziale Ursachen des jüdischen Erfolgs in der Wissenschaft. Juden im Kaiserreich." *Historische Zeitschrift* 45 (1987): 315–42.

vom Bruch, Rüdiger. "Professoren im Deutschen Kaiserreich." In *Gelehrtenpolitik, Sozialwissenschaften und akademische Diskurse in Deutschland im 19. und 20. Jahrhundert*. Ed. Björn Hofmeister and Hans-Christoph Liess, 11–25. Stuttgart: Franz Steiner Verlag, 2006.

Wagner, Siegfried. "Wie aus der Einsteinschen Fabrik Münchens Endzeitsynagoge wurde." *Tribüne* 28 (1989): 167–74.

Walk, Josef. "Das 'Deutsche Komitee Pro Palästina' 1926–1933." *Bulletin des Leo Baeck Instituts* 15, no. 52 (1976): 162–93.

Weiss, Yfaat. "Identity and Essentialism: Race, Culture, and the Jews in the Late Nineteenth and Early Twentieth Centuries." In *German History from the Margins*, ed. Neil Gregor, Nils Roemer, and Mark Rosemen, 49–69. Bloomington: Indiana University Press, 2006.

———. "'Ostjudentum' als Konzept und 'Ostjuden' als Präsenz im deutschen Zionismus." In *Janusfiguren: "Jüdische Heimstätte," Exil und Nation im deutschen Zionismus*, ed. Andrea Schatz and Christian Wiese, 149–65. Berlin: Metropol Verlag, 2006.

———. "'Wir Westjuden haben jüdisches Stammesbewusstsein, die Ostjuden jüdisches Volksbewusstsein': Der deutsch-jüdische Blick auf das polnische Judentum in den beiden ersten Jahrzehnten des 20. Jahrhunderts." *Archiv für Sozialgeschichte* 37 (1997): 157–78.

Weizmann, Chaim. *The Letters and Papers of Chaim Weizmann*. Vol. 10, Series A. July 1920–December 1921. Edited by Bernard Wasserstein and Joel S. Fishman. New Brunswick, NJ: Transition Books; Jerusalem: Israel Universities Press, 1977.

———. *The Letters and Papers of Chaim Weizmann*. Vol. 11. Series A. January 1922–July 1923. Edited by Bernard Wasserstein and Joel S. Fishman. New Brunswick, NJ: Transaction Books; Jerusalem: Israel Universities Press, 1977.

———. *Trial and Error: The Autobiography of Chaim Weizmann*. New York: Harper & Brothers, 1949.

Wheatcroft, Geoffrey. *The Controversy of Zion: Jewish Nationalism, the Jewish State, and the Unresolved Jewish Dilemma*. Reading, MA: Addison-Wesley, 1996.

Wiese, Christian. "'Doppelgesichtigkeit des Nationalismus': Die Ambivalenz zionistischer Identität bei Robert Weltsch und Hans Kohn." In *Janusfiguren: "Jüdische Heimstätte," Exil und Nation im deutschen Zionismus*, ed. Andrea Schatz and Christian Wiese, 213–61. Berlin: Metropol Verlag, 2006.

Winteler-Einstein, Maja. "Albert Einstein—Beitrag für sein Lebensbild," 1924. Typescript, [Besso Nachlaß, Basel].

———. "Albert Einstein—Beitrag für sein Lebensbild." Excerpt. In *The Collected Papers of Albert Einstein*. Vol. 1, *The Early Years: 1879–1902*. Edited by John Stachel et al., xlviii–lxvi. Princeton, NJ: Princeton University Press, 1987.

Wistrich, Robert S. "The Clash of Ideologies in Jewish Vienna (1880–1918): The Strange Odyssey of Nathan Birnbaum." *Leo Baeck Institute Yearbook* 33 (1988): 201–30.

———. "Theodor Herzl: Zionist Icon, Myth-Maker, and Social Utopian." In *The Shaping of Israeli Identity: Myth, Memory and Trauma*, ed. Robert Wistrich and David Ohana, 1–37. London: Frank Cass, 1995.

Woolf, Harry, ed. *Some Strangeness in the Proportion: A Centennial Symposium to Celebrate the Achievements of Albert Einstein*. Reading, MA: Addison-Wesley, 1980.

Yapp, Malcolm E. "Some European Travelers in the Middle East." *Middle Eastern Studies* 39, no. 2 (2003): 211–27.

Zeitlin, Aharon. "On the Notion of Time in Art." *Hatekufa* 19 (1923): 469–76.

Zimmermann, Moshe. "Biography as a Historical Monograph." *Tel Aviver Jahrbuch für deutsche Geschichte* 20 (1991): 449–57.

Zimmermann, Moshe. "Das Gesellschaftsbild der deutschen Zionisten vor dem 1. Weltkrieg." *Trumah* 1 (1987): 139–58.

———. "Deutsche Geschichte in Israel." *Geschichte und Gesellschaft* 15 (1989): 423–40.

———. *Deutsch-jüdische Vergangenheit: Der Judenhass als Herausforderung*. Paderborn: Schöningh, 2005.

———. *Die deutschen Juden. 1914-1945*. Munich: R. Oldenbourg, 1997.

———. "Hannah Arendt, the Early "Post-Zionist." In *Hannah Arendt in Jerusalem*, ed. Steven E. Aschheim, 181–93. Berkeley and Los Angeles: University of California Press, 2001

———. "Jewish History and Jewish Historiography: A Challenge to Contemporary German Historiography." *Leo Baeck Institute Yearbook* 35 (1990): 35–52.

———. "Jewish Nationalism and Zionism in German-Jewish Students' Organizations." *Leo Baeck Institute Yearbook* 27 (1982): 129–53.

Zionist Organization(?), ed. *The Proposed Hebrew University on Mount Scopus Jerusalem Palestine*. London(?): Zionist Organization(?), 1924(?).

INDEX

Page numbers in italics refer to illustrations.
Albert Einstein is abbreviated to "AE" in subentries.

EINSTEIN, ALBERT,
and Palestine: (*cont.*)
AE on Jewish laborers in, 152, 158–
159, 163; AE on landscapes of, 157,
178; lectures in, 149, 151–152, 155–
156, 159, 168, 171, 173, 178; meetings
with Arabs, 151, 153, 155; named
honorary citizen of Tel Aviv, 151; AE
as national icon in, 170, 173; national
importance of AE tour of, 169–170,
172; AE on "Oriental" quality of, 162,
177; planned visit to, 130, 138, 141–
143; preparations for tour of, 143,
167–168; propaganda quality of tour,
170; reception of AE by Jewish community in, 168, 170, 172–174; reception of relativity in, 174; receptions
for AE in, 147–149, 151, 153, 169,
173–174, secondary importance of P.
for AE, 165, 177; as spiritual center,
165, 258; tour of, 139–180, 253; AE's
travel diary, 144–145; visits Ben Shemen, 152; visits Degania, 154–155;
visits Haifa, 152–153; visits Jericho,
146; visits Jerusalem, 146–149, 151,
155, 157; visits Migdal, 154, *154*, 155,
156; visits Mivke Israel, 152; visits
Nahalal, 153–154; visits Nazareth,
154–155; visits Rishon LeZion, 152;
visits the Technion, 153; visits Tel
Aviv, *150*, 151–152; visits Tiberias,
154, 155; warning re future of, 226;
Zionist agenda of AE's tour in, 168–
169, 171, 177 (*see also* Palestine);
on perception by Jews of himself,
205–206;
political awareness of, 40;
political involvement of, 242;
possibility of AE leaving Berlin: AE on,
49; rumors of, 86;
possibility of AE leaving Germany, 141;
on Prague, 34–35;
on Prague Jews, 34, 81, 254;
public affairs: AE's initial interest in, 35;
publications of: on special relativity, 31;
radicalism, limits to AE's, 44, 158;
on relationship to the state, 40, 49, 272;
relationship with his sons, 257; AE on,
39–43, 49–50, 78, 113;
religiosity of, 18–20, 254;
rejection of religious definition of Jewishness, 74–75;

role in Zionist movement of, 5, 58–60,
65, 123, 263–264;
role of science for, 20–21, 27, 30, 38, 41–
43, 79, 242, 257;
schooling of: in Munich, 9, 13–17, *17*; at
Aarau, 23–25;
science: AE's initial interest in, 19–21;
on science: sacred qualities of, 311n52;
on science in Germany, 88;
on scientific output, 78;
search for employment, 28–29, 31;
secularism of, 117, 137, 157;
on Serbians, 35;
signs Zionist invitation, 50;
supports Jewish settlement in Peru, 228;
Swiss citizenship of, 23, 26, 31, 44, 48,
82, 95, 255, 293n25;
at Swiss Patent Office, 29–31;
on the Technion, 66, 206;
tolerance, limits to AE's, 164;
trips: to the Far East, 139–142; and
Japan, 140–141, 145; and Java, 299n9;
and Port Said, 145–146, 157; and Singapore, 144; and Spain, 157;
and United Kingdom: AE invited to, 95;
AE visits, 127;
and United States: AE on, 129–130; AE's
arrival in the U.S., 107; AE's departure for, 102; AE's departure from,
127; invited to join tour of, 89–90,
94–96, 98–99; Jews in, 93, 95, 123,
126–127, 129, 137; motivations for
tour of, 132; AE's planned lecture tour
in, 87–88, 90–92, 107–108, 113, 122,
131; preparations for tour of, 97, 99–
103, 106–107; purpose of tour of, 92–
93; receptions for AE in New York,
109; relative failure or success of tour
of, 125–127, 134–136; AE on success
of tour of, 123, 126–128, 130, 137;
AE's ties with scientists of, 87, 108–
109, 135; universities in, 96, 104–107,
121–123, 135, 137; Zionist tour of,
86–138, 179, 253, 258–259;
on Weimar Republic, 49;
on Western Jews, 5, 74, 83–84;
on World War I, 36–39, 41–44;
on Zionism, 44–46, 52, 66, 84, 209–210,
215, 256, 259, 266; AE on benefits of,
127–128, 149, 165–166, 177–178, 258;
AE on benefits for Arabs in Palestine,
221; as a means to overcome alien-

Ussishkin, Menachem, 99, *103, 104,* 107, 124, 139, 146–148, *148,* 149, 169; on transfer of Arabs, 209

Versailles Peace Treaty, 95
Volcani, Yitzhak, 151

Waldeyer-Hartz, Wilhelm von, 43, 285n123
Warburg family, 240
Warburg, Felix, 91, 99, 101, 185, 189, 191, 198, 200, 203, 231–233, 242, 246
Warburg, Max M., 88, 101
Warburg, Otto, 69, 182–183, 194, 232
Warburg, Otto Heinrich, 43
Warburg, Paul M., 88, 91, 99, 101, 118
Washington University, 105–106
Wassermann, Oskar, 167, 211
Weber, Heinrich, 28–29
Weigert, Charlotte, 70
Weimar Republic, proclamation of, 46, 79. *See also* AE on
Weizmann, Chaim, 3, 7, 211, 273; AE on, 193, 200, 206–208, AE's contacts with, 4, 60, 64, 140, 262; and AE's Palestine tour, 141–143, 153, 164, 166, 169; and AE's United States tour, 86–91, 93–95, 97–103, *103, 104,* 104–107, 109–116, 118–120, 122–126, 128, 130–137, 179, 253, 264; and the Arab-Jewish conflict in Palestine, 217–218, 226–227; and the Hebrew University, 144, 181–182, 185–186, 188–190, 193–197, 199–200, 202–207, 230–234, 239, 241–243, 245–246, 250; relationship with AE, 249, 263, 267
Weizmann, Vera, 102, *104*
Weizmann-Tchemerinsky, Rachel-Lea, 153
Weltsch, Robert, 217–218, 222, 226, 263
Wilhelm II, 37, 46, 79
Wilamowitz-Moellendorf, Ulrich von, 96
Wilson, Woodrow, 56–57
Winteler, Jost, 24

Winteler, Marie, 27
Winteler, Pauline, 24, 27
Winteler-Einstein, Maja, 10, 25, 27, 30–32
World War I, end of, 46

Yale University, 105
Yellin, David, 148, 169
Yishuv. See Palestine: Jewish community in

Zangger, Heinrich, 39, 42, 68
Zionism: factions within, 271; in Germany, 26, 64, 74, 77, 91, 175; in Prague, 33–34; historiography of, 2, 87, 268
Zionist Federation of Germany, 3–4, 44, 46, 50–52, 54, 61, 73, 77, 79, 98–99, 102, 129, 140, 257, 262; and Arabs, 57; and Palestine, 50–51; generations within, 75–76
Zionist Action Committee, 103
Zionist Commission in Palestine, 115, 119
Zionist Organization, 87, 127, 130, 138, 140, 142, 195, 200, 206, 222, 253; AE on 198, 242; and AE's Palestine tour, 147, 167, 169–170; and AE's United States tour, 92–94, 97–98, 100–103, 114–115, 117, 119, 124, 132–133, 143; Berlin bureau, 60, 65–66, 69; and the Hebrew University, 144, 182, 184, 186, 190, 192, 237, 239, 241, 243, 258; London bureau, 58, 60–66, 69, 73, 86, 98, 183, 187, 214, 236, 248; Palestine Office 142; Zionist Executive, 69, 87, 99, 143, 147, 149, 167–168, 170, 175, 183–184, 186–187, 193–194, 241, 262; University Committee, 66, 119, 133, 248; University Fund, 98, 114–120, 125, 128
Zionist Organization of America, 82, 89, 98–99, 104–107, 109–110, 112, 117, 120, 122, 129–131, 134–135
Zionist student associations, 26
Zlatopolsky, Hillel, 99
Zurich Polytechnic, 23, 25–28, 35, 76, 238